Neuroscience, Consciousness and Spirituality

Studies in Neuroscience, Consciousness and Spirituality

Volume 1

For further volumes:
http://www.springer.com/series/10195

Harald Walach • Stefan Schmidt
Wayne B. Jonas
Editors

Neuroscience, Consciousness and Spirituality

Springer

Editors
Harald Walach
Institute for Transcultural Health Science
European Universtiy Viadrina
Frankfurt (Oder), Grosse Scharrnstr. 59
15230 Frankfurt (Oder)
Germany
walach@europa-uni.de

Wayne B. Jonas
Samueli Institute
King Street 1737
Alexandia, VA 22314
USA
wjonas@siib.org

Stefan Schmidt
Department of Environmental
Health Sciences
University Medical Center Freiburg
Breisacherstr. 115b
79106 Freiburg
Germany

European University Viadrina
Frankfurt (Oder)
Germany
stefan.schmidt@uniklinik-freiburg.de

ISSN 2211-8918 e-ISSN 2211-8926
ISBN 978-94-007-2078-7 e-ISBN 978-94-007-2079-4
DOI 10.1007/978-94-007-2079-4
Springer Dordrecht Heidelberg London New York

Library of Congress Control Number: 2011936020

Printed on acid-free paper

Springer is part of Springer Science+Business Media (www.springer.com)

Foreword of the Series Editors

"Neuroscience, Consciousness and Spirituality" was born out of the vision to build bridges and get different disciplines to talk to each other. We have been observing these disciplines for quite a while, doing empirical research in the field of mindfulness meditation, conceptual, psychological and philosophical issues, as well as spirituality. We were struck by the lack of communication between different pockets of research cultures. We thought that neuroscience researchers could learn from philosophers and from those dealing with issues around spirituality and mystical experience, and vice versa. We felt that the philosophical discourse around the issue of what constitutes consciousness and how it can be explained would benefit from hard neuroscientific data on the one hand and from insights stemming from first-person experience on the other hand, as it is the currency of spiritual traditions.

Science within the comfort zone of unidisciplinarity is always nice and easy, and cosy, too. Stepping beyond is not only challenging, it is nothing short of madness and professional suicide. Yet, we felt it is necessary. Spirituality seems to be a necessary ingredient in the scientific debate. Talking about consciousness without taking into account exceptional experiences and personal accounts of conscious states that are beyond the ordinary is a bit like trying to do physics with the constraint of only studying crystal lattices. That won't yield a valid theory of matter. Neither will philosophising about consciousness without taking into account different aspects, especially extraordinary and even rare states of consciousness. Plasma states of matter are rare and not normally observed in our everyday world. Yet, they teach us a lot about matter. In the same sense, extraordinary states of consciousness as reported in the spiritual literature, by those practicing spiritual methods such as meditation, can teach us more about consciousness than thousands of discussions of what consciousness is like in a normal day in the supermarket.

Meditation research is a kind of focal point that has established itself as a new scientific "hot topic" over the past decade. It is done from various angles: Neuroscientists try to map different meditation states using various imaging methods. Sometimes psychologists join in or neuroscientists also use psychological methods trying to tap into the experience of those having such meditation experiences. William James marked the beginning of the scientific study of psychology by defining it as the science of consciousness, and the beginning of consciousness research by studying spiritual experiences. It is this unifying approach which we are trying to regain by fostering dialogue and discourse across the boundaries of disciplines. It has to be regained, because in the beginning of the scientific study of consciousness, in the legacy of Wilhelm Wundt and Sigmund Freud, the study of such exceptional experiences was explicitly banned and this ban has haunted the field like a posthypnotic command.

Consciousness studies cannot be complete without also facing the philosophical question: What, actually, is the stuff consciousness is made of? Is the current mainstream model that favours some sort of emergentist approach that has consciousness arise as an emergent property of complex neuronal systems sufficient to explain conscious experience, especially if we look at spiritual experiences from a phenomenological point of view? Does this phenomenology suggest otherwise? If so, how could we envisage such a model of consciousness? Can we align it with what we know from neuroscience?

Since these are important questions, we will, every now and then, also digress into the philosophical field and discuss models of consciousness that challenge the mainstream view or bridge gaps. We do not do that out of a spirit of dissidence, but of constructive criticism and dialogue.

We developed these ideas a couple of years ago and found sponsors, the Samueli Institute in Alexandria, VA, USA and the Theophrastus Foundation in Germany, both of which were enamoured with them. So it happened that we were able to invite a small and select group of scholars and scientists to our first meeting in Freiburg in 2008 for an open discussion, the proceedings of which we present here. Some guests elected not to publish their ideas and others have changed them considerably in the face of the discussions. We were able to follow on with a second meeting 2010 with the specific topic of meditation research, and we hope to be able to present this volume soon as well. The third in the series is likely to be a piece of discussion by one of us, Harald Walach, that puts forward the argument that spirituality has to be taken into the realm of discourse within academia to proceed with the program of rational and scientific enlightenment. Only if spiritual enlightenment meets scientific enlightenment can we really progress, we suggest. We hope to be able to proceed with our Freiburg meetings, and we also invite volumes and contributions from the wider scientific community touching upon these issues.

We have no hidden agenda, no tacit creed, no criteria for participation in this discourse other than two very pragmatic ones: The submitted contributions need to be interdisciplinary and touch upon the three topics of neuroscience, consciousness and spirituality or use two of these different disciplines to throw light on the third one in particular. And they need to be of good quality, with stringent argumentation and clear style of writing. All contributions are peer reviewed, and the whole volume will again go through review. So expect good quality work addressing an emerging new topic. We are looking forward to contributions, to discourse and discussions.

Freiburg and Frankfurt (Oder), Germany — Harald Walach and Stefan Schmidt

Contents

Neuroscience, Consciousness, Spirituality – Questions, Problems and Potential Solutions: An Introductory Essay

Harald Walach

Abstract Science and spirituality are often seen as two incompatible approaches to reality. This chapter is designed to start bridging this gap. We define science as a joint effort of humans to understand the world and to prevent error, using our senses and invented instruments enhancing our senses. This we call experience of the world in its material aspects. Spirituality can be understood as an effort to understand the general principles or structure of the world through inner experience. There are a few requirements for such an epistemological framework to function. One is that consciousness is understood as complementary to its material substrate, the brain, and hence as capable in principle of having its own access to reality. The other requirement is that dogmatism, both on part of science and on part of religions is put aside and spirituality is understood as a hitherto neglected area of investigation that needs to become part of science as a method of inner experience. Some historical efforts – Roger Bacon's system in the middle ages or Franz Brentano's attempt at the beginning of the history of scientific psychology – can serve as examples. Preconditions and open questions are discussed to pave the way for a better understanding.

Definitions and Explanations

All definitions are provisional. Already Aristotle taught that a definition comes at the end of a long process of understanding. I submit that all terms used here are only incompletely understood at this time. Hence I use these definitions more as a

H. Walach (✉)
Institute for Transcultural Health Sciences, Viadrina European University, Frankfurt (Oder), Germany

Samueli Institute, European Office, Frankfurt (Oder), Germany
e-mail: walach@europa-uni.de

H. Walach et al. (eds.), *Neuroscience, Consciousness and Spirituality*,
Studies in Neuroscience, Consciousness and Spirituality 1,
DOI 10.1007/978-94-007-2079-4_1,

convention of speaking, to make communication clearer and to convey the sense in which I wish these terms to be understood. The readers are welcome to juxtapose these to their own understanding, bearing in mind that this is what I deem most useful for the time being.

This chapter introduces a few preliminary working definitions for the most important terms, such as science and spirituality. It argues that some narrow conceptions of science that are content-based and see science as a kind of modern religion are not helpful. It moves on by analyzing the preconditions for a dialogue between science and spirituality, outlining commonalities and differences. Historical examples of Roger Bacon and Franz Brentano introduce two prominent attempts and also the problems and tasks associated with this dialogue. We end by giving an outlook, how such an enterprise could progress.

Science

Science is most usefully seen as a communal effort of humanity to describe and understand the world and to prevent error as much as possible through systematic inquiry.

It is a *communal* effort: the agent of science is not the single scientist, but the whole community of scientists, that virtual entity often termed "scientific community". We *describe* the world by observing events and by determining which of those observations can be shared by others. This is often referred to as the empiricism of science or knowledge through experience. We *understand* the world by the joint operations of experiencing it through our senses, trying to figure out potential relationships between the events we observe, and modelling these relationships. The process of *experiencing* uses our sense organs and all those enhancements invented by humans to support our sense organs in observing and making contact with the world. Examples of such enhancements are telescopes, microscopes, particle accelerators, immunological bioassays, structured and unstructured interviews, radiography, etc. This experience is usually directed at the material world, hence "outer experience".

Because the scientific enterprise has been going on for quite some centuries now, we need to respect its history and the systematic inquiry that has established itself. Whether this history is a process of cumulative gain of knowledge (Duhem 1965; Lakatos 1978) or revolutionary rewriting (Kuhn 1955, 1983; Toulmin 1985) need not concern us here. However, it is important to understand science as a historical, communal effort of mankind that has created certain institutions, rituals and processes and thereby has given rise to a complex social system (Canguilhem 1979; Latour and Bastide 1986; Latour 1999). One of the methodological hallmarks of science, as opposed to unsystematic experience of everyday life, is that it is *systematic,* installing processes that prevent error as much as possible. One important element in such a systematic process of collective experience is *intersubjectivity*. This requires that observations, experiences and theoretical structures designed to model relationships between observations are shared by a community of competent observers. In practical terms this means that observations are not just singular, but shared by others, or can in principle be shared by others. In theoretical terms it means that models describing relationships between events or observations and giving explanations are consistent with the rest of the current scientific understanding and inherently plausible. In some old-fashioned theoretical models of science this requirement is termed objectivity. We deliberately refrain from using this term, due to a series of severe theoretical problems which we are not addressing here.

There are different canons of methods for different disciplines, but they all attempt to *prevent error*. In the natural sciences the experimental method or empirical testing have become an important tool, but it is only one example of a method to prevent error. Other examples are peer reviewing processes in scientific publishing, replication procedures in the production of data (Schmidt 2009), triangulation of data with different methods, checking of empirical results against theoretical expectations, to name but a few (Collins and Pinch 1993).

Misconception 1: Science as Content Based

Such an understanding implicitly *excludes* some popular *misconceptions* of science, often also held by scientists themselves, if they do not reflect on the preconditions of their own activities. It excludes a dogmatic misunderstanding of science that defines science through particular contents. There is always the temptation to use the standard of knowledge gained so far and exclude everything that cannot be fit into the current framework theoretically and in principle as "unscientific". The error committed by such a misunderstanding is that it limits the vision and potential reach of science. It can be shown in important historical examples that a new idea or invention was rejected by members of the scientific establishment because it did not fit their current knowledge and what was conceivable from their point of view (Oeser 1979a, b; Laudan 1981). Prominent examples include the heliocentric model of cosmology as opposed to a geocentric one. Another example was the idea that the world is curved and not flat which was already proposed in antiquity, lost along the way, rediscovered by Roger Bacon in the thirteenth century (Bacon 1859, 1897, 1983, 1998)[1] but not generally accepted until the seafaring adventures of Columbus and later Maghellan showed that the world can be indeed circumnavigated. A prominent example from the history of medicine is the resistance that William Harvey met, when he declared that the heart is a pump and makes a noise, the heartbeat, when

[1]Compare Bacon's Opus Majus, (Bacon 1897, Vol 1, IV10. De Figura Mundi p. 152 ff, especially p. 156), where Bacon argues that if the earth were flat crews on deck of a ship should see the harbour at the same time as someone on top of the mast. Also in his "De multiplicatione specierum" (Bacon 1983, 1998, I.6, p. 68): "Sed cum mundus sit spherice figure... since the world has a round form..." Regarding Bacon see below.

moving the blood around (Parisano 1647; Walach 2005).[2] The common concept of circulation at the time was influenced by Aristotle and later Galenos who taught the heart is a convection-warmer, warms the blood which moves to the brain, where it is cooled and flows down again. In such a model there was no place for a pumping heart, nor for the associated beat. Hence people did not "hear" the heart beat and did not accept the associated fact. Similarly, Freud's ideas of unconscious mental processes were rejected by the community of psychologists until in our days cognitive science discovered the importance of implicit processing, re-inventing a cognitive unconsciousness (Koukkou and Bräker 2002).

Defining "scientific" through the body of content of what is currently accepted is dangerous. It automatically and by definition excludes whatever cannot be currently conceived to be compatible with this body of knowledge. The history of science teaches that re-conceptualisations make it often possible to include new discoveries, hitherto thought impossible, into the body of scientific knowledge (Laudan 1977). Hence, I think that voices banning "spirituality" and associated ideas as unscientific, or "esoteric", or worse, are not useful. They set out from the implicit understanding that science is a body of knowledge. Quite to the contrary, I suggest understanding science solely through its application of a systematic methodological approach and the communal effort to secure the knowledge gained through procedures against error. Part of these procedures is surely theorising and solid conceptualisations, connecting new ideas with old, established theories and concepts. Sometimes such established theories and ideas have to be revised to include new discoveries. Sometimes new methodologies lead to discoveries that force us to abandon old ideas.

Misconception 2: Science as Mainstream

Such a methodological definition of science also guards us against another common misconception related to the first one. "Scientific" is often shorthand for the "mainstream" view on how things are or should be. While this is surely a useful guiding principle, it is sometimes wrong. All scientists want to belong to the cherished in-group of the mainstream, being accepted and being well respected. This is the social nature of science as a group enterprise and of us humans as social animals; after all, each and every scientist is foremost a human being, with all the social and individual psychology associated with this. There is another host of examples showing that it is frequently the odd one out, who does not care what others think, who prefers his or her findings and instincts over belonging that makes the new discoveries. To read "scientific" as shorthand for "mainstream" is not useful in this context, nor is it essential for a proper understanding of science.

---

[2]The decisive documentary evidence can be found in a compilation of works of William Harvey and his major opponent, the renowned physician-philosopher Emilio Parisano, who famously declared "there is no one in Venice who can hear a heart beat." See Parisano (1947; V. Tactus 79, p. 101 with Harvey's arguments and V. Contactus 77–79, p. 107 with Parisano's counterarguments). I have translated and rendered this passage in full in Walach (2005).

The fact that science is a communal activity reinforces of course certain majority views as being accepted scientific knowledge, while others are deemed wrong or unacceptable. But it is only when we incorporate the historical perspective and see the shifts of opinion across time that we will find that mainstream opinions change, sometimes allowing minority views into focus and giving room to new ideas. From an operational point of view every finding, every idea and every theory that finds its way onto the arena of public ideas, if scrutinised properly, is potentially scientific. Whether it will make its way into the mainstream reasoning is quite another matter. But it is dangerous and unscientific to exclude this possibility using standards of currently accepted opinion.

Misconception 3: Science as a Worldview

Often, the standard of common knowledge is amalgamated in what is then termed a "scientific worldview". By that we mean that all the knowledge we have and the presuppositions we make give rise to a certain generic view about the world. If analysed, such a view rests on presuppositions that cannot be part of the worldview or methodology, but are necessary to make it work (Collingwood 1998, orig. 1940). In other words, science, as much as any human undertaking, needs to ground its activities in a couple of presuppositions that are themselves not "scientific" but the necessary requirements for science to work. Ever since Wittgenstein, Gödel and Collingwood, to name but a few, we know that there cannot be – in principle – a system that can prove the foundations of its own principles using solely its own methods. It has to always fall back on another system, or accept some premises as granted, but unprovable. Thus, the foundations of science are based on implicit social agreement. This is the postmodern condition science is part of. Some examples for such principles that are accepted as granted are, for instance,

- That the world is open to rational scrutiny (rationalism)
- That it is likely that the most important elements of the world are material (materialism)
- That analysis of systems in terms of smaller constituents is a good guiding principle (atomism and analysis)
- That the world is most usefully analysed in terms of outer relations, to name but a few of the most important implicit presuppositions.

While such a worldview is certainly useful within reason, it does not mean that it is a definitive constituent of science as a social, historical enterprise. It happens to be part of the current common belief system. But identifying science with the "scientific worldview" would be converting science into scientism. Scientism, a term introduced by Husserl (1977), refers to a belief system similar to a religious belief, with the difference that in scientism the entity which is supposed to have an answer to all questions and a solution to all problems is science, while in religion it is a transcendent entity often termed God.

Neither view, I submit, is very useful in the context of science as a methodologically defined enterprise. As such, science does not have all answers, and perhaps never will have them. Nor is it necessary to recur to a potential ideal state, where all will be solved, for science to function.

Another important observation needs to be emphasised here: There is an epistemological asymmetry between statements of existence and non-existence, which already Popper has pointed out (Popper 1976). It is one thing to state that something is the case or exists, and then give procedures to ascertain this. For instance, Harvey could say "the heartbeat exists" and tell people to

(a) Listen to chests of people
(b) Cut open live animals and observe the heart beat.

To make negative statements of non-existence is different. Such statements depend on our current knowledge, on our current theories, and on our current methodologies. Our methodologies could change and make things visible or open to experience that currently are inaccessible. Our theories could change and make things conceivable that currently are incompatible with our theoretical ideas. And likewise our body of knowledge could change and make things possible that previously were deemed impossible. For instance, the statement "propulsion at or near the speed of light is impossible" depends on the universal validity of Einstein's special theory or relativity and on the fact derived thereof that, when speeds approach the speed of light, the energy needed to achieve this is near infinite. Now, if we could potentially engineer the vacuum fluctuations of the zero point field, such energy were available in principle. Up to now, no one has achieved this. But the statement clearly depends on a certain state of knowledge. I am not suggesting that propulsion at the speed of light is in fact possible. I am just saying that using science to state that something is impossible, inexistent or inconceivable is itself against the spirit of scientific enquiry, because it exhibits a lack of understanding of the mechanics and preconditions of science.

Science, thus, is most usefully understood methodologically and not content- based, defined through procedures and the values associated with it, and not through outcomes and accepted knowledge.

Spirituality

I define spirituality as an experiential realisation of connectedness with a reality beyond the immediate goals of the individual (Walach and Reich 2005). It gives rise to a holistic type of knowing that manifests cognitively, emotionally and motivationally. This is why it is termed "experience" in the sense of an inner experience of reality.

I have just defined implicitly what it is to "*experience*", in a psychological sense. It means to realise something not only cognitively, but also emotionally and motivationally, and hence denotes a holistic type of knowing. *Connectedness* refers to the fact that every individual goal and every individual life can only be realised within the whole context of our world, in relation to others and to this world. Thus, *reality beyond the individual* and its immediate goals, refers, ideally, to the totality of our world, and practically to the world around us. Sometimes the term "*transcendent*" is evoked to define spirituality, implying that there is a totally different reality, sometimes termed "God" in theistic traditions, out there that spirituality relates to (e.g. Koenig 2008). I submit that "transcendence" is a constituent of the notion of spirituality. Our definition here uses the term in as broad a sense as possible, and everything that is not of immediate concern to an individual "transcends" it. Although this seems to make the concept of transcendence trivial, this is not the case. It simply acknowledges that everything that is not of immediate concern to an individual, by its very nature, transcends these concerns and the individual. Someone giving up smoking because of concerns for his or her own health, for instance, is acting out of immediate concern for his or her individual being. Someone giving up smoking for the sake of his or her partner, or children, is acting out of concern for others and hence out of a "realisation of connectedness" in our terminology here, realising a goal that "transcends" his or her immediate concerns. Ultimately, such a realisation could have the "totality" as a target. It would be here that classical philosophical or theological definitions of "God" meet with the concept envisaged. Meister Eckhart (1260–1324), the great German mystic and scholastic scholar, phrased it as "esse est deus – the being is God" (Eckhart 1964, p. 38). The Latin term "esse" here denotes the totality of being. Since medieval times, both in the Christian and Judaic tradition, there has been a realisation that such "transcendence" also invokes "immanence", i.e. the possibility to actually realise connectedness with such a "transcendent" reality. It is the term "experience" that allows bridging this apparent gap.[3]

When such a realisation of connectedness is made experientially it becomes more than a cognitive insight. It contains an emotional element and the motivational element to change the course of our actions. Complementary to sense experience or experience of the outer aspects of the world, such an experience is an inner experience. While in scientific experience we use our sense organs or instruments to grasp some aspect of the material reality of our world, in spiritual experience we have no dedicated sense organ, except our consciousness as such. Thus, in consciousness we can realise such connectedness, even universal connectedness, beyond space and time. When this is realised, we speak of an inner experience as a different kind of accessing reality or a holistic kind of knowing.

Here are two mundane examples:

We can, for instance, memorise the Pythagorean principle that, in a triangle with a 90° angle, the square of the hypothenusis, or longest side, is equal to the square of the two other sides. Then we have cognitive knowledge. If, however, a good teacher shows

[3] I apologize to the learned reader for this rash and obviously insufficient treatment of such ancient and complex notions. The reason is simple: to give a proper rendering would require more space than is available in an introductory chapter, and the purpose here is, after all, a clarification of notions, and not the systematic argument that the usage of these notions is or is not compatible with certain traditional streams of thought. I will provide a more systematic treatment of these issues in a forthcoming study.

us the proof of this sentence, using geometrical examples, and we, manipulating the squares finally come to "see" that the sentence is true, we have a clearer understanding, an insight, or an "aha experience", as it is sometimes termed. This encompasses more than just our knowing cognitively. It also speaks to our emotion, the joy of discovering something as true and having an insight into a riddle. Often it also has an action or motivation component. Once we have understood this, we can use it in many circumstances and will have the impulse to apply the Pythagorean theorem.

We all have read novels and poems of love at one time, possibly before we fell in love ourselves. At the time we knew from words – cognitive structures – what love is supposed to be. Only when we fell in love ourselves did we truly "know", since we "experienced" it. This comprises a cognitive, emotional and actional or motivational component. Once we knew we loved, we wanted to do something: be closed to the beloved, declare our feelings, have sex, etc.

In that sense spirituality is about experiencing as a holistic type of knowing. The object of what it is that is being experienced is what I have termed "connectedness with reality beyond individual goals". Such a view is deliberately broad and encompasses all instances of transcendence, from the seemingly most trivial to the most complete kind conceivable.

Sometimes such an experience manifests as the intuitive understanding that my existence is somehow dependent on others, as others are on me. Becoming a parent, for instance, is an obvious example of such an experiential realisation. It is cognitive, emotional and motivational at the same time. It is difficult to communicate this type of experience to someone who has not had it, and there seems to be an implicit and intuitive companionship between those, who have had it without many words necessary. Such an experience of being a parent automatically curbs some of our egocentricity, and we need to put some of our individual goals behind the welfare of our offspring.

While this is a comparatively mundane example, there are many other instances of experiences of connectedness with reality beyond our individual goals that are less ubiquitous and well known. Experiences of being one with the world or with others, experiences of sudden illumination or enlightenment, or revelations or just intuitive grasps of parts of reality are some of them (James 1985). Without going into details about whether and how such different experiences might be distinguishable from each other and how to test claims of veridicality – all important questions which are decisive for a future use of the terminology – at the moment I just want to use them as examples. I point to one commonality: they are all instances of an *inner experience*. They occur as something that is experienced within ourselves: the insight into the proof of Pythagoras' theorem, the realisation that one is in love, the experience of parenthood, a potential enlightenment experience, an intuitive grasp of some parts of the reality, they all have in common that they occur, phenomenologically speaking, within ourselves. There may be certain events leading up to it, but the experience itself is clearly inside us. It is a subjective event. Although the reality we seem to grasp is not material reality, but some inner structure, it nevertheless seems to be reality. At least it normally feels like it. While becoming a parent has certain outer, material concomitants – having fathered or given birth to a child, holding a child physically, being legally responsible, etc. – the *experience* has to do with the inner structure. Understanding Pythagoras' theorem also has to do with the abstract mathematical structure, and not with any outer material realisation. When falling in love, although there is a particular person this is normally related to, the experience itself is about a reality in principle. In that sense, inner experience deals with reality, but with inner aspects or structures.

Most of the time, such inner experiences which we term "spiritual" seem to be pointing towards reality or connectedness beyond the self and its immediate goals. We can also have inner experiences about ourselves and our present situation, for instance in a psychotherapy workshop that teaches us about our behavioural patterns. Those insights are also often termed experiences. They are inner experiences, but they are normally not spiritual, in that they revolve around our selves. It appears that in order to term something "spiritual" we also need the qualifier that the experience points beyond our immediate ego and its goals.

Religion

By religion we refer most usefully to a set of teachings, rituals and ethical-behavioral precepts that have developed historically and culturally over time. Humans have used different religions to regulate their relationship among each other and whatever they thought their relationship to a transcendent reality should be like. Arguably, religions can be seen as the form that was given to certain spiritual experiences of founders by their followers, once the original experience and teachings derived thereof have crystallized over time.

While religion sometimes and ideally is a form that contains spirituality and spiritual experiences, facilitates them and channels insights into action, it is nowadays often used and perceived as an empty shell and as a dogmatic system. In the same sense as dogmatic science or scientism is a misconception, religion without a spiritual core is likewise. While religion has to do with the interpretation and enculturation of spiritual experiences and hence might have a potentially important function in the grounding of such experiences, this is not the scope of the definition of spirituality used here. While it might be necessary for lived spirituality to find a vessel to express itself within a current religious and communal context, this is not the case within the scientific remit. Spirituality as a mode of experiencing is, by its very definition, part of an all-encompassing scientific enterprise. Religion, as a form of representing spiritual experiences, is in the same way culturally, historically and politically relative as the currently accepted scientific worldview is. It might be useful to think of spirituality and religion in terms of content and form, similar to a poem. While in reality, both belong together, it is possible to look for the content and deal with it separately. In the same sense we can abstract spirituality as a mode of experiencing from its normal occurrence within religions (Forman 1998). It may even be that at the end of the process or as a result we will identify important insights that also pertain to religions. But this is neither a goal nor a prerequisite for taking spirituality seriously.

While religions are, by definition, historically, culturally and politically contingent as every expression of human experience is, spirituality addresses the experiential core, and as such can be amenable to scientific analysis and debate. Use grief as an example: the human experience of grieving is certainly ubiquitous to humankind and very likely is part of animal culture as well (Sommer 2008). Everyone who experiences the death of a loved one grieves. But the cultural expression of grief is quite different. While in Japan people wear white when they grieve, in Europe and Western cultures they wear black. In the same sense, the core of spiritual experiences might be quite similar, while their interpretation and cultural-historical framing is different.

Science and Spirituality – Commonalities

Monism and Experience

There are some interesting commonalities and dissonances between science and spirituality. Both start from the intuitive assumption that the underlying reality is one. They start from an implicit monism. While the contemporary scientific worldview assumes this unitary reality to be matter, energy, or information (Zeilinger 1999), out of which everything arises, spiritual traditions are most often neutral in their ontology and refuse to define the "stuff" out of which this final underlying reality is made. In some traditions it is termed "mind", "consciousness", giving rise to an implicitly idealist ontology, although it is important to realise that such terms are often more metaphors than anything else. Also, the method seems to be common to both: Science is ultimately based on experience and its rational conceptual analysis. Spirituality draws on inner experience as its ultimate source of knowledge.[4] While the subject matter of this experience in science is the material world, in spirituality it seems to be the inner structure of the world. The outcome of this process of experiencing in science is a set of observations about the world and theoretical rules or laws. In spirituality it is a body of inner experiences, their interpretations in linguistic terms, such as images, paradoxes and complex linguistic structures, and often also a canon of behavioural rules.

Insight and Inner Experience

However, both seem to have one clear commonality: Both, science and spirituality, are trying to make sense of the world we live in. Both need at some point an inner experience or insight. While spirituality starts with such an inner experience,

[4] There is, of course, the difficult issue how "revelations" fit into this picture. They can be conceptualized as inner experiences objectified.

science needs such an insight into structures to discover a viable theory, at least as an initial starting point of a new theoretical model that can inspire further scientific work. While the factual observations about the world are given in science, the theoretical structures combining them to a viable theory are not obvious. They need to be found or discovered. They lie, as it were, "behind" the facts. This is where insight comes into play. The creative idea that transforms observations of regularity or of structures into laws and theorems is not part of the observations as such, but is extracted by the inventive human mind from the observations. But nothing in the observations dictates the structure. The structure is rather "found", "seen", "discovered", even "invented" by an outstanding scientist. If looked at phenomenologically, the great discoveries of scientific theories and inventions often exhibit the very same properties as spiritual experiences or insights: they cannot be "produced at will", they come all of a sudden, after a prolonged period of inner work and dedication, they are the result of an inner insight into structures, and they very often have a compelling sense of truth, beauty and awe. The process that leads to the invention of a scientific theory has been termed *abduction* by C.S. Peirce (1931).[5] Once such a theory exists, one can use this structure to deduce consequences and test these consequences empirically. The results constitute new facts that either confirm or disconfirm an existing theoretical model. If too many deviant facts are discovered, this is normally the time when new theoretical structures are needed that can explain the previously explained facts and the anomalies discovered since. This discovery of a theoretical structure, abduction, is a moment of scientific creativity and theoretical insight that is, we propose, similar to the moment of spiritual experience and in fact very likely *is* such an experience of a particular kind.

Spirituality normally starts out with such an experience or insight, relating to some important aspect of the world, and then works out the consequences in real life and the interpretations relevant until a new experience supersedes these interpretations and engenders new behavioural rules. Thus, science and spirituality have one common denominator in making sense of this world: immediate inner experience or insight. In science it leads to predictions and empirical investigations and finally modification. In spirituality it leads to interpretations and to behavioural rules or consequences within a certain cultural, social and political context until the experience is superseded by a new one. This ideal structure is depicted in Fig. 1.

It should be noted that these circles describe ideal types. In science, a lot of inductive observation and data collection has to happen before someone even has the material to combine these, using an abductive step of reasoning, into a viable theory. Also, in science the rules of observation, methodological prescriptions for

[5] Vol. 7, p. 218, Scientific Method: "Abduction … is the first step of scientific reasoning, as induction is the concluding step". It is of course evident that a large part of scientific work is carried out without any of this inventive-inductive reasoning, for instance, when someone works within the bounds of given models, just exploring certain consequences or simply observing something and amassing data. While whole individual lives of scientists may be lived without any of such an inner experience of how facts fit together within a potential theoretical model, science as a whole and as a collective process has this feature of abductive reasoning as a prerequisite.

Fig. 1 Idealised circle of abduction, induction, and deduction for science and spirituality

making deductions and empirically testing them are well established. In spirituality, no such commonly accepted rules exist, except in certain subgroups and cultures. I have termed the "immediate experience of reality" in the case of spirituality as "induction", as it is the equivalent to an inductive experience of outer reality in science. However, such a direct experience of reality can never be had or communicated without interpretation, which by necessity happens through linguistic structures. This interpretative step converting experience into an interpretation is structurally similar to the abduction of science postulating a theoretical structure behind the observed and experienced facts. Finally, the deductions in the realm of spirituality, what kinds of behaviour are appropriate, are similar in structure to the deductions used by scientists to develop consequences out of theoretical models for further testing. While in science this step serves to find out, whether predictions of a theory are correct and hence the theory useful, in spirituality this final step is a way of putting spirituality into practice.

Some Historical Notes

Scientia Experimentalis of Roger Bacon

Obviously, this conception hinges on the meaning of the term "experience" and what we are willing to let it stand for. We are here transporting a holistic notion of experience, comprising both experience of the outer, material world and inner experience. It is interesting to observe that such a holistic notion of experience had already been conceived of by Roger Bacon (1214/1215–1292), at the beginning of our modern era of science in the middle ages. Roger Bacon was probably one of the most interesting and influential figures for the future development of science after his great and admired predecessor Robert Grosseteste (Crombie 1953; Hackett 1997; Clegg 2003; Power 2006). He influenced William Ockham and his namesake Francis Bacon, who were instrumental for the further development of science. It is difficult to do him justice in a short section, and we have to be content with some sketchy strokes. He was barred from wider communication by his superiors as a Franciscan friar in 1267. Hence, Bacon resorted to a surreptitious means of communication: he contacted the Pope who asked him to write down for him what his concept of academic learning, teaching and scientific scrutiny are. In response he wrote what he termed his "Opus Majus", or "Larger Piece" (Bacon 1897), followed up by his "Opus Minus" or "Lesser Piece" and an "Opus Tertium" or the "Third Piece" (Bacon 1859). Most important for the understanding of these writings is the fact that they existed in two copies only: one intended for the pope himself, and one was Bacon's own. Contrary to some common misconceptions, these texts contain only sketches, graffiti, as it were, of what Bacon would have actually intended to work out had he been given the chance in what he called his prospective "Opus Principale", his Main Work. This he never wrote for reasons unknown, most likely, because he was prohibited. In those surviving three books he sketches out his vision for a completely new way of conceiving scholarly activity and the scientific enterprise. He termed it "scientia experimentalis", experimental science. Here, for the first time in the history of science in the west, he described a unitary experimental science that would embrace both experience of the material world, together with mathematical analysis, quite as it began to develop 300 years later, together and in conjunction with a science of inner or spiritual experience. The latter was, of course, at his time and in his circumstances, being a Franciscan friar, common talk. What was far from common talk was to conceive of such inner experiences as part of a common scientific enterprise, as Bacon obviously did. It is worthwhile quoting him in full:

> Experience comes in two forms – one through our outer senses. Thus we experience what is in heaven and below... And this is human and philosophical experience... but this experience does not suffice man. For it does not give sure and certain evidence about material things because of its difficulty, and about spiritual things it does not attain anything. Therefore human intellect has to be supported otherwise in the way our holy patriarchs and prophets, who first gave knowledge to the world, have received interior illuminations and did not only remain in the senses [Here he quotes Ptolemy, Centilogium]: "There is a twofold way of reaching knowledge of things: one through philosophical experience, another one through divine inspiration which is much better", as he says. This inner science has seven grades .... The seventh consists in the exstases of spiritual experiences (raptibus) and ways of understanding things in different modalities, about which man is not allowed to speak. And who has experience and training in this field, he can certify himself and others not only about spiritual things but also about human sciences... we need this science, which we call experimental. This I want to explain... (Bacon 1897, orig. 1267, Vol 2, p. 169ff, translation HW)[6]

[6]The original Latin text as edited by Bridges reads. "Sed duplex est experientia; una est per sensus exteriores, et sic experimenta ea, quae in coelo sunt... et haec inferiora... experimur.... Et haec experientia est humana et philosophica, quantum homo potest facere secundum gratiam ei datam; sed haec experientia non sufficit homini, quia non plene certificat de corporalibus propter

It is obvious from the text that Bacon had a holistic type of science in mind, part of which would be what later, in fact, became science: systematic experience of the material world and its analysis, using the tools of mathematical language. Another part was what he termed inner science, scientia interior, or inner experience which the spiritual experience of inner enlightenment or illumination was clearly part of. Both, he thought, belong together and help install knowledge about the world and about how to govern it. It is important here to realise the slight difference of remit in these types of experiences: experience of the world would help understand and control it. Inner experience would help in the governance of the world. This belongs to the moral-political complex of agency and values about which science is conspicuously silent these days and which is so much needed.

Hence the idea comes quite naturally that this inner experience and insight into structures is, what is needed to make science complete and help us with those impasses that knowledge about the world alone cannot solve, as science in its current form provides us with. Surely, this is what Bacon had in mind.

As history unfolded, Bacon's attempts remained a colossal fragment. His books arrived at the papal court, when the pope was already dying. He never read them, nor did his immediate successors. Bacon never got a chance to elaborate on his ideas, and he got not to influence the making of the academic mainstream of his days. Hence we understand the split that happened at that time: while science started to unfold and went its own ways, separating from the ways of the Church and religion, religion and theology went a different way altogether. To cut a very complicated and long story extremely short: the theology of mystical experience, although favourably viewed by some, was considered altogether too dangerous and did not become part of mainstream theology. It survived in the monasteries and some mystical branches, in the underground as it were, but it was neither taken up by official mainstream theology, nor by science for that matter.

Thus, around 1260, a split occurs in the Western history of ideas, science and culture: science starts to develop as an experience of the outer, material world and its systematisation. Inner, spiritual experience is relegated to the hidden and forbidden

sui difficultatem, et de spiritualibus nihil attingit. Ergo oportet quod intellectus hominis aliter juvetur, et ideo sancti patriarchae et prophetae, qui primo dederunt scientias mundo, receperunt illuminationes interiores et non solum stabant in sensu... Nam gratia fidei illuminat multum... secundum quod Ptolemaeus dicit in Centilogio quod duplex est via deveniendi ad notitiam rerum, una per experientiam philosohiae, alia per divinam inspirationem quae longe melior est, ut dicit. Et sunt septem gradus hujus scientiae interioris, unus per illuminationes pure scientiales. Alius gradus consistit in virtutibus... p. 171 Virtus ergo clarificat mentem ut non solum moralia sed etiam scientialia homo facilius comprehendat... Tertius gradus est in septem donis Spiritus Sancti... Quartus est in beatitudinis, quas Dominus in evangeliis determinat. Quintus est in sensibus spiritualibus. Sextus est in fructibus, de quibus est pax Domini quae exsuperat omnem sensum. Septimus consistit in raptibus et modis eorum secundum quod diversi diversimode capiuntur, ut videant multa, quae non licet homini loqui. Et qui in his experientiis vel in pluribus eorum est diligenter exercitatus, ipse potest certificare se et alios non solum de spiritualibus, sed omnibus scientiis humanis.... necessaria est nobis scientia, quae experimentalis vocatur. Et volo eam explanare, non solum ut utilis est philosophiae, sed sapientiae Dei, et totius mundi regimini".

realm of mysticism and starts to only thrive in the underground.[7] As I read history, there was a new, decisive move at healing this split, whether consciously or unconsciously is difficult to tell. This happened, when Franz Brentano set out to install a new science of psychology in 1866, nearly exactly 600 years after Roger Bacon.

Franz Brentano and the Making of Psychology

Franz Brentano (1838–1917) is one of the founding figures of modern psychology (Wehrle 1989; Smith 1994; Tiefensee 1998; Benetka 1999; Bühler 2002). Trained as a theologian and philosopher, in his habilitation 1866 in Würzburg he defended the thesis "methodus philosophiae nullus alius nisi scientiae naturalis – the method of philosophy cannot be anything else than the method of science" (Wehrle 1989, p. 45).[8] Thereby he clearly referred to the method of experience. Later on, when he became chair of philosophy in Vienna he urged the city and the university to start a laboratory of psychology as early as 1874 (Kraus 1919).[9] In his psychological works he worked out, what he meant: inner experience or psychognosis, or inner phenomenology, should be the basis for scientific psychology (Place 2002).

For various reasons Brentano did not really succeed in installing his new methodology. Some had to do with his way of working: he never published his final and decisive ideas (Tiefensee 1998). He was also, likely, tied in to his neo-Aristotelian concept of psychology which did not allow him to radically re-envisage a different type of inner experience.[10] Most importantly, his personal choices – renouncing his priesthood, loving a Jewish heiress in Catholic Vienna, having to emigrate in order to marry her – and the intrigues and political jibes following this led to his giving up of his academic career and withdrawing into private life (Brentano 1895). Thus, his most important influences were indirect, through his students and followers.

[7]One could speculate why this was the case. One reason surely is that inner experiences, by their very nature, are paradoxical and difficult to explicate in language, hence misunderstandings are preprogrammed. Another obvious reason is that mystical experiences nearly in all religions threaten the dominance of the ruling class of priests and in particular threatened the teaching of the preeminence of the Church and its role as sole mediator between man and the divine. It seems that today, with political powers of the Churches practically non-existent in the Western world, this issue can become part of the public discourse again.

[8]This habilitation document is unpublished. A quote and the reference to the archival material can be found in Wehrle's dissertation (1989), p. 45.

[9]Normally 1879, the date the experimental laboratory was founded by Wilhelm Wundt in Leipzig is taken as the decisive date for the founding of scientific psychology.

[10]Brentano followed, by and large, Aristotle and his scholastic training, having been trained as a Catholic priest. He obviously thought that introspection would be enough to establish knowledge, thus applying the Aristotelian model of science to psychological, phenomenological data. It is understandable, why he did this; he had no other scientific models at hand. It is also understandable, why this did not work: there is no reference point in inner experience for checking against a given reality, as in scientific experience.

Probably one of the most important of them was Sigmund Freud, who heard his lectures and very likely took his ideas of the analysis of inner experience from Brentano, without ever acknowledging this (Merlan 1945, 1949). This brought inner experience as an epistemological source of insight back into the arena of science, albeit on a comparatively mundane level. Another of his famous students was Carl Stumpf who became seminal for the whole Gestalt psychology movement which influenced Koffka, Köhler, and Wertheimer in Berlin (Münch 2002). One can make a point that with the heyday of cognitive science those insights have become more influential than ever. In the philosophical arena Edmund Husserl took much of his inspiration to develop phenomenology from his teacher Brentano (Husserl 1919). Phenomenology can be seen as a philosophical, perhaps purified, attempt at using inner experience as epistemological access to the structure of the world.

The Legacy of History

Thus, Brentano's attempt at installing inner experience as a modality of epistemology and knowing had some important reverberations in modern history of science. It seems that most have one thing in common: the recursion on inner experience as a potential source of insight. True, there is no accepted methodology associated with it. Also true, mainstream historiography has it that naïve introspectionism after Narziss Ach, Hugo Münsterberg and others is dead and has led nowhere (Place 2002). So why bother?

We feel that there is a reason to bother, because none of the previous attempts has raised the issue of epistemology in a self-reflective way, except Husserl, who pointedly used the term "epoche" to describe the importance of the methodological difference of phenomenological consciousness from everyday consciousness (Husserl 1977; Adams 2006). But Husserl did not tell us how to do this (Adams 2006), and what actually happens, but was concerned with the gleaning of philosophical truth. Most importantly, none made the connex with spiritual experiences and the issue of consciousness that arises with this.

Thus, we feel that there is perhaps a new, even unique chance to take up this project that has been broached several times and dropped again in the history of our sciences. Discussing the interface between science, as exemplified in neuroscience, spirituality understood as experiential access to reality, and mediated by the emerging science of consciousness might help us out here. In that sense, my attempt is both old and new: I am taking up threads that lie around in history; but I am weaving new patterns and knitting new fabrics.

Clearly, there is a lot do to: We do not have a clear epistemology, let alone methodology of inner experience. We do not know, how to distinguish "true" from "fake" claims. No clear demarcations to guard ourselves against error, as in science, are known here. While within specific traditions there are elaborate models of verification of claims of inner experience, such as the Koan testing in Zen Buddhism, the tradition of discernment of spirits in the Christian tradition, and careful triangulation of experiences against texts in other Buddhist traditions, we have no universally agreed methodology.

We do not know how to analyse corresponding theoretical concepts, using tools similar to the tools of mathematics in science. Some such tools have been developed again within particular traditions, but not across traditions. And many more problems are there to be solved.

Problems, Open Questions, Potential Solutions

Not all is rosy in the picture that I have portrayed. There are some, perhaps irreconcilable difficulties and problems lingering on the way. One obvious philosophical and fundamental problem is that of ontology. While both spiritual traditions and our current scientific worldview affirm a kind of monism regarding the world as such, science is normally quite vocal about a material kind of monism, stipulating that the final "stuff" the world is made of is matter. Some admit that it might be energy. But that is the end of it. Spiritual traditions are mostly adamant in claiming that matter is somehow derivative of the "final stuff" and not final reality itself. Will there be common ground? Perhaps in discussing a transcendental kind of monism which recognises that matter itself is a manifestation of a transcendental ground, recognising that our modern notion of "matter" is so devoid of "stuffness" already that it does not seem to be reasonable to speak even of matter (Zeilinger 1999). Will there be recognition that both, matter and consciousness are notions of concrete, phenomenal reality that will never be the final boundary of ultimate reality? Phenomenal reports of spiritual enlightenment experiences often refer to this ultimate reality as somehow "conscious" or "universal awareness" (Kapleau 1969). Some traditions such as various kinds of Buddhism wisely refuse to say anything positive about it, using metaphorical terms such as Dharma, or "the way" for it, as does Daoism. The same is true for Judaism, which is well known for the prohibition of naming the final reality at all, or Islam, as a matter of fact, which only knows the thousand names of Allah, none of which is sufficient. So there seems to be quite some awareness within spiritual traditions that fixing the nature of what-ever-it-is-that-is-experienced-as-final is a dangerous, even blasphemous, thing to do. Perhaps there is a common ground here?

Whoever has experience in spiritual practice of any kind knows that sometimes anomalous experiences – telepathy, precognition, even psychokinesis – are quite normal occurrences. Some would even hold that such instances of non-locality or immediate connectedness across space and time are at the heart of spirituality. It is important to distinguish here between non-locality as an interpretation for anomalous experiences and a generic connectedness across space and time as being at the heart of spirituality. While the latter is surely a defining element of spirituality, manifesting itself in spiritual experiences of connectedness, instances of anomalous experiences are just examples or signs and signals of such a connectedness. Spiritual traditions normally advise against taking them overly serious.

Nevertheless, spiritual lore is full of such experiences, and from a phenomenological-experiential point of view the well-known mainstream opinion that such things do not exist except as anomalous belief systems and distortion of perception does not make sense at all. On the other hand, the current scientific world view does not allow for such experiences, as they seem to defy locality and a host of other well-known scientific findings (Walach and Schmidt 2005). Will there be a potential scientific world-model that will allow for such experiences without distorting either phenomenology or science? We believe that the model of generalised entanglement, which we present in the context of this book and have published elsewhere (Lucadou et al. 2007), might be an avenue to use. Will it be acceptable? Will it prove to be empirically viable?

Perhaps the most difficult point of departure is the mainstream view on consciousness. It seems straightforward to assume that the mainstream neuroscientific view on consciousness is some sort of materialist-monist view. While there are myriads of subcategories and theories, they all have one simple thing in common: they view consciousness as completely dependent on the brain (Metzinger 2000). Some say, consciousness and the brain are identical. Some say, consciousness is a result of the brain's activity, much as immunity is the result of the immune system's functioning. Some say consciousness has some independence, once established, but it is certainly causally dependent on the brain. There are a few pockets of die-hard dualists who claim that consciousness and the brain are two different things. Some say this on empirical grounds, such as Pim van Lommel who has studied Near-Death Experiences carefully and makes his point in this book. Some claim consciousness is different from matter on theoretical grounds, since quantum theory, the best theory to describe matter, presupposes consciousness in the measurement process (Schwartz et al. 2005). But these voices are hardly the mainstream. On the other hand, it will be difficult to reconcile a physicalist view of consciousness that assumes that consciousness is just an emergent property of a physical system with the phenomenal experiences reported within spiritual traditions.

One way of conceiving of the problem and bridging the gap we have provided previously. It is through the notion of complementarity (Walach and Römer 2000). This idea, originally introduced by Nils Bohr into physics to describe the paradoxical nature of the quantum, means that we need two maximally incompatible descriptions to describe one and the same thing. The thing to describe is a conscious human being. The two incompatible descriptors are material reality and consciousness. One way of coming to grips with the problem is stipulating that both, consciousness and brain, material and mental reality, are two incompatible and irreducible sides of one transcendent reality. If this is so, we have two alternative routes of access to this reality. One is through our material senses and outer experience. Another one is through our "inner sense" or through consciousness. By directing our attention inwards we might potentially create an alternative avenue of access to reality. This is the way spiritual traditions use, when they train their followers in meditation or similar practices.

Thus, the methodology of how to have insights and access to inner experiences and reality seems to be tightly linked with a viable view of what consciousness is.

In our view, complementarity of mind and body is a minimally sufficient notion to guarantee both, a scientifically viable notion of consciousness and a methodologically viable access route to reality: via consciousness as one manifestation of this reality.

This begs the question that such a view is neither compatible with the mainstream reductionist concept of consciousness, nor with the somewhat difficult to maintain view of a dual nature (Beauregard and O'Leary 2007). Although the complementarist stance is by default phenomenologically dualist, it is ontologically monist, combining the best of two worlds.

This might, however, not be sufficient to accommodate the insights of spiritual traditions, I submit. It might be necessary to stipulate, above and beyond what I would like to term Consciousness 1, personal consciousness, some Consciousness 2, superpersonal or transpersonal or spiritual consciousness, which transcends Consciousness 1 (Walach 2007). Such a supra-individual type of consciousness has been stipulated since Plato, by Aristotle and his interpreters, who conceived of the intellectus agens as such (Merlan 1963). Brian Lancaster on Kabbala in this volume makes reference to this tradition. This is also, likely, what Pim van Lommel has in mind, and what many spiritual traditions, such as Vedanta or Tibetan Buddhism, theorise about and others, such as Theravada Buddhism, shoot into the ground of Nothingness. After all, this is ultimately a scientific question, I submit. If the notion is inescapable and defies Ockham's razor and if it is supported by incontrovertible evidence through experience, it will have to be adopted. This is, where inner experience as a new method yet to be developed comes into play. This is surely for the future. The most important thing about all these questions, in our view, is to remain open and un-dogmatic. After all, this is the beauty of science that true science can ask, is allowed to ask and even has to ask all questions, even seemingly silly ones. This is where true spirituality and true science really meet: being radically open about the future, about asking questions, about abandoning trodden paths, and about rejecting dogmatic answers.

References

Adams, W.A. (2006). Transpersonal heterophenomenology? *Journal of Consciousness Studies, 13*(4), 89–93.

Bacon, R. (1859). In J. S. Brewer (Ed.), *Opera quaedam hactenus inedita. Vol 1: Opus tertium, Opus minus, Compendium philosophiae*. London: Longman, Green, Longman & Roberts.

Bacon, R. (1897). *The Opus majus of Roger Bacon*, 2 vols. Oxford: Clarendon.

Bacon, R. (1983, 1998). *Roger Bacon's philosophy of nature. A critical edition, with English translation, introduction, and notes, of De multiplicatione specierum and De Speculis comburentibus*. Oxford: Clarendon.

Beauregard, M., & O'Leary, D. (2007). *The spiritual brain. A neuroscientist's case for the existence of the soul*. New York: Harper One.

Benetka, G. (1999). Die Methode der Philosophie ist keine andere als die der Naturwissenschaft...: Die "empirische Psychologie" Franz Brentanos. In T. Slunecko, O. Vitouch, C. Korunka, H. Bauer, & B. Flatschacher (Eds.), *Psychologie des Bewusstseins – Bewusstsein der Psychologie. Giselher Guttmann zum 65. Geburtstag*. (pp. 157–175). Wien: Wiener Universitätsverlag.

Brentano, F. (1895). *Meine letzten Wünsche für Österreich*. Stuttgart: J. W. Cottasche Buchhandlung.
Bühler, K.-E. (2002). Die Bedeutung der Deskriptiven Psychologie Brentanos für die Psychotherapie. *Brentano Studien, 10*, 151–173.
Canguilhem, G. (1979). *Wissenschaftsgeschichte und Epistemiologie. Gesammelte Aufsätze hrsg. v. Wolf Lepenies*. Frankfurt: Suhrkamp.
Clegg, B. (2003). *The first scientist. A life of Roger Bacon*. London: Constable.
Collingwood, R.G. (1998, orig. 1940). *An essay on metaphysics*. Oxford: Clarendon Press.
Collins, H., & Pinch, T. (1993). *The Golem. What everyone should know about science*. Cambridge: Cambridge University Press.
Crombie, A.C. (1953). *Robert Grosseteste and the origins of experimental science 1100–1700*. Oxford: Clarendon.
Duhem, P.M.M. (1965). *To save the phenomena: An essay on the idea of physical theory from Plato to Galileo*. Chicago: University of Chicago Press.
Eckhart, M. (1964). In K. Weiss (Ed.), *Die deutschen und lateinischen Werke. Bd. 1: Prologi. Expositio Libri Genesis. Libri parabolorum*. Stuttgart: Kohlhammer.
Forman, R.K.C. (Ed.). (1998). *The innate capacity: Mysticism, psychology, and philosophy*. New York/Oxford: Oxford University Press.
Hackett, J. (1997). Roger Bacon: His life, career and works. In J. Hackett (Ed.), *Roger Bacon and the sciences* (pp. 9–23). Leiden: Brill.
Husserl, E. (1919). Erinnerungen an Franz Brentano. In O. Kraus (Ed.), *Franz Brentano, Zur Kenntnis seines Lebens und seiner Lehre* (pp. 151–167). München: Beck.
Husserl, E. (1977). *Die Krisis der europäischen Wissenschaften und die transzendentale Philosophie*. Hamburg: Meiner.
James, W. (1985). *The works of William James. The varieties of religious experience*. Cambridge, MA: Harvard University Press.
Kapleau, P. (1969). *The three pillars of Zen: Teaching, practice, enlightenment*. New York: Harper.
Koenig, H.G. (2008). Concerns about measuring "spirituality" in research. *The Journal of Nervous and Mental Disease, 196*, 349–355.
Koukkou, M., & Bräker, E. (2002). Das Mentale und die psychoanalytische Theorie des Mentalen im Zustands-Wechsel-Modell der Funktionen des menschlichen Gehirns. In P. Giampieri-Deutsch (Ed.), *Psychoanalyse im Dialog der Wissenschaften. Band 1: Europäische Perspektiven* (pp. 143–191). Kohlhammer: Stuttgart.
Kraus, O. (1919). *Franz Brentano: Zur Kenntnis seines Lebens und seiner Lehre*. München: Beck.
Kuhn, T. (1955). *The structure of scientific revolutions*. Chicago: University of Chicago Press.
Kuhn, T.S. (1983). Rationality of theory choice. *Journal of Philosophy, 80*, 563–571.
Lakatos, I. (1978). *The methodology of scientific research programmes. Philosophical papers* (Vol. 1). Cambridge: Cambridge University Press.
Latour, B. (1999). *Pandora's hope: An essay on the reality of science studies*. Cambridge, MA: Harvard University Press.
Latour, B., & Bastide, F. (1986). Writing science – Fact and fiction: The analysis of the process of reality construction through the application of socio-semiotic methods to scientific texts. In M. Callon, J. Law, & A. Rip (Eds.), *Mapping the dynamics of science and technology: Sociology of science in the real world* (pp. 51–66). Houndmills/London: Macmillan.
Laudan, L. (1977). *Progress and its problems: Towards a theory of scientific growth*. Berkeley: University of California Press.
Laudan, L. (1981). *Science and hypothesis*. Dordrecht: Reidel.
Lucadou, Wv, Römer, H., et al. (2007). Synchronistic phenomena as entanglement correlations in generalized quantum theory. *Journal of Consciousness Studies, 14*, 50–74.
Merlan, P. (1945). Brentano and Freud. *Journal of the History of Ideas, 6*, 375–377.
Merlan, P. (1949). Brentano and Freud – A sequel. *Journal of the History of Ideas, 10*, 451–452.
Merlan, P. (1963). *Monopsychism mysticism metaconsciousness. Problems of the soul in the neoaristotelian and neoplatonic tradition*. The Hague: Martinus Nijhoff.
Metzinger, T. (2000). *Neural correlates of consciousness: Empirical and conceptual questions*. Cambridge, MA: MIT Press.

Münch, D. (2002). Erkenntnistheorie und Psychologie. Die Wissenschaftliche Weltanschauung Carl Stumpfs. *Brentano Studien, 10*, 11–66.

Oeser, E. (1979a). *Wissenschaftstheorie als Rekonstruktion der Wissenschaftsgeschichte. Band 2: Experiment, Erklärung, Prognose*. München: Oldenbourg.

Oeser, E. (1979b). *Wissenschaftstheorie als Rekonstruktion der Wissenschaftsgeschichte. Band 1: Metrisierung, Hypothesenbildung, Theoriendynamik*. München: Oldenbourg.

Parisano, E. (1647). *Recentiorum disceptationes de motu cordis, sanguinis et chyli*. Leiden: Ioannis Maire.

Peirce, C.S. (1931). *Collected papers. Ed. Ch. Hartshorne & P. Weiss: Bd. 1–6; Ed. A. Burks: Bd. 7–8*. Cambridge: Harvard University Press.

Place, U.T. (2002). Some thoughts on the work of the Würzburg School and the controversy it provokated, prompted by a visit to Würzburg 10–16 October 1989. *Brentano Studien, 69*, 269–287.

Popper, K.R. (1976). *Logik der Forschung*. Tübingen: J.C.B. Mohr.

Power, A. (2006). A mirror for every age: The reputation of Roger Bacon. *English Historical Review, 121*(492), 657–692.

Schmidt, S. (2009). Shall we really do it again? The powerful concept of replication is neglected in the social sciences. *Review of General Psychology, 13*, 90–100.

Schwartz, J.M., Stapp, H.P., et al. (2005). Quantum physics in neuroscience and psychology: A neurophysiological model of mind-brain interaction. *Philosophical Transactions of the Royal Society B: Biological Sciences, 1458*, 1309–1328.

Smith, B. (1994). *Austrian philosophy. The legacy of Franz Brentano*. Chicago: Open Court.

Sommer, V. (2008). *Darwinisch denken. Horizonte der Evolutionsbiologie*. Stuttgart: Hirzel.

Tiefensee, E. (1998). *Philosophie und Religion bei Franz Brentano (1838–1917)*. Tübingen: Francke.

Toulmin, S. (1985). Conceptual revolutions in science. In R. S. Cohen & M. W. Wartofsky (Eds.), *A portrait of twenty-five years: Boston colloquium for the philosophy of science 1960–1985* (pp. 58–74). Dordrecht: Reidel.

Walach, H. (2005). Generalized entanglement: A new theoretical model for understanding the effects of complementary and alternative medicine. *Journal of Alternative and Complementary Medicine, 11*, 549–559.

Walach, H. (2007). Mind – body – spirituality. *Mind and Matter, 5*, 215–240.

Walach, H., & Reich, K.H. (2005). Science and spirituality: Towards understanding and overcoming a taboo. *Zygon, 40*, 423–441.

Walach, H., & Römer, H. (2000). Complementarity is a useful concept for consciousness studies. A reminder. *Neuroendocrinology Letters, 21*, 221–232.

Walach, H., & Schmidt, S. (2005). Repairing Plato's life boat with Ockham's razor: The important function of research in anomalies for mainstream science. *Journal of Consciousness Studies, 12*(2), 52–70.

Wehrle, J.M. (1989). *Franz Brentano und die Zukunft der Philosophie. Studien zur Wissenschaftsgeschichte und Wissenschaftssystematik im 19. Jahrhundert*. Amsterdam: Rodopi.

Zeilinger, A. (1999). A foundational principle for quantum mechanics. *Foundation of Physics, 29*, 631–643.

Mindfulness in East and West – Is It the Same?

Stefan Schmidt

Abstract The transference of the practice and concept of *mindfulness* from Eastern Buddhist into modern Western society is reconsidered. The underlying question is whether mindfulness as practiced in the ancient Buddhist tradition is still compatible with its modern expressions growing more and more popular in the Western world. Definitions and contexts within the Eastern tradition and the Western approach which is more scientifically dominated are compared, and the process of transference and secularization is addressed. Also, reasons for the popularity of mindfulness and meditation in the West based on societal developments are identified. This analysis comes to the conclusion that there are huge differences between the Western and Eastern approaches and contexts. Thus, it is more appropriate to see the heterogeneous, and mostly secular, practices in the West as a newly emerging culture of mindfulness which has not been there before. Nevertheless, this movement itself entails a transcultural aspect connecting East and West.

Introduction

The notion of *mindfulness* enjoyed a steep ascent within the Western hemisphere in the last 20–30 years. Mindfulness based approaches and interventions are applied within many different areas starting from mind-body and behavioral medicine and stretching as far as the art therapy (Monti et al. 2006) or coaching (Passmore and Marianetti 2007). Research on mindfulness is a hot topic within the medical and

S. Schmidt (✉)
Department of Environmental Health Sciences, Center for Meditation, Mindfulness and Neuroscience Research, University Medical Center Freiburg, Freiburg, Germany

European University Viadrina, Frankfurt (Oder), Germany
e-mail: stefan.schmidt@uniklinik-freiburg.de

H. Walach et al. (eds.), *Neuroscience, Consciousness and Spirituality*, Studies in Neuroscience, Consciousness and Spirituality 1,
DOI 10.1007/978-94-007-2079-4_2,

psychological sciences but also in neuroscience. Thus, the question of what we are really talking about when applying the notion of mindfulness, is of some importance. What we see today is that the meaning of mindfulness is more and more diluted the more popular mindfulness becomes. Thus, it makes sense to go back to the roots and to closely observe the process of the transition of this conception from its historic religious context in the East to the modern West.

Mindfulness has its origin in the Buddhist teachings in the ancient East. After having been a concept of importance without much change for 2500 years in the Eastern Buddhist realms, the concept was introduced into the modern Western culture. There it is incorporated into and applied within quite a different cultural context, often in a secularized version. This chapter addresses the question, whether after such a considerable change of context the notion, concept and practice is still the same. In order to do so, Eastern and Western definitions of mindfulness will be sketched and compared and also the Eastern and Western contexts, in which mindfulness is practiced will be described and contrasted. Next follows a section which tries to answer why meditation and Buddhism is so popular in the West. Finally, all pieces are brought together and the proposition is made that, inspired by ancient Eastern sources and driven by the needs created out of the modern Western society a new culture of mindfulness is emerging, which in this form is completely new.

Definition: Mindfulness in the East

The oldest written references for the notion of Mindfulness, *sati* in the Pali language, can be found in the so called *Pali Canon* of the Theravada Buddhist branch. Theravada (literally teaching of the elders) is the oldest Buddhist school, which is today still practiced in Sri Lanka, Burma (or Myanmar), Laos, Cambodia and Thailand. All other Buddhist traditions, such as Tibetan or Zen have their origin in this tradition. It is said that in the first century BC Buddhist monks wrote down the talks and teachings of Gautama Buddha, who lived approximately in the fifth century BC. These texts, which were transmitted orally before, form the oldest written account of the Buddhist teachings (i.e. the *Pali canon*, see also www.accesstoinsight.org/canon). For the study of the conception of mindfulness there are mainly two talks (Pali: sutta) of importance: The *Satipaṭṭhāna Sutta* (see e.g. Analayo 2004; Nyanaponika 1983) and the *Ānāpānasati Sutta* (see Rosenberg and Guy 2004). A detailed description and interpretation of these two suttas is not within the scope of this chapter and the interested reader may be directed to two excellent books fulfilling this task, Analayo (2004) and Rosenberg and Guy (2004). Both suttas describe solely a meditation practice but not a concept. Yet what mindfulness or *sati* entails expressed in abstract terms, can be inferred from this practice.

According to Analayo (2004), a Theravadin monk and scholar, the word *sati* has its origin in the verb *sarati* which means 'to remember' (p. 46). However, it can be shown that *sati* is not meant as memory but as *awareness of the present moment* which, in turn, will facilitate memory. Present moment awareness and memory complement each other: "…sati seems to combine both present moment awareness and remembering what the Buddha has taught" (p. 48). In order to achieve this, the mind in the state of *sati* needs to be "wide awake in regard to the present moment" (p. 48). Here the notion of "breadth of the mind" (compared to a narrow focus state) or wide awareness is emphasized. Another monk and scholar from the Theravada tradition, Nyānaponika (1983), describes sati as "bare attention". The term "bare" here refers to the fact that the observer tries only to perceive the object of observation, rather than to interact with it by e.g. appraisal, judgment, taking position or by changing it through volitional acting. Salzberg (2008, p. 135) expresses the same, but puts it the other way round:

> Mindfulness is a quality of relationship to the object of awareness. Just having an experience, say hearing a sound, is not really being mindful. Knowing a sound without grasping, aversion, or delusion is being mindful.

So, if *sati* is practiced in meditation[1] it can be described as a state of awareness of the present moment with a certain breadth of the mind in which one tries to observe without interfering. *Sati* is also often described by images and similes through the *pali canon* and these images emphasize different functions of *sati*, such as the qualities of alertness, relaxation, detachment, non-reactiveness, or undirectedness. An excellent collection of these images can also be found in Analayo (2004, p. 53ff).

While all these inferred conceptualizations of *sati* bring forward the idea of a fixed theoretical notion, one has to take care not to forget that mindfulness or *sati* is based on experience. For its full understanding one needs to pursue introspection practices resulting in a first person experience. A natural consequence of this fact is that sati is not a static concept, but changes with the meditator's gaining more experience. Furthermore it has to be noted that speaking, writing and theorizing about *sati* is always an incomplete approach. Direct experiences can only be communicated incompletely by language, since in principal every experience has more facets than language can express.

The Ancient Eastern Context of Mindfulness

The crucial role of *sati* for the Buddhist teachings can be inferred from two citations from the *Satipaṭṭhāna Sutta*. In the beginning it is said:

> "Monks, this is the direct path for the purifications of beings, for the surmounting of sorrow and lamentation, for the disappearance of *dhukkha*[2] and the discontent, for acquiring the

---

[1] Sati refers not only to a passive meditative state. Right mindfulness or *sammā sati* is also related to acting in accordance with certain ethical guidelines and the Buddhist principles. Therefore *sati* has to be combined with *sampajāna* (clearly knowing) and *ātāpī* (diligence). These two other notions are often aligned with *sati* and have to be considered to understand the conception of mindfulness to the full.

[2] *dhukka* is most often translated with 'suffering', although this translation does not capture the full notion this word has within the *Pali* language.

true method, for the realization of *Nibbāna*,[3] namely, the four *satipaṭṭhānas*" (translated into English by Analayo 2004, p. 3).

And at the end it is said:

"Monks (...) if anyone should develop these four *satipaṭṭhānas* in such a way for seven days, one of two fruits could be expected for him, either final knowledge here and now, or, if there is a trace of clinging left, non-returning" (translated into English by Analayo 2004, p. 13).

These two citations clearly stress the unique importance of *sati* for the Buddhist practice. It is only through the practice and cultivation of mindfulness that the ultimate goal of liberation can be reached. According to the Buddhist view, the ongoing cultivation of mindfulness will lead to insights into important fundamental truths, and it is this personally experienced insight which will finally lead to the ultimate goal of liberation. Thus, mindfulness meditation is often also termed *Vipassanā* (i.e. insight) meditation.

But such a cultivation of mindfulness is not a solitary procedure. It is of course embedded in a wider context of other meditative practices and ethical guidelines (Allmen 2007). The core of the Buddhist teaching are the *Four Noble Truths* which express, in a simplified version that all human suffering can be ended by following the guidelines and practices of the so called *Noble Eightfold Path*[4] which is the consequence of the fourth and last *Noble Truth*. The practice of *sammā sati* (right mindfulness) is one of these eight aspects of the *Noble Eightfold Path*. This Eightfold Path is the backbone of a spiritual path leading to personal transformation. Thus, it is obvious that the practice of *sammā sati* or right mindfulness cannot be separated from the other seven aspects. These other aspects include other meditative practices (*samādhi*), wisdom (*paññā*) and also a set of ethical guidelines (*sīlas*), which are different for lay persons and ordained ones (see also Fig. 1).

For illustration two examples shall be mentioned shortly here. *Sammā diṭṭhi* (right view) subsumes amongst others the belief in reincarnation, which is often a difficult point for persons from the Western hemisphere interested in Buddhism. Another aspect called *sammā kammanta* or right action is understood as refraining from killing or injuring living beings, refraining from stealing, refraining from certain forms of sexual behavior (misconduct) and refraining from intoxicants, such as e.g. alcohol. Furthermore the meditation practice is also intimately connected to the aim of cultivating the four so called *bramavihāras* or divine abodes, even if these are not directly mentioned in the Eightfold Path. They include loving kindness (*mettā*), compassion (*karunā*), sympathetic joy (*muditā*) and equanimity (*uppekhā*) and reflect another aspect of how to interact socially, and how to relate to the world while practicing insight meditation (*vipassanā*).

[3] *Nibbāna* can be translated as 'blowing out' or 'ultimate liberation'.

[4] It may be noted that *following* here does not imply a blind and unreflected adherence to certain religious views. Rather the Buddhist teachings stress the importance of being 'empirical' within the practice and to hold the resulting own experience always higher than any statements by written texts or religious leaders.

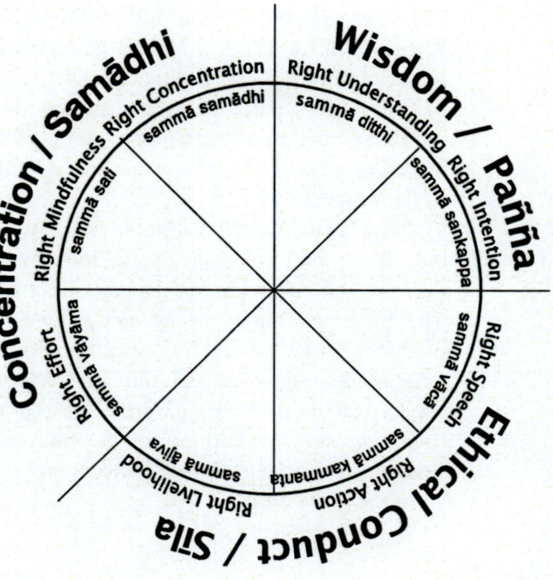

Fig. 1 The Noble Eightfold Path (Adapted from Allmen 2007)

In summary, the practice of mindfulness in the original Buddhist context is not just a solitary meditation technique performed to enjoy a period of silence or self-exploration, but part of a larger spiritual path. The main motive and intention to follow this path is to embark on a process of personal transformation leading to compassion for all living beings and has the ultimate goal of liberation (either in this or in other lives). This path encompasses many other practices, views and ethical guidelines mentioned above.

Mindfulness in the West

While mindfulness as a formal concept taught by meditative training was developed by the Buddhist tradition, it can be found implicitly in almost all spiritual traditions, Eastern ones as well as Western ones, e.g. in Christian mystical teachings (see Buchheld and Walach 2004). The practice of Buddhist inspired mindfulness meditation has meanwhile also found its place in our modern Western society. It was introduced by several sources and I would like to mention the most important ones:

1. The foundation of the *Insight Meditation Society* (IMS) in Barre, Massachusetts, USA by Jack Kornfield, Joseph Goldstein and Sharon Salzberg in 1974. They were travelling and working for the Peace Corps in the early 1970s in the Far East and came into contact with Buddhist teachings. IMS offers meditation retreats in the tradition of Theravada Buddhism and brought the *Vipassanā* practice with some of its original religious context into the United States.

2. The development of *Mindfulness Based Stress Reduction* (MBSR) in 1979 by Jon Kabat-Zinn (1990). MBSR is a structured 8-week course teaching several forms of mindfulness meditation as well as yoga to persons seeking coping strategies for stress, pain or chronic diseases. The MBSR program is in its orientation secular, non-religious and non-esoteric. Here only the techniques of mindfulness are taught, but the Buddhist context as outlined in the section above, is not part of the program.
3. The 10-day *vipassanā* meditation retreats taught by S. N. Goenka and his followers. This organization has meditation centers all over the world, where interested persons can participate in 10-day silent retreats to learn *vipassanā* meditation. The understanding of *vipassanā* by Goenka is that this form of meditation can be practiced without any religious orientation and is independent of faith. He mainly stresses this non-sectarian and secular interpretation and teaches only selected aspects of the practices described in the ancient literature (i.e. mindfulness of breathing and mindfulness of the body sensations or body-scan). During the retreats participants are asked to comply with several ethical guidelines taken from the *pali canon*.

Definition: Mindfulness in the West

From these and also other sources the concept of mindfulness has spread out in several different areas of modern Western society. And it is obvious that the way by which the idea of mindfulness was brought into the West is also reflected in the interpretation of the concept, e.g. either as a secular attitudinal quality or a spiritual Buddhist practice. Further distribution and adoption processes followed and finally the notion of mindfulness turned into a fuzzy cloud combining and mixing all kinds of ideas and meanings. Thus, without a proper definition of the specific context the term is used in, it will remain unclear, what is meant when talking about 'mindfulness'. Mindfulness may refer to (i) a formal meditation procedure, more precisely termed *mindfulness meditation*, (ii) to a theoretical concept from the Buddhist teachings, (iii) to a certain attitude towards one's own experience and actions in daily life (which could also be described as *informal mindfulness*), (iv) to a psychological concept derived from the Buddhist teachings but expressed in terms of Western psychological science, (v) to a different psychological concept by the same name defined by Ellen Langer (1989) and finally (vi) to the noun related to the adjective 'mindful' and its everyday life meaning. The last two aspects are of course by definition not influenced by the Eastern sources as described above and thus won't be followed up here.

The notions (i) and (ii) were already described in the first part of this chapter. The third aspect could be best characterized by a formulation more familiar to the Western mind and context. Jon Kabat-Zinn e.g. describes mindfulness "(...) as moment-to-moment, non-judgemental awareness, cultivated by paying attention in a specific way, that is, in the present moment, and as non-reactively and as nonjudgmentally and openheartedly as possible" (Kabat-Zinn 2005, p. 108). He furthermore defines certain qualities, which describe the internal attitude of this special way to pay attention and thus this attitude is indirectly part of the definition. According to Jon Kabat-Zinn (1990, p. 33ff), these are *non-judging, patience, beginner's mind, trust, non-striving, acceptance* and *letting go*. Shapiro and Schwartz (1999) extend this list by adding the following additional qualities: *gentleness, generosity, empathy, gratitude* and *loving kindness*. While this description will have the function to explain the concept in words to the interested Westerner, it is of course too imprecise for scientific research. Thus, there are several attempts to translate the notion into a scientific language and furthermore to connect and link such a concept to already existing psychological concepts. Bishop et al. (2004) held a consensus meeting in Toronto to develop a testable operational definition. According to this approach, mindfulness can be described by a combination of two components: (i) *self-regulation of attention*. This component characterizes the effort to maintain the focus of attention in the present moment. The second component (ii) is described as *orientation to experience* and is furthermore characterized by an attitude of curiosity, openness and acceptance. These or similar basic approaches towards a scientific definition can often be found, when *present moment experience* and *acceptance* are characterized as core features of mindfulness. Shapiro et al. (2006) developed a model to explain the positive effects of mindfulness based interventions in clinical studies (see e.g. Grossman et al. 2004). They propose that mindfulness can be explained by the three axioms of (i) *intention*, stressing the specific purpose of the practice, (ii) *attention*, stressing the volitional self-regulation aspect of one's attention towards the present moment and, (iii) *attitude*, stressing the qualities with which this purposeful attention is practiced. The authors emphasize that these axioms are "... interwoven aspects of a single cyclic process..." (Shapiro et al. 2006, p. 375).

Scientific definitions are often not verbally expressed but practically realized by instruments, which are designed to measure the concept under consideration. Thus, it is also important to look at the notions and presuppositions underlying questionnaires which claim to measure mindfulness. Meanwhile, there are eight such instruments published (Baer et al. 2004, 2006; Brown and Ryan 2003; Cardaciotto 2005; Chadwick et al. 2008; Feldman et al. 2007; Lau et al. 2006; Walach et al. 2006) with at least one more in the process of development. Unfortunately these instruments overlap only partially and thus result more in a dilution of the concept rather than in a clarification. There is also a strong argument by Grossman (2008) that for several reasons it may not be possible to measure mindfulness by a questionnaire approach in principle. Since these questionnaires are also given to people who had no prior experience with the idea of mindfulness (see e.g. Creswell et al. 2007) the resulting score of 'mindfulness' may have almost nothing to do with the original Buddhist concept or the practice within mindfulness meditation. If given to practitioners, there is the possibility of measurement error, as such individuals know, what the "correct answer" is supposed to be and can therefore either under- or overestimate their "true" mindfulness.

Furthermore, all these concepts and definitions are made from a third person perspective in accordance with the Western scientific approach. But the Buddhist practice gives always priority to the *experience* of the meditator, i.e. the first person perspective. Such a personal experience is, as already mentioned, changing with experience and cannot be directly shared with others. Thus, mindfulness, if taken seriously, resists a scientific definition and remains elusive to a certain degree.

The Modern Western Context of Mindfulness

In our Western society mindfulness is practiced, cultivated and applied in a much more diverse context than in the East. Whoever practices mindfulness meditation or other techniques to develop mindfulness will not solely seek the ultimate liberation within an ancient religious or philosophical system, but may have a wide set of motivations to do so, many of them being secular. An incomplete list of such motivations or intentions may contain topics such as:

- Coping with stress
- Coping with illness
- Self-regulation
- Self-exploration
- Self-experience
- Interest in psychology or Eastern philosophy
- Interest in spirituality
- Self-transformation
- Following a spiritual path

In order to further explore the question, whether a practice of mindfulness based on the motivations listed above can be compared to its Eastern form as described in section "The Ancient Eastern Context of Mindfulness", it is also important to study the cultural background for the growing interest in this Eastern concept and the related meditation procedures.

Why is Mindfulness and Buddhism so Popular in the West?

First of all, it is important to consider that there are different types of sources of Buddhism in the West. Jan Nattier (1995) differentiates between *Import, Export* and *Baggage Buddhism* to characterize the different developments leading to Buddhist activities in the West. Baggage (or *Ethnic*) Buddhism refers to immigrants from Buddhist societies, who bring their religion and religious practice as a baggage into the country they are migrating to. Export (or *Evangelical*) Buddhism refers to a more missionary approach initiated by certain groups in Eastern Buddhist societies, who actively export their ideas into the Western society. Finally, Import Buddhism refers to a process that Buddhist ideas and concepts are actively sought by members of Western society in the East and then in a next step are 'imported' into our society (see the example of Jack Kornfield, Sharon Salzberg and Joseph Goldstein as described above). It is mainly this latter group, which is responsible for the fast growing interest in the concept of mindfulness and mindfulness meditation. Nattier stresses an important point here, as he emphasizes that "only a member of the elite level of society can start an Import Buddhist group" (Nattier 1995, p. 43) and thus the resulting groups are far from ordinary regarding their education and profession. Nattier terms this as *Elite Buddhism* and what is said next here about meditation, Buddhism and mindfulness in the West refers most likely only to this elite aspect of society, which is an important issue to keep in mind.

So, what drives the interest of people towards mindfulness? There are several motives:

1. Individualization of religion and spirituality
   It is no secret that our Western culture emphasizes a strong individualization. To be different from others and to have one's own individual profile was a driving source of our cultural development over the last 100 years. This idea stretches further and further and has meanwhile also reached the areas of belief, religion and spirituality. While people in earlier times were socialized into the religious context, belief and systems of their family, and then usually stayed within this context, the situation nowadays is somewhat different. Many people develop either for themselves or within their family or peer group new religious or spiritual biographies. And, unless in the earlier days, they now have the chance to individually select from a huge market of offered religious spiritual teachings and services of all kinds (Huber 2007).
2. Fundamental attitude to cope with modern life
   Our modern society is changing at a fast pace and the demands to cope just with daily life are constantly altering and at the same time are getting higher and higher. This, in my view, has mainly to do with modern information and communication possibilities and technologies and the resulting faster timing of interaction processes. For instance, we are faced with instant information about almost every catastrophe in the world. At the same time, we are also asked to respond by continuously reduced time intervals between communications, e.g. by email, cell phone, text messages and many other new technologies. The speed at which these interactions are taking place is rising with every new communication medium. Other factors making daily life more complex and demanding are increased mobility and travel as well as the breakdown of stable structures providing guidance in life such as families and other social units. The already mentioned pressure to individualize in almost all aspects of life shifts the responsibility for many decisions from the social unit to the individual. This shift affects decisions of major importance (e.g. choice of profession, to develop an individual professional biography) as well as only minor ones (e.g. to individually select one's telephone provider or health insurance), but they all add up to an overload on the single individual. This information bombardment and overload of responsibility creates the wish in many persons to find an inner guide or attitude how to position oneself towards the demands of modern society. Such an attitude should be ideally independent of content of the demands and teach generic principles.

The idea to encounter the world, including oneself, with a stance of mindfulness seems to be an obviously helpful solution fitting the current needs and requests.

3. Needs for periods of timeout and retreat
Related to this issue is also the idea of an intentional and systematic interruption of this information and communication stream by deliberate time-outs or periods of retreat. This can be done on a daily basis by a (regular) meditation interval or on a larger scale by a whole 'day of mindfulness' or more formal 'retreats' for several days in special locations. Common to all these interruptions is, amongst other aspects, the idea of (noble) silence, non-responsiveness and withdrawal. Many people feel such a break as a necessary condition not to get lost in an externally oriented endless time stream of more or less automatic responses to information and communication requests. Of course, there are many other procedures fulfilling similar demands (e.g. not reading or answering the email for a day, switching off mobile phones, or not owning one in the first place, etc.) and the idea of interruption is not necessarily tied to meditation and silence retreats. But the latter ones can be considered as the most effective and intense interruptions, a fact which holds some irony in itself. It may be also interesting to add that the Christian influence on our society supported and generated such a break by indicating the Sunday as a day of rest for a long time period throughout history.[5] But the protestant reformation brought a new attitude towards the role of work in one's life and resulted in a softening of this precept (Idler and George 1998; Wetzel 2004). This change may be the reason, why concepts such as 24/7 availability are (next to its economic impact) held at high esteem within certain branches of Western society.
4. Introspection and self-actualization
Meditation as well as informal mindfulness practice offer another aspect fitting well with the current cultural demands and developments. This is introspection and self-actualization. The emphasis on individualization as described above can only be fulfilled successfully, if one has access to one's own internal world (i.e. internal self representations). On the other hand, the growing information and communication load have a strong drive towards the outer world. We also know e.g. from the theory of volition by Kuhl (see e.g. Fuhrman and Kuhl 1998) that stress and negative affect result in an inhibition of holistic processing, especially of the access to self-representations. In these cases one tends to switch to a decontextualized and analytical processing mode with a focus on isolated details. Presumably everybody knows such a situation, where under stress only aspects singled out from a context are perceived and the bigger picture and especially aspects of the self (e.g. why am I doing this? What has this to do with me?) are lost. A reconnection with the self, also called self-actualization, is often helpful in these situations. Practicing meditation, especially with its impact on non-reactiveness and interruption of behavioral patterns, results in introspective insights and also in self-actualization.

[5] The demands of the Jewish Shabbath can be considered as an enforced and earlier version of the same concept and of course the idea of periodic rest periods can be also found in virtually every culture.

It may be interesting, and quite ironic, to note here that according to this analysis many of the motives, why Westerners turn towards Eastern meditation techniques and teachings, are created out of the intention to individualize and to function more effectively in a society which rewards individualization. But at the same time the Buddhist philosophy neglects the existence of a time-invariant self-concept as a mere illusion. One of the central insights of the Buddhist teachings is called anattā (no-self) and expresses the idea that there is no independent self which stays the same over time. According to this concept, one has to release the idea of a distinctive ego and personal individuality in order to reach the ultimate goal. Thus, whoever sets out on this path and then decides to consider and analyze its concepts seriously will suddenly find a surprising fruit. At a first glance, it may not fit personal intentions of this person, but it may make sense from a larger view.

5. A systematic approach
In addition to what was said above, the Eastern spiritual approaches have something to offer, which is very hard to find in the Jewish and Christian traditions as practiced today. The Eastern teachings give detailed and systematic descriptions on how to develop and practice certain capacities, such as e.g. mindfulness or loving kindness. While almost all religious and spiritual systems stress the importance of love and orienting towards the present experience, the Eastern traditions also describe, how these aspects can be practically and concretely developed. With their teachings, they provide different meditation techniques with detailed guidelines describing how one can start to develop these capacities. In this sense, they are practical and experience based at a fundamental level in a way which is hard to find in Western religions, except perhaps in monastic contexts or esoteric teachings. And this is exactly what many people are looking for these days. They don't just want to believe what they are told, they want to experience for themselves. In the Christian tradition the idea of an experienced spiritual connection with God was often at the core of rejuvenating efforts, such as during the Franciscan movement in the middle ages or during the reformation, but neglected again and replaced by a rationalistic-theological approach which is predominating today. Experience based approaches can mainly be found in the mystical traditions which never reached standard theological teaching.
6. Self-determination, belief and empirical approaches
Finally, there is an important aspect related to the concepts of sin and the grace of God in Christian theology which is entirely missing in Eastern philosophies.[6] According to most Christian theologies, especially the dominating Catholic theology, humans are regarded as sinners starting from the first sin with Adam and Eve. Living a life according to Christian principles in devotion to God may help to redeem from these sins but there is no guarantee. This is because the absolution of sins as well as redemption is in the grace of God, which will be given as a gift to

[6] The Christian doctrine of original sin is, in teaching and structure, very similar to the first Noble Truth that all life is suffering. It would be very interesting and enlightening to study these parallels in more depth.

mankind. This leaves the final decision on the destiny of the Christian follower out of his or her hand. In this aspect one has no full control over one's own future destiny, but is dependent on some higher power or rather on one's faith in this power.

In Eastern philosophies, here mainly Buddhism, the situation is different and this difference is a crucial one. Reaching the ultimate goal of liberation is a logical and somewhat guaranteed consequence of one's practice and efforts (see the above quote from the Satipaṭṭhāna Sutta). The only limitation here is that the goal may not be reached in this life but in later ones. If one compares these two positions one can see that the follower within Buddhism is given a larger amount of control and thus self-determination than in Christianity. This position of an individual as an autonomous, self-determined person is more in accordance with the modern Western humanistic tradition after the Enlightenment than the Christian concept. This can also be seen in the consequences of daily spiritual practice. In the Christian tradition prayer is at the centre. There are different forms of prayer amongst other intercessory, petitioning, gratefulness and contemplative prayer. Maybe the most common is petitioning prayer in which one asks (amongst others) for strength in belief and for the grace of God. In the centre of the Buddhist practice are different forms of meditations and common to all of them is that they are empirical in their orientation. 'Empirical' here is understood as an exploration of the functioning and contents of one's own mind in order to find insights and 'ultimate truth', which can be seen as a self-determinated process.

The Emergence of a New Culture of Mindfulness

Coming back to our comparison of Eastern and Western contexts for meditation and mindfulness practice the differences mentioned so far are summarized in Table 1.

One can see that the contextual embedding is quite different and the crucial question is, whether the old Eastern concept of mindfulness is still the same when placed in such a different environment. Asking mindfulness practitioners in the West most of them would answer "yes". They have the impression that in principle they are practicing the same as the monks in a Theravada Monastery in Sri Lanka or Thailand, maybe only a bit less intense.

Table 1 Comparison of the embedding of mindfulness in traditional Eastern and modern Western contexts

Context	Traditional East	Modern West
Intention	Transformation, liberation, compassion for all beings	Many secular and non-secular ones
Religion	Constitutive	Subordinated
Placement in society	Collective system	Individualized, privatized

I propose that this is not the case. What is actually practiced and what is achieved by a continuous practice is mainly based on the intention and motivation which leads to this practice. A wish for self-regulation or coping with chronic pain is quite different from embarking on a spiritual path to achieve self-transformation. This could be also demonstrated empirically in a study by Shapiro (1992). He interviewed long-term Vipassanā meditators (on average 4.27 years of meditation practice) and asked them, amongst others, for their intentions and motivation when starting meditation. They also had to list the effects meditation had on their life. The results showed that 67% of the meditators "stated positive effects which were congruent with their reasons for beginning" (Shapiro 1992, p. 29f). Or as the author phrases it "what you get is related to what you want" (Shapiro 1992, p. 29). This may not only be true for meditators, but is most likely a reflection of a larger principle stating the intimate relationship between intention and outcome, which can e.g. also be seen in placebo research (see e.g. Moerman and Jonas 2002). Shapiro could also show that for his sample the goals were different in relation to the length of meditation practice. He interprets this as a shift along a continuum from self-regulation via self-exploration to self-liberation. In accordance with this finding he also reports some indications that length of meditation practice is related with religious orientation.

Also in the above mentioned theoretical notion of mindfulness according to Shapiro et al. (2006) one of the three axioms of mindfulness is *intention* or the purpose for practicing mindfulness. This notion shows that the intention is a formative and thus crucial part of any practice in mindfulness.

The last two sections made clear that the intentions for practicing mindfulness in our Western society are manifold and mainly driven by our cultural situation. This brings new elements into the old Eastern techniques, and as *intention* is crucial, the transference from East to West results in a marked change. Also related to this is the fact that a large part of mindfulness practice in the West is conducted in a secularized form, with MBSR being only one of many examples. Another important point in the translation process is language. It turns out that for many of the *Pali* words there is no exact translation into modern languages. Scholars knowledgeable in the field mainly agree that words like *dukkha* (nearest applied translation: suffering, but see also Mikulas 2007) or *dhamma* (nearest applied translation: teachings of the Buddha but also used as justice, law of nature, tradition, mental object and many more) have no direct counterparts in Western languages and can only be grasped indirectly. On the other hand, it is surprising that there is no word for meditation in the *Pali* language. The closest notion here may be *bhāvanā* which translates to contemplation or 'unfolding' (of the mind). And these difficulties are not only known within the Buddhist context. The problem of mixing and also shifting meanings, when e.g. translating the Chinese 'Chi' into the English 'energy', as this is often done, may just serve as another example.

One can see from the above analysis that by the translation process from the ancient East to the modern West the concept of mindfulness was altered through secularisation, translation and cultural context with the latter one mainly expressed through changes in intention and motivation. My proposition here is that these Western forms of mindfulness are different from its Eastern origin. What is currently practiced in our culture is clearly something new which has not been there before. The description of such a practice as an 'Eastern tradition in the West' with the implicit idea that it is just the same, is misleading. Rather a new culture of mindfulness emerges. The specific cultural context of our modern society finds its imprint in the single biographies of the members of our society and this imprint creates in some of them the wish to practice mindfulness (meditation). Eastern technique and Western intentions melt into various mindfulness activities placed in our Western society which form a new movement. Thus, we are not witnessing, as many think, the process of Eastern meditation taking over the West. Rather these techniques are applied in different contexts and with different motivations in a way that changes the results of their practice in a non-trivial way. Mindfulness is becoming a regulating force in many cases, or a counter movement for demands of society which people find ever harder to cope with.

The Transcultural Perspective

The fact that this movement is based on techniques from another culture and another epoch brings a *transcultural* perspective to the practice of mindfulness. Most people starting a regular mindfulness practice are not only interested in the meditation techniques itself. Sooner or later they come into contact with the origins of these methods and start reading about Buddhist philosophy or listen to *dhamma* talks (i.e. Buddhist teachings). Shapiro (1992) proposed that there is an inherent tendency once one has embarked on a meditation practice to develop a more religious orientation after certain time and to shift intentions, although this may not be true for everybody. Indeed, it seems reasonable to assume that while people may have started to meditate in order to find a break in a busy day, and to have a time interval, where they can temporarily slip out of the continuous doing mode of a busy modern life, they may develop a deeper interest in the origin of these techniques after some time. In addition to the wish for a regular break comes now a motivation to explore the own mind or to embark on some form of self-exploration. If Shapiro's results are taken seriously and if we generalize them to a larger group, then we can assume that people who stick with their meditation or mindfulness practice for a longer period may be in a process of changing their intentions in a way which comes gradually closer to the Eastern origin. It can be hypothesized furthermore that with a long-standing meditation practice peculiarities stemming out of one's own individual biography will have a reduced impact on the experience, while at the same time more universal principles of the human condition are moving into the centre of the practice. In this sense, long term meditators socialized in Eastern and Western society are likely to have a more similar practice than novices from these two cultures. Of course, this assumption has so far not been tested empirically, but if it is true, then we can see that the newly emerged Western mindfulness culture is aiming at least in some cases back towards its origin (see also Walsh and Shapiro 2006). This is what can be considered the true transcultural aspect of such a process. The cultural exchange is going into both directions. While the knowledge and wisdom of the ancient East is transformed in the West to a new culture of mindfulness, this new culture reaches in return back to the East. But at the same time the modern Eastern cultures are of course heavily influenced by values and living styles of the Western hemisphere and here e.g. traditional Buddhist cultures are confronted with Western concepts of self, individuality and person.

Thus, we are currently witnessing a process of mutual exchange, assimilation and transformation of spiritual approaches between East and West. It is quite likely that based on these transcultural processes the traditional religious systems will experience some radical reformations and transformations, which are driven by the developments and needs of our modern culture. The growing interest in mindfulness and meditation in the West might just be a first glimpse of such a change.

References

Allmen, Fv. (2007). *Buddhismus (Lehren – Praxis – Meditation)*. Stuttgart: Theseus Verlag.

Analayo. (2004). *Satipatthana: The direct path to realization*. Cambridge: Windhorse.

Baer, R. A., Smith, G. T., & Allen, K. B. (2004). Assessment of mindfulness by report. The Kentucky inventory of mindfulness skills. *Assessment, 11*, 191–206.

Baer, R. A., Smith, G. T., Hopkins, J., Krietemeyer, J., & Toney, L. (2006). Using self-report assessment methods to explore facets of mindfulness. *Assessment, 13*, 27–45.

Bishop, S. R., Lau, M., Shapiro, S., Carlson, L. E., Anderson, N. D., Carmody, J., Segal, Z. V., Abbey, S., Speca, M., Velting, D., & Devins, G. (2004). Mindfulness: A proposed operational definition. *Clinical Psychology: Science and Practice, 11*, 230–241.

Brown, K. W., & Ryan, R. M. (2003). The benefit of being present: Mindfulness and its role in psychological well-being. *Journal of Personality and Social Psychology, 84*, 822–848.

Buchheld, N., & Walach, H. (2004a). Die historischen Wurzeln der Achtsamkeitsmditation – ein Exkurs in Buddhismus und christliche Mystik. In T. Heidenreich & J. Michalak (Eds.), *Achtsamkeit und Akzeptanz in der Psychotherapie* (pp. 25–46). Tübingen: DGVT-Verlag.

Cardaciotto, L. (2005). *Assessing mindfulness: The development of a bi-dimensional measure of awareness and acceptance*. Cambridge: ProQuest Company.

Chadwick, P., Hember, M., Symes, J., Peters, E., Kuipers, E., & Dagnan, D. (2008). Responding mindfully to unpleasant thoughts and images: Reliability and validity of the Southampton mindfulness questionnaire (SMQ). *British Journal of Clinical Psychology, 47*, 451–455.

Creswell, D., Way, B. M., Eisenberger, N. I., & Lieberman, M. D. (2007). Neural correlates of dispositional mindfulness during affect labeling. *Psychosomatic Medicine, 69*, 560–565.

Feldman, G., Hayes, A., Kumar, S., Greeson, J., & Laurenceau, J. P. (2007). Mindfulness and emotion regulation: The development and initial validation of the Cognitive and Affective Mindfulness Scale-Revised (CAMS-R). *Journal of Psychopathology and Behavioral Assessment, 29*, 177–190.

Fuhrman, A., & Kuhl, J. (1998). Maintaining a healthy diet: Effects of personality and self-reward versus self punishment on commitment to and enactment of self-chosen and assigned goals. *Psychology and Health, 13*, 651–686.

Grossman, P. (2008). On measuring mindfulness in psychosomatic and psychological research. *Journal of Psychosomatic Research, 64*, 405–408.

Grossman, P., Niemann, L., Schmidt, S., & Walach, H. (2004). Mindfulness-based stress reduction and health benefits: A meta-analysis. *Journal of Psychosomatic Research, 57*, 35–43.

Huber, S. (2007). Are religious beliefs relevant in daily life? In H. Streib (Ed.), *Religion inside and outside traditional institutions* (pp. 211–230). Leiden: Brill.

Idler, E. L., & George, L. K. (1998). What sociology can help us understand about religion and mental health. In H. G. Koenig (Ed.), *Handbook of religion and mental health* (pp. 51–62). New York: Academic.

Kabat-Zinn, J. (1990). *Full catastrophe living: using the wisdom of your body and mind to face stress, pain, and illness.* New York: Bantam Dell.

Kabat-Zinn, J. (2005). *Coming to our senses. Healing ourselves and the world though mindfulness.* New York: Hyperion.

Langer, E. J. (1989). *Mindfulness.* Reading: Addison-Wesley.

Lau, M. A., Bishop, S. R., Segal, Z. V., Buis, T., Anderson, N. D., Carlson, L., Shapiro, S., Carmody, J., Abbey, S., & Devins, G. (2006). The Toronto mindfulness scale: Development and validation. *Journal of Clinical Psychology, 62*, 1445–1467.

Mikulas, W. L. (2007). Buddhism & Western psychology. *Journal of Consciousness Studies, 14*, 4–49.

Moerman, D. E., & Jonas, W. B. (2002). Deconstructing the placebo effect and finding the meaning response. *Annals of Internal Medicine, 136*, 471–476.

Monti, D. A., Peterson, C., Kunkel, E. J., Hauck, W. W., Pequignot, E., Rhodes, L., & Brainard, G. C. (2006). A randomized, controlled trial of mindfulness-based art therapy (MBAT) for women with cancer. *Psycho-Oncology, 15*, 363–373.

Nattier, J. (1995). Visible and invisible: Jan Nattier on the politics of representation in Buddhist America. *Tricycle, 5*, 42–49.

Nyanaponika, T. (1983). *The heart of Buddhist meditation.* London: Rider.

Passmore, J., & Marianetti, O. (2007). The role of mindfulness in coaching. *The Coaching Psychologist, 3*, 131–137.

Rosenberg, L., & Guy, D. (2004). *Breath by breath. The liberating practice of insight meditation.* Boston: Shambala Publications.

Salzberg, S. (2008). Interview with Sharon Salzberg. In R. Shankman (Ed.), *The experience of samadhi. An in-depth exploration of Buddhist meditation* (pp. 130–135). Boston: Shambala.

Shapiro, D. H. (1992). A preliminary study of long-term meditators: Goals, effects, religious orientation, cognitions. *Journal of Transpersonal Psychology, 24*, 23–39.

Shapiro, S. L., & Schwartz, G. E. (1999). Intentional systemic mindfulness: An integrative model for self-regulation and health. *Advances in Mind-Body Medicine, 15*, 128–134.

Shapiro, S. L., Carlson, L. E., Astin, J. A., & Freedman, B. (2006). Mechanisms of mindfulness. *Journal of Clinical Psychology, 62*, 373–386.

Walach, H., Buchheld, N., Buttenmüller, V., Kleinknecht, N., & Schmidt, S. (2006). Measuring mindfulness – the Freiburg Mindfulness Inventory (FMI). *Personality and Individual Differences, 40*, 1543–1555.

Walsh, R., & Shapiro, S. L. (2006). The meeting of meditative disciplines and Western psychology: A mutually enriching dialogue. *American Psychologist, 61*, 227–239.

Wetzel, S. (2004). *Arbeit & Muße. Leben & Sinn. Thesen und Übungen. Eine Hommage an Hannah Arendt.* Berlin: edition tara libre.

Setting Our Own Terms: How We Used Ritual to Become Human

Matt J. Rossano

Abstract Archeological evidence of the sophisticated cognitive attributes thought to define humanity – such as symbolism, language, theory of mind, and a spiritual sense – is, by and large, late-emerging (after 50,000 years before present [ybp]), postdating the emergence of anatomically modern humans (AMH). This suggests that the relevant selection pressures for these abilities did not emerge until after the arrival of the fully human body and brain. I argue that this stands to reason to reason since the selection pressure responsible for the emergence of uniquely human cognition was human-made. Human culture created human cognition. The key facet of that culture was ritual. Ritual selection pressure filtered *Homo sapiens sapiens* for the very cognitive attributes that made us what we are today.

100,000 Years Before Present

Imagine you were transported back 100,000 years ago and happened upon a group of our ancestors. Would you be surprised to see them gathered before a blazing campfire – singing, chanting, dancing or simply sitting transfixed before the flames as an elder told a tale? So natural are these "campfire" activities that we may miss their potential evolutionary significance. Only humans gather communally to engage in ritualized activities that required focused attention. To one degree or another, these ritual activities incorporate many of the same cognitive, behavioral, and spiritual elements commonly found in meditative practices today. Often they involve an altered state of consciousness producing important psycho/physical health effects. The thesis of this chapter is that ritual activities of this type were not merely

M.J. Rossano (✉)
Department of Psychology, Southeastern Louisiana University, Hammond, LA, USA
e-mail: mrossano@selu.edu

H. Walach et al. (eds.), *Neuroscience, Consciousness and Spirituality*,
Studies in Neuroscience, Consciousness and Spirituality 1,
DOI 10.1007/978-94-007-2079-4_3,

incidental to human evolution – they were fundamental to it. Rituals of focused attention created the selective environment from which uniquely human cognition emerged.

Uniquely Human Cognition

What is uniquely human cognition? While many cognitive traits including language, recursion, symbolism, and theory of mind have been proposed as the 'human Rubicon', it appears that no single attribute categorically defines human cognition. Instead, humans seem to have all these attributes to a greater extent or in a more sophisticated form compared to other species. Thus, the critical question changes to one of identifying the necessary prerequisite(s) for developing these cognitive attributes to the unprecedented level of complexity seen only in our species. On this question, the answer seems to be working memory capacity.

Coolidge and Wynn (Wynn and Coolidge 2007) have built a compelling case that the emergence of uniquely human cognition resulted from a slight but significant increase in working memory capacity. This increase made anatomically modern humans (AMH) better able to hold information in mind, especially information about behavioral procedures and intended goals, in spite of competing signals or response competition (Kane and Engle 2002). Thus, when confronting cognitive challenges, AMH were better equipped to resist mental sets and other prior habits of thought and behavior. This ability was essential for exploring novel relationships, engaging in cognitive innovation, and ultimately creating and using symbols (Wynn and Coolidge 2007). It may also have provided the foundation for the uniquely human capacity for complex culture and theory of mind (Tomasello et al. 2005).

Why Us?

Saying that working memory capacity lies at the heart of modern cognition simply pushes the question of origins back another level. What selection pressure produced enhanced working memory capacity exclusively in our ancestors? This question is an especially vexing one given that very little in the archeological record distinguishes AMH from other archaic hominid forms prior to the Upper Paleolithic. Tool kits are largely comparable, and both Neanderthals and AMH collected natural pigments, built fires, and engaged in large mammal hunting. In fact, recent studies have shown that Neanderthals were highly skilled hunters and foragers, whose abilities compared favorably with Cro-Magnons and contemporary hunter-gatherers (Adler et al. 2006; Sorensen and Leonard 2001).

Thus, it is hard to argue that the cognitive demands of hunting, tool-making, or of survival in harsh climates differentiated *Homo sapiens* from Neanderthals. If these activities created selection pressure for enhanced working memory capacity and symbolism then these traits would have arisen in Neanderthals as well. But what if the difference was cultural? Unique facets of human culture might have specifically targeted focused attention and working memory. This could also explain the rapidity with which modern cognition emerged. Culture is known to produce rapid evolutionary change – the emergence of lactose tolerance in adults, for example, became widespread among those populations who adopted dairying in matter of only a few thousand years (Burger et al. 2007). Part of reason for this rapid change is that culture can often produce Baldwinian selective forces (to be discussed in more detail later) that serve to augment natural selective forces.

A Tough Neighborhood

The world into which AMH arrived was an unforgiving one indeed. Ice-core data indicate that rapid climate changes, sometimes occurring within decades, were not uncommon from about 100,000 to 10,000 ybp (Alley 2000, pp. 118–126). These shifts would have produced periods of drought and deprivation, stressing our ancestors' survival capacity to the limit. Furthermore, around 70,000 ybp a massive volcanic eruption on Sumatra Island may have further exacerbated already arduous conditions (Ambrose 1998a). While the impact of this eruption (the Mt. Toba eruption) is debated (Petraglia et al. 2007), genetic evidence confirms that humanity passed through a population bottleneck at this time, with numbers dropping to near extinction levels (Ambrose 1998a; Behar et al. 2008). Those of our ancestors who managed to survive were those who derived a 'social solution' to recurrent patterns of resource stress.

The Social Solution to Resource Stress

The !Kung San of southern Africa are traditional hunter-gatherers living in the harsh habitat of the Kalahari desert. (Note: the "!" refers to the "click" sound in the San's native language.) Critical to their success is a system of inter-group gift exchange called hxaro. This exchange helps to build a relationship of trust and cooperation among different bands, producing further exchanges of material goods and vital bits of information such as where game or water were last cited. Body ornaments, such as shell beads worn as necklaces, are commonly exchanged gifts in hxaro (Weissner 1982).

The first evidence of shell beads dates back to about 100,000 ybp or slightly older (Vanhaeren et al. 2006). Beads from Blombos Cave, South Africa and Oued Djebbana, Algeria, have been dated to around 75,000 ybp (Henshilwood et al. 2004; Vanhaeren et al. 2006), while those from Enkapune Ya Muto, Kenya are more recent (about 40,000 ybp; Ambrose 1998b). Thus from about 100,000 to 40,000 ybp, AMH were making shell beads suitable for purposes similar to the hxaro practice of the !Kung. The fact that each find is composed of beads of a single type suggests that a particular value was associated with them supporting the notion that they were used as gifts. Furthermore, many shells were found at sites remote from their sea-shore origin suggesting that they were transported there, possible by trade networks.

Along with beads, evidence of expanded trade networks in Africa can also be found in the Howiesons Poort and Still Bay (South Africa) tool industries, which are dated to around 60–70,000 ybp. These industries contain fine-grained microliths made from non-local, 'exotic' raw materials. An analysis by Ambrose and Lorenz (Ambrose and Lorenz 1990) supports the notion that these industries may very well represent the emergence of inter-group exchange networks in response to increasing resource scarcity. This process began what Ambrose (Ambrose 2002, p. 22) has termed the 'troop to tribe' transition in human evolution. Increasingly, survival meant interacting with groups on the perimeter of one's range; groups often composed of more distantly related kin and outright strangers. While these expanding alliances provided access to more widely dispersed resources, they also stressed social/cognitive and communicative capacities. Evidence of trading networks and generally greater social complexity are also present in the Upper Paleolithic archeological record for Cro-Magnons, but not contemporaneous Neanderthals (Bar-Yosef 2000; Feblot-Augustins 1999; Gamble 1999; Hayden 2003; Hoffecker 2002).

Good Rituals Make Good Friends

In response to rapid climate changes (possibly accentuated by the Toba eruption) and the ensuing resource stresses they entailed, our ancestors did something that no other hominid replicated – they created wide-ranging inter-group reciprocal trade alliances. Increasingly, the social world of our ancestors was expanding to incorporate regular interactions with more distantly related kin and outright strangers. But how does one go about forging relationships with wary outsiders? The answer is one with a deep evolutionary history: ritual.

Ritual behavior is widespread across the animal kingdom, especially where cautious communication is required (de Waal 1990; Guthrie 2005, p. 68). In this context, ritual is defined as an attention-getting, formalized, and invariantly ordered sequence of behaviors designed to convey a particular meaning (Bell 1997, pp. 138–169; Rappaport 1999, p. 24). For example, a common ritual used for social bonding among male baboons is called 'scrotum-grasping'. Two males wishing to signal friendship will momentarily allow each other to hold their testicles (Smuts and Watanabe 1990; Whitham and Maestripieri 2003). This ritual is especially effective given that grabbing and ripping at the genitals is common when primates fight. Thus, the 'scrotum-grasp' can be understood as a ritualized version of this fighting action. However, the 'scrotum-grasp' is a formalized or more restricted form of the action (i.e. a momentary grasp rather than aggressive 'grabbing and ripping'). The act itself is undoubtedly attention-getting (it's hard to ignore someone handling your genitals); and it follows a rule-governed, relatively invariant sequence: While making affiliative gestures such as lip-smacking and flattening of the ears, one baboon strides up to another using a rapid, straight-legged gait. The other responds in like fashion, and they each present their hind-quarters for a quick genital squeeze. It is noteworthy that male alliances seem to require successful execution of this ritual. Younger males usually fail to complete this greeting ritual and are less likely to form social alliances than older males (Smuts and Watanabe 1990).

The power of ritual is located in its ability to direct attention away from pre-potent defensive responses so that social interactions can be extended, thus allowing social bonding emotions an opportunity to take hold. For example, dominant female monkeys use certain vocalizations, grunts and gurneys, when approaching subordinates to signal them of non-threatening intentions, forestalling the subordinate's natural tendency to flee (Silk 2001). A successful approach can lead to another common social ritual among primates: grooming. Grooming causes the release of endogenous brain opiates helping to bring about a mental state conducive to affiliation (Keverne et al. 1989). In this sequence then, one can see how successfully executed ritual can focus attention on a relevant signal (the approach grunt), inhibit defensive emotions (fright in the subordinate) and allow time for social bonding emotions (associated with grooming) to operate.

As highly social creatures, primates possess a wealth of social rituals designed to build trust, promote group harmony, and reinforce social relations (de Waal 1990; Goodall 1986; van Roosmalen and Klein 1988, p. 515). For example, when chimpanzee, bonobo, and spider monkey foraging parties reunite, they engage in ritualized acts of welcoming and social re-affirmation including mutual embracing, kissing, group pant-hooting, and grooming. Gelada baboons use rhythmic back-and-forth approach vocalizations to signal benign intent during close-quarter feeding sessions. These vocalizations allow two baboons to peacefully feed near one another without threat (Richman 1987). Finally among chimpanzees, reconciliation between combatants is signaled by submissive bows, plaintiff vocalizations, and the hand-out begging gesture (on the part of the loser) followed by embraces and kisses (from the winner; de Waal 1990). The many social rituals present among our primate cousins indicates that our hominid ancestors were pre-adapted for using ritual as a means of social bonding and could call upon a rich repertoire of them in their everyday social lives.

Ritual Behavior and Working Memory

Recent neuroscience research has linked ritual behaviour and working memory, especially in the context of the inhibition of pre-potent responses. Areas of the frontal lobe, especially the dorsolateral prefrontal cortex (dlPFC) and anterior cingulate cortex (ACC) are central to working memory, focused attention, and the ability to direct willful actions (Curtis and D'Esposito 2003; Ingvar 1994; Kelly et al. 2006). Studies specifically addressing the issue of inhibitory control have also implicated the dlFPC and ACC (Cunningham et al. 2004; Knoch and Fehr 2007).

For example, Beauregard, Levesque, and Bourgouin (Beauregard et al. 2001) monitored brain activity while subjects viewed erotic films. Not unexpectedly, they found that the films increased activity in areas of the brain known to be associated with sexual arousal such as the amygdala and hypothalamus. However, subjects who were given specific instructions to inhibit any sexual response were found to have no increased activity in these subcortical regions but significantly increased activity in the dlPFC and ACC. This was interpreted as an example of top-down inhibitory control over a naturally elicited response. These studies are consistent with a growing body of neuroscience and neuropsychological literature showing that the dlFPC (especially on the right side) is critical to the ability to filter out competing signals, inhibit immediate emotion-based responses, and exercise conscious self-control (Knoch and Fehr 2007; Sanfey et al. 2003; Stuss et al. 2002).

Further research (Kelly et al. 2006) has found that increases in working memory capacity allow greater resources to be dedicated to inhibitory processes. This improves the efficacy of those processes making it more likely that the subject will be capable of maintaining attentional focus on current task demands. The effect of practice therefore is to increasingly automatize the controlled aspects of a task, freeing up more cognitive resources for inhibitory control.

Ritual behavior directly relates to the willful direction of action and the suppression of pre-potent responses. As mentioned earlier, ritual's attention-directing quality makes it effective in focusing attention on a selected signal while directing attention away from defensive reactions (recall the approach grunts of dominant monkeys). Furthermore, the repetitive elements of ritual provide opportunities for practice effects, whereby working memory capacity can be 'freed up' for greater inhibitory control.

In our ancestral past the ability to exert inhibitory control would very likely have been stressed to unprecedented levels. Evidence from traditional societies indicates that social rituals designed to build group solidarity and establish inter-group alliances are extremely demanding in terms of self-control, focused attention, and inhibition of pre-potent responses. Furthermore, those who successfully complete these demanding rituals tend to gain fitness advantages in the form of greater access to resources (via reciprocal arrangements), enhanced status, and psychophysical health benefits.

Social Rituals Among Traditional Societies

Across a range of traditional societies, three types of social rituals are common for enhancing within-group social cohesion and building between-group alliances: rituals of trust-building and reconciliation, rituals of initiation, and shamanistic rituals of community and individual healing. The degree to which these rituals can be unalterably projected into our ancestral past is unclear. However, they provide the best starting point for understanding past rituals, and a consistent feature of them is physical and psychological rigor.

Rituals of Trust-Building and Reconciliation

As inter-group interactions became more frequent in the late Pleistocene, rituals for establishing inter-group trust, while maintaining intra-group cohesion, very likely rose in salience and importance. Examples of these rituals from traditional societies show that they frequently 'ritualize' the expression of the explosive emotions that must be contained if trust and reconciliation are to be achieved. By exhibiting these dangerous emotions in ritual form while controlling their effects, participants signal their willingness and ability to let longer-term group-level interests direct their actions rather than short-sighted, self-interested inclinations.

Disputes among the Ammassalik of Greenland are often addressed using a traditional 'drum match,' where the aggrieved parties drum and sing about how the other has injured them (Mirsky 1937). Tradition governs nearly every element of the match including the tone, expression, and movement of the participants. This, however, does not eliminate the tension inherent in the ritual. As they face, the singer uses mocking tones to detail the other's personal and familial faults. Even as the confrontation escalates with the singer occasionally butting heads with the listener, the listener remains frustratingly indifferent to the singer's taunts and accusations. When the singer is done, the roles reverse. Matches are rarely settled in one round, but may be continued for months or years.

An even 'edgier' example is the peace-making ritual of the Yanamamo, a traditional people of Amazonian jungle (Chagnon 1968). The party requesting a truce invites its enemies to a ceremonial feast. As their adversaries arrive, the host warriors recline unarmed in hammocks. With weapons drawn, the 'guests' taunt their hosts with insults and intimidating gestures. But the hosts remain calm and unaffected by the threats. In time, hosts and guests trade places and the threats and insults begin anew. Only when each are satisfied as to the other's peaceful intentions does the feast begin, which includes the exchanging of gifts, the forging of new alliances and the arranging of marriages.

Our late Paleolithic ancestors' rituals of trust building and reconciliation may not have been as elaborate as these. However, even the most mundane ritual of this type requires some level of self-control. A handshake is only modestly removed from a swinging fist. Those of our ancestors unable to inhibit their aggressive or defensive inclinations long enough to allow for ritual-based trust and reconciliation to take hold very likely found themselves social outcasts, separated from the reciprocal benefits of within and between group alliances.

Rituals of Initiation

Adolescent rites of passage occur in over 70% of traditional societies studied (Alcorta 2006; Lutkehaus and Roscoe 1995).The severity of these initiations varies and tends to increase where ecological or external threats are greater (Hayden 2003, pp. 104–105; Sosis 2006, p. 82) Among aboriginal societies in Australia, for example, the most severe initiation rites are found among tribes living in the driest, harshest conditions.

Rapid climate changes very likely magnified by the effects of the Toba eruption may very well have placed unprecedented resource and social stress on ancestral groups. Maintaining social stability during these periods and establishing inter-group relations would have undoubtedly led to heightened social tensions. Though neighboring groups would have been essential for material trade and information exchange, the xenophobic nature of humans in general and of tight-knit traditional societies in particular, would have made these interactions a constant source of tension and unease (Richerson and Boyd 2001). Group interactions almost always entail an elevated degree of group competition. Thus, it is not unreasonable to conclude that initiation ceremonies may have either arisen or intensified in the late Pleistocene as the social world became more complex.

Adolescent rites of passage can be trying events, often requiring a young person to endure isolation, deprivation, physical pain, and psychological stress. For example, female initiation ceremonies among many traditional societies in southern Africa involve forced seclusion, bloodletting, genital cutting, and rigorous training in ceremonial dances (Knight et al. 1995; Power 1998). Deprivation, beatings, exhaustive physical exertion, exposure to harsh elements, genital mutilation, ritual scarring, tooth removal, and forced dancing and chanting are among the torturous trials included in many male initiation ceremonies among Australian aborigines, native Americans, New Guinea tribes, Pacific Islanders, and many African tribes (Catlin 1867; Glucklich 2001; Whitehouse 1996). Possibly the most dramatic of these initiations was the famous Mandan (Native American) Sun Dance ceremony where new warriors were suspended from the top beam of a large ceremonial enclosure with ropes attached to skewers embedded in their chests. They might remain there for hours or days as dancing and chanting went on below them.

With regard to modern cognition, the important point is that the capacity to endure such rituals required a degree of mental control over reflexive responses that only humans have mastered. It is hard to know how severe our Pleistocene ancestors' earliest initiation rituals may have been. But current ethnographic models indicate that to some degree they would have required initiates to inhibit natural pre-potent responses in order to signal their commitment to the tribe. Furthermore, those initiates best equipped to pass these tests very likely achieved higher status within the tribe and with it greater reproductive success.

Shamanistic Healing Rituals

There is considerable evidence that shamanism (broadly defined) is humanity's oldest form of religion (Guenther 1999; Lee and Daly 1999; Winkelman 1990). It is ubiquitous, found in nearly all traditional societies (Townsend 1999; Vitebsky 2000). Furthermore, many paleoanthropologists argue that some Upper Paleolithic cave art and artifacts reflect shamanistic rituals and/or experiences (Eliade 1972; Hayden 2003; Lewis-Williams 2002; Winkelman 2002). These two qualities, ubiquity and antiquity, suggest that the roots of shamanism run deep in human history. Two recent finds suggest that shamanism may actually pre-date the Upper Paleolithic.

A stone slab dated to around 35,000 ybp from Fumane cave in Italy appears to depict a human form with the antlered headgear typical of a shaman (Balter 2000). In 2006, archeologists discovered a ritually-modified snake-rock, dated to around 70,000 ybp in a deep cave site in the Tsodilo Hills of Botswana (Minkel 2006). The setting, along with the intentional enhancements to rock's exterior strongly suggested use of the site for the consciousness altering rituals associated with shamanism.

The shaman serves as the community's spiritual emissary using ritually-induced trance to commune and communicate with supernatural powers in order to cure illness, manipulate natural forces, and reduce suffering and social strife. As the spirit world's messenger, the shaman plays a critical role in binding supernatural authority to social norms, thereby strengthening community and discouraging deviance. Shamanistic rituals typically involve sensory deprivation, the ingestion of psychoactive substances, rhythmic drumming, dancing, and chanting often by hypnotic firelight, all designed to produce an altered state of consciousness. These conditions typically evoke intense emotions and the release of brain opiates both of which can have powerful social bonding effects (Frecska and Kulscar 1989).

Along with its role in strengthening social cohesion, McClenon (2002) argues that in our evolutionary past, shamanism would have been our ancestors' primary means of healing. Research confirms that ritual healing practices involving altered states of consciousness can be effective for maladies where a significant psychological factor is present (see section below "Ritual Healing Theory"). The Kalahari !Kung conduct 'healing dances' about every 2 weeks, where shaman healers dance about frenetically, laying hands on and transmitting "healing power" to all present. These dances are considered essential to the health and vitality of the !Kung, both individually and as a community (Katz 1982).

It is not hard to imagine our late Pleistocene ancestors engaging in similar rituals around a blazing campfire. At times these rituals may only have involved group chanting, dancing, or hypnotic silence before the flames (the benefits of which should not be casually dismissed). At other times they may have involved intensely dramatic shamanistic ceremonies where soul flight, supernatural encounters, and 'miraculous' healings took place. Shamanistic healing rituals such as those of the !Kung, always involve techniques designed to bring about a health-enhancing altered state of consciousness. In our ancestral past those most able to achieve this state would have had a selective advantage over others by virtue of its positive physical and psychological effects.

Shamanistic Healing: Ritual Healing Theory

Traditional healing practices involving shamanistic rituals and altered states of consciousness are ubiquitous among traditional societies. Shamanistic healing rituals may have been especially important in selecting for the enhanced working memory capacity necessary for modern cognition. McClenon (1997, 2002) has marshaled considerable evidence indicating that those of our ancestors who were most susceptible to the beneficial physical and psychological effects of shamanistic rituals had a selective advantage over others in surviving illness or injury, overcoming debilitating emotional states, and enduring the rigors of childbirth. This 'ritual healing' theory is based on a number of converging lines of evidence, including: (McClenon 2002 or Rossano 2010, Chap. 4 for references)

1. The universality (or near universality) of ritual healing practices across traditional societies.
2. The fact that ritual healing always involves hypnotic processes and altered states of consciousness.
3. Evidence showing that hypnotizability or the ability to achieve a mental state highly prone to suggestion is measurable, variable, and has heritable components.
4. **The finding that ritual healing is often highly effective for a range of maladies** where psychological factors are involved such as chronic pain, burns, bleeding, headaches, skin disorders, gastrointestinal disorders, and the discomforts and complications of childbirth. Relatedly, the fact that meditative practices affect levels of beta endorphins, serotonin, and melatonin, all of which are implicated in immune system function, pain reduction, and subjective well-being.
5. The evidence from comparative and archeological studies indicating the existence of ritual, altered states of consciousness, and care of the sick among our primate cousins and hominid ancestors.
6. The fact that the earliest medical texts (from Mesopotamian and Egyptian civilizations) closely connect healing with religious ritual.
7. **The finding that anomalous events associated with ritual, such as 'miraculous'** healing, are effective in inducing supernatural beliefs. Thus, healing rituals would have reinforced supernatural beliefs among our ancestors and encouraged their expansion.

Meditative practices have also recently been found to promote the anti-**inflammatory response responsible for restoring immune system balance; thereby** reducing symptoms of autoimmune disorders such as rheumatoid arthritis, colitis, and septic shock (Peng et al. 2004; van Dixhoorn and White 2005; see Tracey 2007 for review).

What this evidence indicates is that ritual healing can be effective health care. Thus, shamanistic healing rituals would have disproportionally enhanced the health of those whose brains permitted the deepest immersion in the rituals. What type of brain would this have been? The next section reviews evidence indicating that it would have been a brain with increased working memory capacity.

Shamanism, Neuroscience, and Working Memory

Shamanistic rituals and working memory are connected in that the techniques used for altering consciousness are known to activate areas of the brain associated with working memory and focused attention. Recent neuroimaging and EEG studies examining a wide range of meditative practices show consistent activation in the dlPFC and ACC (Azari et al. 2001; Beauregard and Paquette 2006; Lazar et al. 2000, 2005; Lou et al. 1999; Newberg et al. 2001), Furthermore, meditative practices producing altered states can lead to long term changes in both brain structure and attentional capacities, and they typically increase levels of arginine vasopressin (AVP) which has important functions in learning and memory (Carter et al. 2005; Lazar et al. 2005; Newberg 2006).

When our ancestors were engaging in rituals around a campfire, focusing their attention on the flames or chanting a repeated phrase to the incessant rhythm of a pounding drum, they were very likely taxing the very brain areas involved in attention and working memory. Those whose brains were most 'ritually capable' by virtue of increased working memory and attentional control capacity would also have been the ones to reap the greatest health and fitness benefits. Furthermore, unlike hunting, tool-making or other cognitively challenging activities, rituals around campfires could have easily involved children, thus affecting their brain ontogeny.

The Fortuitous Mutation

To construct and sustain an increasingly complex social world, our ancestors would have needed ever-more demanding social rituals. These rituals would have taxed working memory providing a fitness advantage to those with greater working memory capacity and attentional control. To become an advantageous heritable trait, enhanced working memory must reduce to a genetic change passed from parent to offspring. Stanford archeologist Richard Klein has argued that ultimately the difference between us and other hominids came down to a fortuitous genetic mutation that reorganized brain structure and function resulting in a critical cognitive advantage – greater working memory capacity (Klein and Edgar 2002).

As an explanatory mechanism, a fortuitous mutation would seem to require no deeper causal force. Mutations, it has generally been thought, are more or less inevitable and largely random. However, recent work in evolutionary developmental biology has provided a clearer framework for understanding how 'random' genetic mutations are translated into non-random phenotypic variations (Jablonka and Lamb 1995; Kirschner and Gerhart 2005). This work shows that mutations may be far less random than originally thought and that evolved developmental processes place constraints on how genetic mutations get expressed in the phenotype. The type of selection process whereby the probability of adaptive phenotypic changes closely tailored to selective conditions is enhanced is called Baldwinian selection. There are good reasons to suspect that Baldwinian processes were at work in the emergence of uniquely human cognition.

A Baldwinian Process

The Baldwin effect provides a non-Lamarckian way for acquired characteristics to become genetically heritable. If an acquired trait (brought about either by learning or physiological adaptation) provides a fitness advantage, then any genetic disposition that makes that trait easier to acquire can be positively selected. Over time, little or no environmental exposure may be required for the trait to emerge, indicating that it has become genetically encoded (Weber and Depew 2003).

The classic example of this was provided by Waddington (Waddington 1942), who exposed pupal fruit flies (*Drosophila melanogaster*) to heat shock. Some of the heat-shocked pupae later developed into flies without the typical cross-vein pattern on their wings. Waddington bred the no-cross-vein flies and once again exposed their pupal offspring to heat shock. After successive breedings, Waddington found that the no-cross-vein trait would emerge in nearly 100% of the offspring even in the absence of heat shock. In other words, an initially environmentally induced trait (no-cross-veins) eventually became encoded and transmitted genetically.

Research with human raised apes has shown that when atypical environmental demands are present, novel cognitive capacities such as symbolism, syntax, and deferred imitation can arise (Bjorklund and Rosenberg 2005; Tomasello and Call 1997). This could provide a potential model for how hominids acquired increasingly complex cognitive skills. These skills may first have appeared as novel acquired traits induced by atypical environmental demands. Then, as those demands persisted, a Baldwinian process could have led to the traits becoming genetically heritable and stabilized. Interestingly, over the course of hominid evolution, the atypical environmental demands were increasingly products of hominids themselves.

Jablonka and Lamb (Jablonka and Lamb 2005, pp. 158–176) have recently provided a specific mechanism for how Baldwinian effects may occur. They cite considerable research indicating that a substantial amount of genetic variation in a developing phenotype becomes masked as selection 'canalizes' ontogenesis down particular adaptive pathways. Belyaev's (Belyaev 1979) work on the domestication of silver foxes provides an example of this.

Belyaev found that as he selected foxes for tameness, a variety of other phenotypic variation emerged including: the length of the reproductive season, the droopiness of the ears, the lengths of tails and legs, the spotting on the fur, and even the shape of the skull. For Belayaev this cluster of variability emerged too quickly to be solely the result of genetic mutations. Instead, echoing Waddington (Waddington 1942), he argued that they resulted from epigenetic changes brought on by environmental stress. By this, he was referring to changes in gene regulation – how the effects of genes are switched on and off during the course of development. Thus, (put crudely) genes that had previously been 'turned off' in the foxes were being 'turned on' and vice versa. An example of this in humans might be the presence of atavisms (such as tails or webbed feet). The genes coding for this information are still present but the effects have been turned off over the course of our evolution. Thus, the normal course of human development (i.e. the canalized ontogenetic pathway) producing the typical human phenotype masks this genetic variability – presumably certain environmental stressors could (and occasionally do) unmask it.

According to Jablonka and Lamb (Jablonka and Lamb 2005) environmental stressors can push development off of its canalized pathway revealing previously masked genetic variation on which selection can operate. Any mutation that facilitates ontogenetic canalization to a more fit phenotype would then also be positively selected. Through this process, what initially emerged as a somatic change can become genetically heritable.

Jablonka and Lamb's model raises the possibility that about 70,000 years ago, increasingly demanding social rituals disturbed human ontogenesis enough to throw it off its canalized pathway, revealing new variance in working memory capacity. Those with greater working memory capacity accrued a fitness advantage by virtue of greater access to resources (through reciprocal social arrangements constructed via rituals of peacemaking and reconciliation), enhanced social status (through better performance in initiation rituals), and psychophysical health benefits (from shamanistic healing rituals). Modern cognition emerged as a byproduct of the mental requirements for successful ritual performance – sustaining attentional focus, inhibiting pre-potent responses and retaining a calm equanimity in the face of distracting, even threatening, signals. In short, ritual made us human.

References

Adler, D.S., Bar-Oz, G., Belfer-Cohen, A., & Bar-Yosef, O. (2006). Ahead of the game: Middle and Upper Paleolithic hunting behaviors in the southern Caucasus. *Current Anthropology, 47*, 89–118.

Alcorta, C.S. (2006). Religion and the life course: Is adolescence an "experience expectant" period for religious transmission? In P. McNamara (Ed.), *Where God and science meet* (Vol. 2, pp. 55–79). Bridgeport: Praeger Publishing.

Alley, R.B. (2000). *The two-mile time machine: Ice cores, abrupt climate change and our future*. Princeton: Princeton University Press.

Ambrose, S.H. (1998a). Late Pleistocene population bottlenecks, volcanic winter, and the differentiation of modern humans. *Journal of Human Evolution, 34*, 623–651.

Ambrose, S.H. (1998b). Chronology of the later stone age and food production in East Africa. *Journal of Archaeological Science, 25*, 377–392.

Ambrose, S.H. (2002). Small things remembered: Origins of early microlithic industries in Subsaharan Africa. In R. Elston & S. Kuhn (Eds.), *Thinking small: Global perspectives on microlithic technologies* (pp. 9–29). Arlington: American Anthropological Association. Archaeological papers of the American Anthropological Association #12, Washington, DC.

Ambrose, S.H., & Lorenz, C.G. (1990). Social and ecological models for the Middle Stone Age in Southern Africa. In P. Mellars (Ed.), *The emergence of modern humans* (pp. 3–33). Edinburgh: University of Edinburgh Press.

Azari, N.P., Nickel, J.P., Wunderlich, G., Niedeggen, M., Hefter, H., Tellmann, L., et al. (2001). Neural correlates of religious experience. *European Journal of Neuroscience, 13*, 1649–1652.

Balter, M. (2000). Paintings in Italian cave may be oldest yet. *Science, 290*, 419–421.

Bar-Yosef, O. (2000). A Mediterranean perspective on the Middle/Upper Paleolithic Revolution. In C.B. Stringer, R.N.E. Barton, & J.C. Finlayson (Eds.), *Neanderthals on the edge* (pp. 9–18). Oxford: Oxbow.

Beauregard, M., & Paquette, V. (2006). Neural correlates of mystical experience in Carmelite nuns. *Neuroscience Letters, 405*, 186–190.
Beauregard, M., Levesque, J., & Bourgouin, P. (2001). Neurocorrelates of conscious self-regulation of emotion. *Journal of Neuroscience, 21*(RC 165), 1–6.
Behar, D., Villems, M.R., Soodyall, H., Blue-Smith, J., Pereira, L., et al. (2008). The dawn of human matrilineal diversity. *The American Journal of Human Genetics, 82*, 1130–1140.
Bell, C. (1997). *Ritual: Perspectives and dimensions*. Oxford: Oxford University Press.
Belyaev, D.K. (1979). Destabilizing selection as a factor in domestication. *Journal of Heredity, 70*, 301–308.
Bjorklund, D.F., & Rosenberg, J. S. (2005). The role of developmental plasticity in the evolution of human cognition. In B. J. Ellis & D. F. Bjorklund (Eds.), *Origins of the social mind* (pp. 45–75). New York: Guilford.
Burger, J., Kirchner, M., Bramanti, B., Haak, W., & Thomas, M.G. (2007). Absence of the lactase-persistence-associated allele in early Neolithic Europeans. *Proceedings of the National Academy of Sciences, 104*, 3736–3741.
Carter, O.L., Presti, D.E., Callistemon, C., Ungerer, Y., Lui, G.B., & Pettigrew, J.D. (2005). Meditation alters perceptual rivalry in Buddhist monks. *Current Biology, 15*, R412.
Catlin, G. (1867). *O-kee-pa: A religious ceremony and other customs of the Mandans*. London: Trubner.
Chagnon, N. (1968). *Yanamamo: The fierce people*. New York: Holt, Rinehart, Winston.
Cunningham, W.A., Johnson, M.K., Raye, C.L., Gatenby, J.C., Gore, J.C., & Banaji, M. (2004). Separable neural components in the processing of black and white faces. *Psychological Science, 15*, 806–813.
Curtis, C.E., & D'Esposito, M. (2003). Persistent activity in the prefrontal cortex during working memory. *Trends in Cognitive Science, 7*, 415–423.
de Waal, F.B.M. (1990). *Peacemaking among primates*. Cambridge: Harvard University Press.
Eliade, M. (1972). *Shamanism: Archaic techniques of ecstasy*. London: Routledge and Kegan Paul.
Feblot-Augustins, J. (1999). Raw material transport patterns and settlement systems in the European Lower and Middle Paleolithic: Continuity, change, and variability. In W. Roebroeks & C. Gamble (Eds.), *The Middle Paleolithic occupation of Europe* (pp. 193–214). Leiden: University of Leiden Press.
Frecska, E., & Kulscar, Z. (1989). Social bonding in the modulation of the physiology of ritual dance. *Ethos, 17*, 70–87.
Gamble, C. (1999). *The Upper Paleolithic societies of Europe*. Cambridge: Cambridge University Press.
Glucklich, A. (2001). *Sacred pain*. New York: Oxford University Press.
Goodall, J. (1986). *The chimpanzees of Gombe*. Cambridge: Harvard University Press.
Guenther, M. (1999). From totemism to shamanism: Hunter–gatherer contributions to world mythology and spirituality. In R.B. Lee & R. Daly (Eds.), *Cambridge encyclopedia of hunters and gatherers* (pp. 426–433). Cambridge: Cambridge University Press.
Guthrie, R.D. (2005). *The nature of Paleolithic art*. Chicago: University of Chicago Press.
Hayden, B. (2003). *Shamans, sorcerers and saints: A prehistory of religion*. Washington, DC: Smithsonian Institution Books.
Henshilwood, C., d'Errico, F., Vanhaeren, M., van Niekerk, K., & Jacobs, Z. (2004). Middle Stone Age shell beads from South Africa. *Nature, 304*, 404.
Hoffecker, J.F. (2002). *Desolate landscapes: Ice-age settlement in Europe*. New Brunswick: Rutgers University Press.
Ingvar, D.H. (1994). The will of the brain: Cerebral correlates of willful acts. *Journal of Theoretical Biology, 171*, 7–12.
Jablonka, E., & Lamb, M.J. (1995). *Epigenetic inheritance and evolution*. Oxford: Oxford University Press.
Jablonka, E., & Lamb, M.J. (2005). *Evolution in four dimensions*. Cambridge: MIT Press.

Kane, M.J., & Engle, R.W. (2002). The role of prefrontal cortex in working-memory capacity, executive attention, and general fluid intelligence: An individual-differences perspective. *Psychonomic Bulletin & Review, 9*, 637–671.

Katz, R. (1982). *Boiling energy: Community healing among the Kalahari !Kung*. Cambridge: Harvard University Press.

Kelly, A.M., Hester, R., Foxe, J.J., Shpaner, M., & Garavan, H. (2006). Flexible cognitive control: Effects of individual differences and brief practice on a complex cognitive task. *NeuroImage, 31*, 866–886.

Keverne, E.B., Martinez, N.D., & Tuite, B. (1989). Beta-endorphin concentrations in cerebrospinal fluid of moneys influenced by grooming relationships. *Psychoneuroendocrinology, 14*, 155–161.

Kirschner, M.W., & Gerhart, J.C. (2005). *The plausibility of life*. New Haven: Yale University Press.

Klein, R.G., & Edgar, B. (2002). *The dawn of human culture*. New York: Wiley.

Knight, C.D., Power, C., & Watts, I. (1995). The human symbolic revolution: A Darwinian account. *Cambridge Archeological Journal, 5*, 75–114.

Knoch, D., & Fehr, E. (2007). Resisting the power of temptations: The right prefrontal cortex and self-control. *Annals of the New York Academy of Science, 1104*, 123–134.

Lazar, S.W., Bush, G., Gollub, R.L., Fricchione, G.L., Khalsa, G., & Benson, H. (2000). Functional brain mapping of the relaxation response and meditation. *NeuroReport, 11*, 1581–1585.

Lazar, S.W., Kerr, C.E., Wasserman, R.W., Gray, J.R., Greve, D.N., Treadway, M.T., Mcgarvey, M., Quinn, B.T., Dusek, J.A., Benson, H., Rauch, S.L., Moore, C.I., & Fischl, B. (2005). Meditation experience is associated with increased cortical thickness. *NeuroReport, 16*, 1893–1897.

Lee, R.B., & Daly, R. (1999). Introduction: Foragers and others. In R.B. Lee & R. Daly (Eds.), *The Cambridge encyclopedia of hunters and gatherers* (pp. 1–22). Cambridge: Cambridge University Press.

Lewis-Williams, D. (2002). *The mind in the cave*. London: Thames & Hudson.

Lou, H.C., Kjaer, T.W., Friberg, L., Wildschiodtz, G., Holm, S., & Nowak, M. (1999). A 15O-H20 PET study of meditation and the resting state of normal consciousness. *Human Brain Mapping, 7*, 98–105.

Lutkehaus, N.C., & Roscoe, P.B. (1995). *Gender rituals: Female initiation in Melanesia*. London: Routledge.

McClenon, J. (1997). Shamanic healing, human evolution and the origin of religion. *Journal for the Scientific Study of Religion, 36*, 345–354.

McClenon, J. (2002). *Wondrous healing: Shamanism, human evolution and the origin of religion*. DeKalb: Northern Illinois University Press.

Minkel, J.R. (2006). Offerings to a stone snake provide the earliest evidence of religion. *Scientific American* online http://www.sciam.com/article.cfm?articleID=3FE89A86-E7F2-99DF-366D045A5BF3EAB1. Accessed 21 May 2007.

Mirsky, J. (1937). The Eskimo of Greenland. In M. Mead (Ed.), *Cooperation and competition among primitive peoples* (pp. 51–86). New York: McGraw-Hill.

Newberg, A.B. (2006). Religious and spiritual practices: A neurochemical perspective. In P. McNamara (Ed.), *Where God and science meet* (Vol. 2, pp. 15–31). Westport: Praeger.

Newberg, A., Alavi, A., Baime, M., Pourdehnad, M., Santanna, J., & d'Aquili, E. (2001). The measurement of regional cerebral blood flow during the complex cognitive task of meditation: A preliminary SPECT study. *Psychiatry Research: Neuroimaging Section, 106*, 113–122.

Peng, C.K., et al. (2004). Heart rate dynamics during three forms of meditation. *International Journal of Cardiology, 95*, 19–27.

Petraglia, M., Korisettar, R., Boivin, N., Clarkson, C., Ditchfield, P., et al. (2007). Middle Paleolithic assemblages from the Indian subcontinent before and after the Toba super-eruption. *Science, 317*, 114–116.

Power, C. (1998). Old wives' tales: The gossip hypothesis and the reliability of cheap signals. In J.R. Hurford, M. Studdert-Kennedy, & C. Knight (Eds.), *Approaches to the evolution of language: Social and cognitive bases* (pp. 111–129). Cambridge: Cambridge University Press.

Rappaport, R.A. (1999). *Ritual and religion and the making of humanity*. Cambridge: Cambridge University Press.
Richerson, P.J., & Boyd, R. (2001). The evolution of subjective commitments to groups: A tribal instincts hypothesis. In R. Neese (Ed.), *Evolution and the capacity for commitment* (pp. 186–220). New York: Russell Sage.
Richman, B. (1987). Rhythm and melody in gelada vocal exchanges. *Primates, 28*, 199–223.
Rossano, M. J. (2010). *Supernatural selection: How religion evolved*. New York: Oxford University Press.
Sanfey, A.G., Rilling, J.K., Aronson, J.A., Nystrom, L.E., & Cohen, J.D. (2003). The neural basis of economic decision-making in the ultimatum game. *Science, 300*, 1755–1758.
Silk, J.B. (2001). Grunts, girneys, and good intentions: The origins of strategic commitment in nonhuman primates. In R.M. Neese (Ed.), *Evolution and the capacity for commitment* (pp. 138–158). New York: Russell Sage.
Smuts, B.B., & Watanabe, J.M. (1990). Social relationships and ritualized greetings in adult male baboons (*Papio cynocephalus anubis*). *International Journal of Primatology, 11*, 147–172.
Sorensen, M.V., & Leonard, W.R. (2001). Neandertal energetics and foraging efficiency. *Journal of Human Evolution, 40*, 483–495.
Sosis, R. (2006). Religious behaviours, badges, and bans: Signally theory and the evolution of religion. In P. McNamara (Ed.), *Where God and science meet* (Vol. 1, pp. 61–86). Bridgeport: Praeger Publishing.
Stuss, D.T., Binns, M.A., Murphy, K.J., & Alexander, M. P. (2002). Dissociations within the anterior attentional system: Effects of task complexity and irrelevant information on reaction-time speed and accuracy. *Neuropsychology, 16*, 500–513.
Tomasello, M., & Call, J. (1997). *Primate cognition*. New York: Oxford University Press.
Tomasello, M., Carpenter, M., Call, J., Behne, T., & Moll, H. (2005). Understanding and sharing intentions: The origins of cultural cognition. *The Behavioral and Brain Sciences, 28*, 675–735.
Townsend, J.B. (1999). Shamanism. In S. D. Glazier (Ed.), *Anthropology of religion* (pp. 429–469). Westport: Praeger.
Tracey, K.J. (2007). Physiology and immunology of the cholinergic antiinflammatory pathway. *Journal of Clinical Investigation, 117*, 289–296.
van Dixhoorn, J., & White, A. (2005). Relaxation therapy for rehabilitation and prevention in ischaemic heart disease: A systematic review and meta-analysis. *European Journal of Cardiovascular Prevention and Rehabilitation, 12*, 193–202.
van Roosmalen, M.G.M., & Klein, L.L. (1988). The spider monkeys, genus *Ateles*. In R. Mittermeier, A.B. Rylands, A.F. Coimbra-Filho, & G.A.B. da Fonesca (Eds.), *Ecology and behavior of neotropic primates* (pp. 445–537). Washington, DC: World Wildlife Fund.
Vanhaeren, M., d'Errico, F., Stringer, C., James, S. L., Todd, J.A., & Mienis, H.K. (2006). Middle Paleolithic shell beads in Israel and Algeria. *Science, 312*, 1785–1788.
Vitebsky, P. (2000). Shamanism. In G. Harvey (Ed.), *Indigenous religions* (pp. 55–67). London: Cassell.
Waddington, C.H. (1942). Canalization of development and the inheritance of acquired characteristics. *Nature, 150*, 563–565.
Weber, B.H., & Depew, D.J. (2003). *Evolution and learning: The Baldwin effect reconsidered*. Cambridge: MIT Press.
Whitehouse, H. (1996). Rites of terror: Emotion, metaphor and memory in Melanesian cults. *The Journal of the Royal Anthropological Institute, 2*, 703–715.
Whitham, J.C., & Maestripieri, D. (2003). Primate rituals: The function of greetings between male guinea baboons. *Ethology, 109*, 847–859.
Wiessner, P. (1982). Risk, reciprocity, and social influences on !Kung San economics. In E. Leacock & R. Lee (Eds.), *Politics and history in band societies* (pp. 61–84). Cambridge: Cambridge University Press.
Winkelman, M. (1990). Shamans and other "magico-religious" healers: A cross-cultural study of their origins, nature, and social transformation. *Ethos, 18*, 308–352.

Winkelman, M. (2002). Shamanism and cognitive evolution. *Cambridge Archaeological Journal, 12*, 71–101.

Wynn, T., & Coolidge, F.L. (2007). Did a small but significant change in working memory capacity empower modern thinking? In P. Mellars, K. Boyle, O. Bar-Yosef, & S. Stringer (Eds.), *Rethinking the human evolution: New behavioural and biological perspectives on the origin and dispersal of modern humans* (pp. 79–90). Cambridge: Cambridge University McDonald Institute Monographs.

Neuroscience and Spirituality – Findings and Consequences

Mario Beauregard

Abstract In this chapter we first define religious, spiritual and mystical experiences (RSMEs). We then review clinical data about patients with epilepsy suggesting a role for the temporal lobe and the limbic system in RSMEs. The possibility of experimentally inducing such experiences by stimulating the temporal lobe with weak electromagnetic currents is examined. The limbic-marker hypothesis is also presented. The findings of brain imaging studies of RSMEs carried out during the last decade are then examined. Next, these findings and the phenomenology of RSMEs in regard to the mind-brain problem are discussed. Finally, we terminate the chapter with a few concluding remarks.

Abbreviations

BA	Brodmann area
BOLD	Blood oxygen level dependent
CAT	Computed axial tomography
EEG	Electroencephalography

M. Beauregard (✉)
Mind/Brain Research Lab (MBRL), Centre de Recherche en Neuropsychologie et Cognition (CERNEC), Département de Psychologie, Université de Montréal, Montreal, QC, Canada

Centre de Recherche en Sciences Neurologiques, Université de Montréal, Montreal, QC, Canada

Département de Radiologie, Université de Montréal, Montreal, QC, Canada

Centre de recherche du Centre hospitalier de l'Université de Montréal, Montreal, QC, Canada
e-mail: mario.beauregard@umontreal.ca

H. Walach et al. (eds.), *Neuroscience, Consciousness and Spirituality*, Studies in Neuroscience, Consciousness and Spirituality 1,
DOI 10.1007/978-94-007-2079-4_4,

FMRI	Functional magnetic resonance imaging
NDE	Near-death experience
PET	Positron emission tomography
rCBF	regional cerebral blood flow
RSMEs	Religious spiritual and mystical experiences
SPECT	Photon emission computed tomography
TLE	Temporal-lobe epilepsy

Introduction

The past decade has seen the emergence of the neuroscience of spirituality. The central objective of this domain of research is to use neuroscience methods (e.g., brain imaging, stimulation, psychopharmacological, electrophysiological recordings) to explore the neural mechanisms supporting religious/spiritual/mystical experiences (RSMEs). These experiences relate to a fundamental dimension of human existence and are frequently reported across all cultures (Hardy 1975; Hay 1990). One of the basic assumptions of this emergent field is that RSMEs are brain-mediated (i.e., they have neurophysiological correlates) as are all other aspects of human experience. About this issue, it is important to fully appreciate that elucidating the neural substrates of RSMEs does not diminish or depreciate their meaning and value.

Religious experiences arise from following a religious tradition and involve a contact with the divine or a religious figure. Spiritual experiences are subjective experiences that do not arise from following a religious tradition. These experiences, however, can also bring the experiencer into contact with the divine or a transcendent reality. Mysticism refers to the pursuit of an altered state of consciousness that enables the mystic to commune with, or identify with a divinity or ultimate reality through an immediate, direct, intuitive knowledge and experience. James (1902) has proposed that ineffability, noetic quality, transiency and passivity are the most important features of mystical experiences. Other characteristics attributed to mystical experiences include feelings of unity, peace and bliss, numinosity, sense of incommunicability of the experience, loss of ego, an altered perception of space and time, and profound transformative changes (i.e., changes in one's worldview, belief system, relationships, and sense of self) (Stace 1960; Waldron 1998).

In the second section of this chapter, we review data suggesting a role for the temporal lobe and the limbic system in RSMEs. In the following section, we examine the findings of brain imaging studies of RSMEs conducted to date (due to space limitations, the results from neuroimaging studies of various types of meditation will not be reviewed here). In the fourth section, we discuss these findings and the phenomenology of RSMEs with respect to the mind-brain problem. In the last section, we present a few concluding remarks about the data and issues discussed in this chapter.

Role of the Temporal Lobe and the Limbic System in Religious/Spiritual/Mystical Experiences

Temporal Lobe Epilepsy

Clinical observations suggest an association between temporal-lobe epilepsy (TLE) and RSMEs during (ictal), after (postictal), and in between (interictal) seizures (Devinsky and Lai 2008). Howden (1872–1873) first observed a man who had a religious conversion after a generalized seizure in which he experienced being transported to "Heaven." Afterwards, Mabille (1899) described a patient who, following a seizure, reported that God had given him a mission to bring law to the world. A few years later, Spratling (1904) reported a religious aura or a premonitory period of hours or several days associated with religiosity in 52 of 1,325 patients with epilepsy (4%). Furthermore, Boven (1919) reported the case of a 14-year-old boy who after a seizure "saw the good God and the angels, and heard a celestial fanfare of music."

More recently, Dewhurst and Beard (1970) reported six patients with TLE who underwent sudden and often lasting religious conversions in the postictal period. Some of these patients had prior or active psychiatric disorders. There was an obvious temporal relationship between conversion and first seizure or increased seizure frequency in five patients.

Studies have shown that between 0.4% and 3.1% of TLE patients had ictal RSMEs while postictal RSMEs occured in 2.2% of patients with TLE. Ictal RSMEs occur most often in patients with right TLE whereas there is a predominance of postictal and interictal RSMEs in TLE patients with bilateral seizure foci. Of note, many of the epilepsy-related religious conversion experiences occur postictally (Devinsky and Lai 2008).

From an experiential perspective, ictal religious experiences during seizures can be accompanied by intense emotions of God's presence, the sense of being connected to the infinite (Alajouanine 1963), hallucinations of God's voice (Hansen and Brodtkorb 2003), the visual hallucination of a religious figure (Karagulla and Robertson 1955), or repetition of a religious phrase (Ozkara et al. 2004). It has been suggested that some of the greatest religious figures in history (e.g., Saint Paul, Muhammad, Joan of Arc, Joseph Smith) were probably suffering form TLE (Saver and Rabin 1997).

Naito and Matsui (1988) described an elderly woman whose seizures were characterized by joyful visions of God. Interictal electroencephalography (EEG) revealed spike discharges in the left anterior and middle temporal areas during sleep. Morgan (1990) described a patient whose seizures were associated with feelings of ineffable contentment and fulfillment; visualizing a bright light recognized as the source of knowledge; and sometimes visualizing a bearded young man resembling Jesus Christ. A computed axial tomography (CAT) scan displayed a right anterior temporal astrocytoma. Following anterior temporal lobectomy, ecstatic seizures vanished.

Ogata and Miyakawa (1998) examined 234 Japanese epileptic patients for ictus-related religious experiences. Three (1.3%) patients were found to have had such experiences. All three cases had TLE with post-ictal psychosis, and interictal experiences with hyperreligiosity. Patients who had ictus-related or interictal religious experiences did not believe only in Buddhism (a traditional religion in Japan), but rather in a combination of Buddhism and Shintoism, new Christian sects, contemporary Japanese religions and/or other folk beliefs. This suggests that these experiences were related not only to the personality characteristic of TLE, but also to the social circumstances and conditions under which such experiences occur in contemporary Japan. These findings indicate that manifestations of religious experience in post-ictal psychosis were influenced by some psychosocial factors.

Interictal Personality Syndrome of TLE

Waxman and Geschwind (1975) suggested that hyperreligiosity is a core feature of a distinctive interictal personality syndrome of TLE (also called the Geschwind syndrome). A heightened state of religious conviction, an increased sense of personal destiny, intense philosophical and cosmological concerns and strong moral beliefs usually characterize interictal religiosity. The putative temporal-lobe personality type is also characterized by hypermoralism, deepened affects, humorlessness, aggressive irritability, and hypergraphia.

Support for this hypothetical syndrome was provided by Bear and Fedio (1977) who found that religiosity trait scores were significantly higher in TLE patients than in healthy control subjects. In keeping with this, Roberts and Guberman (1989) found that 60% of 57 consecutive patients with epilepsy had excessive interests in religion. Subsequent studies using religion questionnaires, however, failed to find any differences regarding interictal religiosity between patients with TLE versus idiopathic generalized epilepsy, or between patients with epilepsy and normal control subjects (Willmore et al 1980; Tucker et al 1987). It has been proposed that differences in religiosity measures and in control group selection account for some of the discrepancy among studies (Saver and Rabin 1997).

Interestingly, Wuerfel et al. (2004) used magnetic resonance imaging (MRI) to investigate mesial temporal structures in 33 patients with refractory partial epilepsy, comparing 22 patients without and 11 patients with hyper-religiosity. High ratings on the religiosity scale were correlated with a significantly smaller hippocampus in the right hemisphere. The hippocampal atrophy may reflect the duration and severity of hyperreligiosity. This does not mean that it is the critical cerebral structure for religious experience (Devinsky and Lai 2008).

The Limbic-Marker Hypothesis

Saver and Rabin (1997) have theorized that temporolimbic discharges underlie each of the core features of RSMEs (e.g., the noetic and the ineffable; the sense of having touched the ultimate ground of reality; the sense of incommunicability of the experience; the experience of unity, timelessness and spacelessness; feelings of positive affect, peace and joy). The limbic system integrates external stimuli with internal drives and is part of a distributed neural circuit that marks the valence (positive or negative) of stimuli and experiences (Damasio et al 1991). Saver and Rabin (1997) posited that temporolimbic discharges may mark experiences as (1) depersonalized or derealized, (2) crucially important and self-referent, (3) harmonious-indicative of a connection or unity between disparate elements, and (4) ecstatic-profoundly joyous.

According to the limbic-marker hypothesis, the perceptual and cognitive contents of a RSME are comparable to those of ordinary experience, except that they are tagged by the limbic system as of deep importance, as united into a whole, and/or as joyous. Therefore, descriptions of the contents of the RSME resemble descriptions of the contents of ordinary experience, and the feelings associated with them cannot be captured fully in words. As in the case of strong emotions, these limbic markers can be named but cannot be communicated in their full visceral intensity, resulting in a report of ineffability.

The temporal lobe and the limbic system may not be the only cerebral structures involved in RSMEs. About this question, Devinsky and Lai (2008) hypothesized that alterations in frontal functions in the right hemisphere may contribute to increased religious interests and beliefs as a personality trait. This hypothesis is based on the finding that dramatic changes in self, defined as a change in political, social, or religious views can be seen in patients with a dysfunction affecting selectively the right frontal lobe (Miller et al. 2001).

Stimulation of the Temporal Lobe

Persinger (1983) speculated that RSMEs are evoked by transient, electrical microseizures within deep structures of the temporal lobe, and that it is possible to experimentally induce RSMEs by stimulating the temporal lobe with weak electromagnetic currents. Persinger and Healey (2002) tested this hypothesis by exposing 48 university students to weak (100 nanoTesla to 1 microTesla) complex, pulsed electromagnetic fields. These fields were applied in one of three ways: over the right temporoparietal region, over the left temporoparietal region, or equally across the temporoparietal region of both hemispheres of the brain (one treatment per group). Fields were applied for 20 min while subjects were wearing opaque goggles in a very quiet room. A fourth group was exposed to a sham field condition – that is, subjects were not exposed to an electromagnetic field, although all subjects were told that they might be. Beforehand, the Hypnosis Induction Profile (Spiegel et al. 1976) was administered to subjects (psychology students), to test for suggestibility.

Two-thirds of the subjects reported a sensed presence under the influence of the electromagnetic fields. But 33% of the control (sham-field) group reported a sensed presence too. In other words, Persinger and Healey (2002) found that twice as many subjects reported a sensed presence under the influence of the electromagnetic field as those who reported one without an electromagnetic field. About half of these subjects stated that they felt "someone else" in the chamber. Another approximate half of the group described a sentient being who moved when they tried to "focus attention" upon the presence. About one-third of subjects attributed the presence to a deceased member of the family or to some cultural equivalent of a "spirit guide." In the study, those who had received stimulation over the right hemisphere or both hemispheres reported more unusual phenomena than those who had received stimulation over the left hemisphere.

Persinger and Healey (2002) concluded two things: that the experience of a sensed presence can be manipulated by experiment, and that such an experience may be the source for phenomena attributed to visitations by spiritual entities.

A research team at Uppsala University in Sweden, headed by Pehr Granqvist (Granqvist et al. 2005), mirrored Persinger's experiment by testing 89 undergraduate students, some of whom were exposed to the electromagnetic fields and some of whom were not. Using Persinger's equipment, the Swedish researchers could not reproduce his key results. They attributed their findings to the fact that they ensured that neither the participants nor the experimenters interacting with them had any idea who was being exposed to the electromagnetic fields.

Granqvist and colleagues made sure that their experiment was double blind by using two experimenters for each trial. The first experimenter, who was not told about the purpose of the study, interacted with the subjects. The second experimenter switched electromagnetic fields off or on without advising either the first experimenter or the subject. So if the subject had not already been advised that a RSME was likely at Granqvist's laboratory, the study experimenters were not in a position to provide that clue.

Study participants included undergraduate theology students as well as psychology students. Neither group were asked for prior information on spiritual or paranormal experiences, nor was any participant told that there was a sham-field (control) condition. Rather, participants were told only that the study investigated the influence of weak electromagnetic fields on experiences and feeling states. Personality characteristics that might predispose a person to report an unusual experience were used as predictors for which subjects would report one. These characteristics included absorption (the ability to become completely absorbed in an experience), signs of abnormal temporal-lobe activity, and a "New Age" lifestyle orientation.

No evidence was found for a "sensed presence" effect of weak electromagnetic fields. The characteristic that significantly predicted the outcomes was personality.

Of the three subjects who reported strong spiritual experiences, two were members of the control group. Of the 22 who reported "subtle" experiences, 11 were members of the control group. Those subjects who were rated as highly suggestible on the basis of a questionnaire filled out after they completed the study reported paranormal experiences whether the electromagnetic field was on or off while they were wearing the stimulation helmet. Granqvist and colleagues also noted that they had found it difficult to evaluate the reliability of Persinger's findings, because no information on experimental randomization or blindness was provided, which left his results open to the possibility that psychological suggestion was the best explanation.

Brain Imaging Studies of Religious/Spiritual/Mystical Experiences

The first brain imaging study of a religious experience was conducted by Azari et al. (2001). These researchers studied a group of six self-identified religious subjects, who attributed their religious experience to biblical Psalm 23. These subjects, who were members of a 'Free Evangelical Fundamentalist Community' in Germany, all reported having had a conversion experience (related to the first verse of biblical Psalm 23, which states "The LORD is my shepherd; I shall not be in want"), and interpreted biblical text literally as the word of God. Religious subjects were compared to six non-religious individuals. The texts used for the different tasks were 'religious' (first verse of biblical Psalm 23), 'happy' (a well-known German children's nursery rhyme) and 'neutral' (instructions on using a phone card from the Düsseldorf telephone book).

Subjects were scanned with positron emission tomography (PET) during various conditions: reading silently or reciting biblical Psalm 23; reading silently or reciting the children's nursery rhyme; reading silently the set of instructions; and while lying quietly. The PET images revealed a significant activation of the right dorsolateral prefrontal cortex in the religious subjects during the religious state as compared with non-religious subjects. During the religious state, the religious subjects showed additional loci of activation, including the dorsomedial frontal cortex and the right precuneus. Limbic areas did not show regional cerebral blood flow (rCBF) changes.

According to Azari and colleagues (2001), these results strongly support the view that religious experience is a cognitive attributional phenomenon, mediated by a pre-established neural circuit, involving dorsolateral prefrontal, dorsomedial frontal and medial parietal cortex. Religious attributions are based on religious schemata which consist in organized knowledge about religion and religious issues, and include reinforced structures for inferring religiously related causality of experienced events (Spilka and McIntosh 1995). Azari and co-workers (2001) proposed that the dorsolateral prefrontal and medial parietal cortices were probably involved in the subject's own religious schemata whereas the dorsomedial frontal cortex would be implicated in the felt immediacy of religious experience.

Newberg et al. (2003) used single photon emission computed tomography (SPECT) to scan three Franciscan nuns while they performed a "centering prayer" to open themselves to the presence of God. This prayer involved the internal repetition of a particular phrase. Compared to baseline, the prayer condition scan showed increased rCBF in the prefrontal cortex, inferior parietal lobes, and inferior frontal lobes. There was a strong inverse correlation between the rCBF changes in the prefrontal cortex and in the ipsilateral superior parietal lobule. Newberg et al. (2003) hypothesized that increased frontal rCBF reflected focused concentration whereas increased rCBF in the superior parietal lobule was related to an altered sense of space experienced by the nuns during prayer. In this pilot study, there was no attempt to analyze and quantify in a rigorous and systematic manner the nuns' subjective experiences during their "centering prayer." In other words, Newberg and colleagues could not determine whether focusing attention on a phrase from a prayer over a period of time really led the nuns to feel the presence of God.

Newberg et al. (2006) also used SPECT to investigate changes in cerebral activity during glossolalia ("speaking in tongues"). This unusual mental state is associated with specific religious traditions. Glossolalia is one of the "gifts of the Spirit" according to Saint Paul and, hence, some fundamentalist religious traditions see it as a sign of being visited by the Spirit. This is due to the Pentecost experience, where, according to the Acts of the Apostles, the Apostles "spoke in the tongues" of all those present, i.e. made themselves understood to everybody, whereby later on just babbling something became synonymous with glossolalia. In this state, the individual seems to be speaking in an incomprehensible language over which he/she claims to have no voluntary control. Yet, the individual perceives glossolalia to have great personal and religious meaning. In their study, Newberg and colleagues examined five practitioners (women) of glossolalia. Participants described themselves as Christians in a Charismatic or Pentecostal tradition who had practiced glossolalia for more than 5 years. Structured clinical interviews excluded current psychiatric conditions. Glossolalia was compared to a religious singing state since the latter is similar except that it involves actual language (English). Earphones were used to play music to sing and to perform glossolalia (the same music was used for both conditions). Several significant rCBF differences were noted between the glossolalia and singing state. During glossolalia (compared to the religious singing state), significant decreases were found in the prefrontal cortices, left caudate and left temporal pole. Decreased activity in the prefrontal lobe is consistent with the participants' description of a lack of volitional control over the performance of glossolalia. Newberg et al. (2006) proposed that the decrease in the left caudate may relate to the altered emotional activity during glossolalia.

Recently, we sought to identify the neural correlates of a mystical experience (as understood in the Christian sense) in a group of contemplative nuns using functional magnetic resonance imaging (fMRI) (Beauregard and Paquette 2006). Fifteen Carmelite nuns took part in the study. Blood oxygen level dependent (BOLD) signal changes were measured during a Mystical condition, a Control condition, and a Baseline condition. In the Mystical condition, subjects were asked to remember and relive the most intense mystical experience ever felt in their lives as a member of the Carmelite Order.

This strategy was adopted given that the nuns told us before the onset of the study that "God can't be summoned at will." In the Control condition, subjects were instructed to remember and relive the most intense state of union with another human ever felt in their lives while being affiliated with the Carmelite Order. The week preceding the experiment, subjects were requested to practice these two tasks. The Baseline condition was a normal restful state. Immediately at the end of the scan, the intensity of the subjective experience during the Control and Mystical conditions was measured using numerical rating scales ranging from 0 (no experience of union) to 5 (most intense experience of union ever felt): self-report data referred solely to the experiences lived during these two conditions, not to the original experiences recalled to self-induce the Control and Mystical states. The phenomenology of the mystical experience during the Mystical condition was assessed with 15 items of the Mysticism Scale (Hood 1975). This scale, which comprises 32 items, aims at measuring reported mystical experience (for each participant, scores of 15 or above were considered significant for a given item). In addition, qualitative interviews were conducted after the experiment to obtain additional information regarding the nature of the subjective experiences during the Control and Mystical conditions.

As regards the phenomenology of the subjective experience during the Mystical condition, summed scores of 15 or above were noted for three items of the Mysticism Scale: (1) "I have had an experience in which something greater than my self seemed to absorb me" (average score: 15); (2) "I have experienced profound joy" (average score: 22); (3) "I have had an experience which I knew to be sacred" (average score: 20). During the qualitative interviews conducted at the end of the experiment, several subjects mentioned that during the Mystical condition they felt the presence of God, His unconditional and infinite love, as well as plenitude and peace. All subjects reported that from a first-person perspective, the experiences lived during the Mystical condition were different than those used to self-induce a mystical state. Subjects also reported the presence of visual and motor imagery during both the Mystical and Control conditions. In addition, the subjects experienced a feeling of unconditional love during the Control condition.

The Mystical versus Baseline contrast produced significant loci of BOLD activation in the right medial orbitofrontal cortex (Brodmann area [BA] 11), right middle temporal cortex (BA 21), right inferior parietal lobule (BA 40) and superior parietal lobule (BA7), right caudate, left medial prefrontal cortex (BA 10), left dorsal anterior cingulate cortex (BA 32), left inferior parietal lobule (BA 7), left insula (BA 13), left caudate, and left brainstem. A few loci of activation were also seen in the extra-striate visual cortex.

Based on the studies indicating a relationship between RSMEs and the temporal lobe, we posited that the right middle temporal activation noted during the Mystical condition was related with the subjective impression of contacting a spiritual reality. We also proposed that the caudate activations reflected feelings of joy and unconditional love since the caudate nucleus has been systematically activated in previous functional brain imaging studies implicating positive emotions such as happiness (Damasio et al. 2000), romantic love (Bartels and Zeki 2000), and maternal love (Bartels and Zeki 2004). Concerning the brainstem, there is some empirical support for the view that certain brainstem nuclei map the organism's internal state during emotion (Damasio 1999). Given this it is conceivable that the activation in the left brainstem was linked to the somatovisceral changes associated with the feelings of joy and unconditional love. As for the insula, this cerebral structure is richly interconnected with regions involved in autonomic regulation (Cechetto 1994). It contains a topographical representation of inputs from visceral, olfactory, gustatory, visual, auditory and somatosensory areas and is proposed to integrate representations of external sensory experience and internal somatic state (Augustine 1996). The insula has been seen activated in several studies of emotional processing and appears to support a representation of somatic and visceral responses accessible to consciousness (Critchley et al. 2004; Damasio 1999). It is plausible that the left insular activation (BA 13) noted in our study was related to the representation of the somatovisceral reactions associated with the feelings of joy and unconditional love.

In addition, we suggested that the left medial prefrontal cortical activation (BA 10) was linked with conscious awareness of those feelings. Indeed, the results of functional neuroimaging studies indicate that the medial prefrontal cortex is involved in the metacognitive representation of one's own emotional state (Lane and Nadel 2000). This prefrontal area receives sensory information from the body and the external environment via the orbitofrontal cortex and is heavily interconnected with limbic structures such as the amygdala, ventral striatum, hypothalamus, midbrain periaqueductal gray region, and brainstem nuclei (Barbas 1993; Carmichael and Price 1995). In other respects, brain imaging findings (Lane et al. 1997, 1998) support the view that the activation of the left dorsal anterior cingulate cortex (BA 32) reflected that aspect of emotional awareness associated with the interoceptive detection of emotional signals during the Mystical condition. This cortical region projects strongly to the visceral regulation areas in the hypothalamus and midbrain periaqueductal gray (Ongur et al 2003). Regarding the medial orbitofrontal cortex, there is mounting evidence that this prefrontal cortical region codes for subjective pleasantness (Kringelbach et al. 2003). The medial orbitofrontal cortex has been found activated with regard to the pleasantness of the taste or smell of stimuli (Araujo et al. 2003; Rolls et al. 2003) or music (Blood and Zatorre 2001). It has reciprocal connections with the cingulate and insular cortices (Carmichael and Price 1995; Cavada et al. 2000). The right medial orbitofrontal cortical activation (BA 11) noted in the Mystical condition was perhaps related to the fact that the experiences lived during the mystical state were considered by the subjects emotionally pleasant.

Given that the right superior parietal lobule is involved in the spatial perception of self (Neggers et al. 2006), it is conceivable that the activation of this parietal region (BA 7) reflected a modification of the body schema associated with the impression that something greater than the subjects seemed to absorb them. Moreover, there is evidence that the left inferior parietal lobule is part of a neural system implicated in the processing of visuospatial representation of bodies (Felician et al. 2003). Therefore, the left inferior parietal lobule activation in the Mystical condition was perhaps related to an alteration of the body schema. In keeping with this, there is some evidence indicating that the right inferior parietal lobule is crucial in bodily consciousness and the process of self/other distinction (Ruby and Decety 2003).

However, the inferior parietal lobule plays an important role in motor imagery (Decety 1996). It is thus plausible that the activations in the right (BA 40) and left (BA 7) inferior parietal lobules were related to the motor imagery experienced during the Mystical condition. Last, regarding the loci of activation found in the extra-striate visual cortex during this condition, it has been previously shown (Ganis et al. 2004) that this region of the brain is implicated in visual mental imagery. It is likely that the BOLD activation in visual cortical areas was related to the visual mental imagery reported by the nuns.

These results suggest that several brain regions and networks mediate the various aspects of RSMEs. This conclusion should not come as a surprise given that these experiences are complex and multidimensional, that is, they implicate changes in perception, self-awareness, cognition and emotion.

Neuroscience, Religious/Spiritual/Mystical Experiences and the Mind-Brain Problem

Physicalism is the mainstream metaphysical view of modern neuroscience with respect to the mind-body problem, i.e., the explanation of the relationship that exists between mental processes and bodily processes. According to this view, consciousness and mental events (e.g., thoughts, emotions, desires) can be reduced to their neural correlates, i.e., the brain electrical and chemical processes whose presence necessarily and regularly correlates with these mental events. Physicalist philosophers and neuroscientists believe that mental events are equivalent to brain processes. About this issue, it is important to bear in mind that neural correlates do not yield a causal explanation of mental events, i.e., they cannot explain how neural processes become mental events. Indeed correlation does not entail causation. And the external reality of "God" or ultimate reality can neither be confirmed nor disconfirmed by neural correlates.

Newberg and colleagues (Newberg et al. 2001) submitted that the most important criterion for judging what is real is the subjective vivid sense of reality. They argued that individuals usually refer to dreams as less real than waking (baseline) reality when they are recalled within baseline reality. In contrast, RSMEs (e.g., "cosmic consciousness" states, religious visions, near-death experiences) appear more real to the experiencers than waking (baseline) reality when they are recalled from baseline reality.

A major problem with this criterion is its subjectivity. This problem is well illustrated by the fact that individuals suffering from psychosis are unable to distinguish personal subjective experience from the reality of the external world. They experience hallucinations and/or delusions as being very real. From a neuroscientific perspective, a more satisfactory approach to evaluate the "objective" ontological reality of RSMEs is to determine whether it is possible for a human being to have a spiritual experience during a state of clinical death, i.e., when her/his brain is not functioning. In this state, vital signs have ceased: the heart is in ventricular fibrillation, there is a total lack of electrical activity on the cortex of the brain (flat EEG), and brain-stem activity is abolished (loss of the corneal reflex, fixed and dilated pupils, and loss of the gag reflex).

The thought-provoking case of a patient who apparently underwent a profound spiritual experience while her brain was not functioning has been reported by cardiologist Michael Sabom (1998). In 1991, 35-year-old Atlanta-based singer and songwriter Pam Reynolds began to suffer dizziness, loss of speech, and difficulty moving. A CAT scan revealed that she had a giant basilar artery aneurysm (a grossly swollen blood vessel in the brain stem). If it burst, it would kill her. But attempting to drain and repair it might kill her too. Her doctor offered no chance of survival using conventional procedures. Reynolds heard about neurosurgeon Robert Spetzler, at the Barrow Neurological Institute in Phoenix, Arizona. He was a specialist and pioneer in a rare and dangerous technique called hypothermic cardiac arrest, or "Operation Standstill." He would take her body down to a temperature so low that she was clinically dead, but then bring her back to a normal temperature before irreversible damage set in. At a low temperature, the swollen vessels that burst at the high temperatures needed to sustain human life become soft. Then they can be operated upon with less risk. Also, the cooled brain can survive longer without oxygen, though it obviously cannot function in that state. So for all practical purposes, Reynolds would actually be clinically dead during the surgery. But if she didn't agree to it, she would soon be dead anyway with no hope of return. So she consented.

As the surgery began, her heart and breathing ceased, the blood was completely drained from her head and her EEG brain waves flattened into total silence (indicating no cerebral activity – during a cardiac arrest, the brain's electrical activity disappears after about 10 s – Clute and Levy 1990). Her brain stem became unresponsive (her eyes had been taped shut and her ears had been blocked by molded ear speakers), and her temperature fell to 15°C. When all of Reynolds's vital signs were stopped, the surgeon began to cut through her skull with a surgical saw. She reported later that at that point, she felt herself "pop" outside her body and hover above the operating table. From her out-of-body position, she could see the doctors working on her lifeless body. She described, with considerable accuracy for a person who knew nothing of surgical practice, the Midas Rex bone saw used to open skulls. Reynolds also heard and reported later what was happening during the operation and what the nurses in the operating room had said. At a certain point, she became conscious of floating out of the operating room and traveling down a tunnel with a light. Deceased relatives and friends were waiting at the end of this tunnel, including her long-dead grandmother. She entered the presence of a brilliant, wonderfully warm and loving Light and sensed that her soul was part of God and that everything in existence was created from the Light (the breathing of God) (Sabom 1998).

The anecdotal case of Pam Reynolds strongly challenges the physicalist doctrine in regard to the mind-brain problem. This case suggests that mental processes and events (consciousness, perception, cognition, emotion) can be experienced at the moment that the brain seemingly no longer functions (as evidenced by a flat EEG) during a period of clinical death. This case also suggests that RSMEs can occur when the brain is not functioning, that is, these experiences are not necessarily delusions created by a defective brain. In other words, it would be possible for humans to experience a transcendent reality during an altered state of consciousness in which perception, cognition, identity and emotion function independently from the brain. This raises the possibility that when a spiritual experience happens while the brain is fully functional, the neural correlates of this experience indicate that the brain is de facto connecting with a transcendent level of reality.

It should be noted that since Pam Reynolds did not die, there were likely residual brain processes not detectable by EEG that persisted during the clinical death period at sufficient levels so as to permit return to normal brain functioning after the standstill operation. Yet it is difficult to see how the brain could generate higher mental functions in absence of cortical and brainstem activity. Scientific research is clearly needed to investigate the possibility that a functioning brain may not be essential to higher mental functions and spiritual experiences. It is noteworthy that NDEs are reported by 10–18% of cardiac arrest survivors (Parnia et al. 2001; van Lommel et al. 2001; Greyson 2003, see chapter by van Lommel in this volume).

More than a century ago, William James (1898) proposed that the brain may serve a permissive/transmissive/expressive function rather than a productive one, in terms of the mental events and experiences it allows (just as a prism – which is not the source of the light – changes incoming white light to form the colored spectrum). Following James, Henri Bergson (1914) and Aldous Huxley (1954) posited that the brain acts as a filter or reducing valve by blocking out much of, and allowing registration and expression of only a narrow band of, perceivable reality. Bergson and Huxley believed that over the course of evolution, the brain has been trained to eliminate most of those perceptions that do not directly aid our everyday survival. This outlook implies that the brain normally limits the human capacity to have a RSME. A significant alteration of the electrochemical activity of the brain would be necessary for the occurence of a RSME (Beauregard and O'Leary 2007).

Concluding Remarks

Taken together, the clinical studies of epileptic patients suggest that the temporal lobe and the limbic system can be involved in the experiential aspect of RSMEs. However, the relationship between these brain regions and RSMEs is still poorly understood since (1) most people who have these experiences are not epileptics; and (2) very few epileptics report RSMEs during seizures. In addition, the experimental induction of such experiences by stimulating the temporal lobes with weak electromagnetic currents does not appear easily achievable when psychological suggestibility is controlled using a randomized, double-blind, placebo approach.

Contrary to the assertion that neural discharges in the temporal lobe and limbic system underlie each of the main features of RSMEs (Saver and Rabin 1997), brain imaging studies conducted during the last decade indicate that several brain regions and networks support the diverse aspects of these experiences (perception, cognition, emotion, etc.). The distinct results in these studies are mainly related to differences between tasks and spiritual experiences/states. For instance, Azari et al. (2001) used a rather cognitive task whereas an important emotional dimension characterized the task/state in our fMRI study of Carmelite nuns (Beauregard and Paquette 2006). Not surprisingly, the neural correlates found in these two neuroimaging studies were quite different.

Finally, the case of Pam Reynolds and many cases of NDEs during cardiac arrest (Parnia et al. 2001; van Lommel et al. 2001; Greyson 2003) stand against the physicalist credo as regards RSMEs and the mind-brain problem. Collectively, these cases point out the possibility that RSMEs can happen when the brain is seemingly not functioning (i.e., there is no cerebral activity detectable by EEG). In this context, **it is conceivable that the neural correlates of RSMEs reflect the actual connection of the brain with a spiritual level of reality. Solid scientific research is required** to tackle this fascinating issue. One way to address this question is to conduct an experiment aiming to test the veridicity of out-of-body perceptions with randomly changing pictures presented (on a video screen) in the operating room during hypothermic cardiac arrest.

References

Alajouanine, T. (1963). Dostoiewski's epilepsy. *Brain, 86*, 209–218.

Augustine, J.R. (1996). Circuitry and functional aspects of the insular lobe in primates including humans. *Brain Research Reviews, 22*(3), 229–244.

Azari, N.P., Nickel, J., Wunderlich, G., Niedeggen, M., Hefter, H., Tellmann, L., et al. (2001). Neural correlates of religious experience. *European Journal of Neuroscience, 13*(8), 1649–1652.

Barbas, H. (1993). Organization of cortical afferent input to the orbitofrontal area in the rhesus monkey. *Neuroscience, 56*(4), 841–864.

Bartels, A., & Zeki, S. (2000). The neural basis of romantic love. *NeuroReport, 11*(17), 3829–3834.

Bartels, A., & Zeki, S. (2004). The neural correlates of maternal and romantic love. *NeuroImage, 21*(3), 1155–1166.

Bear, D., & Fedio, P. (1977). Quantitative analysis of interictal behavior in temporal lobe epilepsy. *Archives of Neurology, 34*(8), 454–467.

Beauregard, M., & Paquette, V. (2006). Neural correlates of a mystical experience in Carmelite nuns. *Neuroscience Letters, 405*(3), 186–190.

Beauregard, M., & O'Leary, D. (2007). *The spiritual brain*. New York: Harper Collins.

Bergson, H. (1914). Presidential address. *Proceedings of the Society for Psychical Research, 27*, 157–175.

Blood, A., & Zatorre, R. (2001). Intensely pleasurable responses to music correlate with activity in brain regions implicated in reward and emotion. *Proceedings of the National Academy of Sciences USA, 98*(20), 11818–11823.

Boven, W. (1919). Religiosité et épilepsie. *Schweiz Archives of Neurological Psychiatry, 4*, 153–169.

Carmichael, S.T., & Price, J.L. (1995). Limbic connections of the orbital and medial prefrontal cortex in macaque monkeys. *The Journal of Comparative Neurology, 363*(4), 615–641.

Cavada, C., Company, T., Tejedor, J., Cruz-Rizzolo, R.J., & Reinoso-Suarez, F. (2000). The anatomical connections of the macaque monkey orbitofrontal cortex, a review. *Cerebral Cortex, 10*(3), 220–242.

Cechetto, D.F. (1994). Identification of a cortical site for stress-induced cardiovascular dysfunction. *Integrative Physiological and Behavioral Science, 29*(4), 362–373.

Clute, H.L., & Levy, W. J. (1990). Electroencephalographic changes during brief cardiac arrest in humans. *Anesthesiology, 73*(5), 821–825.

Critchley, H.D., Wien, S., Rotshtein, P., Ohman, A., & Dolan, R.J. (2004). Neural systems supporting interoceptive awareness. *Nature Neuroscience, 7*(2), 189–195.

Damasio, A.R. (1999). *The feeling of what happens: Body and emotion in the making of consciousness*. New York: Harcourt Brace.

Damasio, A.R., Tranel, D., & Damasio, H. (1991). Somatic markers and the guidance of behaviour. In H. Levin, H. Eisenberg, & A. Benton (Eds.), *Frontal lobe function and dysfunction* (pp. 217–228). New York: Oxford University Press.

Damasio, A.R., Grabowski, T.J., Bechara, A., Damasio, H., Ponto, L.L., Parvizi, J., et al. (2000). Subcortical and cortical brain activity during the feeling of self-generated emotions. *Nature Neuroscience, 3*(10), 1049–1056.

de Araujo, I.E., Rolls, E.T., Kringelbach, M.L., McGlone, F., & Phillips, N. (2003). Taste-olfactory convergence, and the representation of the pleasantness of flavour, in the human brain. *European Journal of Neuroscience, 18*(7), 2059–2068.

Decety, J. (1996). Do imagined and executed actions share the same neural substrate? *Cognitive Brain Research, 3*(2), 87–93.

Devinsky, O., & Lai, G. (2008). Spirituality and religion in epilepsy. *Epilepsy & Behavior, 12*(4), 636–643.

Dewhurst, K., & Beard, A.W. (1970). Sudden religious conversions in temporal lobe epilepsy. *The British Journal of Psychiatry, 117*(540), 497–507.

Felician, O., Ceccaldi, M., Didic, M., Thinus-Blanc, C., & Poncet, M. (2003). Pointing to body parts: A double dissociation study. *Neuropsychologia, 41*(10), 1307–1316.

Ganis, G., Thompson, W.L., & Kosslyn, S. M. (2004). Brain areas underlying visual mental imagery and visual perception: An fMRI study. *Cognitive Brain Research, 20*(2), 226–241.

Granqvist, P., Fredrikson, M., Unge, P., Hagenfeldt, A., Valind, S., Larhammar, D., et al. (2005). Sensed presence and mystical experiences are predicted by suggestibility, not by the application of transcranial weak complex magnetic fields. *Neuroscience Letters, 379*(1), 1–6.

Greyson, B. (2003). Incidence and correlates of near-death experiences in a cardiac care unit. *General Hospital Psychiatry, 25*(4), 269–276.

Hansen, B.A., & Brodtkorb, E. (2003). Partial epilepsy with "ecstatic" seizures. *Epilepsy & Behavior, 4*(6), 667–673.

Hardy, A. (1975). *The biology of God*. New York: Taplinger.

Hay, D. (1990). *Religious experience today: Studying the facts*. London: Mowbray.

Hood, R.W., Jr. (1975). The construction and preliminary validation of a measure of reported mystical experience. *Journal for the Scientific Study of Religion, 14*(1), 21–41.

Howden, J.C. (1872–1873). The religious sentiments in epileptics. *The Journal of Mental Science, 18*, 491–497.

Huxley, A. (1954). *The doors of perception*. New York: Haper & Row.

James, W. (1902). *The varieties of religious experience: A study in human nature*. New York: Longmans, Green and Co.

James, W. (1960). Human immortality: Two supposed objections to the doctrine. In G. Murphy & R.O. Ballou (Eds.), *William James on psychical research* (pp. 279–308). New York: Viking. Original work delivered as a lecture (1898).

Karagulla, S., & Robertson, E. E. (1955). Psychical phenomena in temporal lobe epilepsy and the psychoses. *British Medical Journal, 1*(4916), 748–752.

Kringelbach, M.L., O'Doherty, J., Rolls, E. T., & Andrews, C. (2003). Activation of the human orbitofrontal cortex to a liquid food stimulus is correlated with its subjective pleasantness. *Cerebral Cortex, 13*(10), 1064–1071.

Lane, R.D., & Nadel, L. (2000). *Cognitive neuroscience of emotion*. New York: Oxford University Press.

Lane, R.D., Fink, G.R., Chau, P.M. L., & Dolan, R.J. (1997). Neural activation during selective attention to subjective emotional responses. *NeuroReport, 8*(18), 3969–3972.

Lane, R.D., Reiman, E.M., Axelrod, B., Yun, L.S., Holmes, A., & Schwartz, G.E. (1998). Neural correlates of levels of emotional awareness. Evidence of an interaction between emotion and attention in the anterior cingulate cortex. *Journal of Cognitive Neuroscience, 10*(4), 525–535.

Mabille, H. (1899). Hallucinations religieuses dans l'épilepsie. *Annales Médicopsychologiques, 9–10*, 76–81.

Miller, B.L., Seeley, W.W., Mychack, P., Rosen, H.J., Mena, I., & Boone, K. (2001). Neuroanatomy of the self: Evidence from patients with frontotemporal dementia. *Neurology, 57*(5), 817–821.

Morgan, H. (1990). Dostoevsky's epilepsy: A case report and comparison. *Surgical Neurology, 33*(6), 413–416.

Naito, H., & Matsui, N. (1988). Temporal lobe epilepsy with ictal ecstatic state and interictal behavior of hypergraphia. *Journal of Nervous and Mental Diseases, 176*(2), 123–124.

Neggers, S.F., van der Lubbe, R.H., Ramsey, N.F., & Postma, A. (2006). Interactions between ego- and allocentric neuronal representations of space. *NeuroImage, 31*(1), 320–331.

Newberg, A., d'Aquili, E., & Rause, V. (2001). *Why God won't go away*. New York: Ballantine Books.

Newberg, A., Pourdehnad, M., Alavi, A., & d'Aquili, E.G. (2003). Cerebral blood flow during meditative prayer: Preliminary findings and methodological issues. *Perceptual and Motor Skills, 97*(2), 625–630.

Newberg, A.B., Wintering, N.A., Morgan, D., & Waldman, M.R. (2006). The measurement of regional cerebral blood flow during glossolalia: A preliminary SPECT study. *Psychiatry Research, 148*(1), 67–71.

Ogata, A., & Miyakawa, T. (1998). Religious experiences in epileptic patients with a focus on ictus-related episodes. *Psychiatry and Clinical Neuroscience, 52*(3), 321–325.

Ongur, D., Ferry, A.T., & Price, J.L. (2003). Architectonic subdivision of the human orbital and medial prefrontal cortex. *The Journal of Comparative Neurology, 460*(3), 425–449.

Ozkara, C., Sary, H., Hanoglu, L., Yeni, N., Aydogdu, I., & Ozyurt, E. (2004). Ictal kissing and religious speech in a patient with right temporal lobe epilepsy. *Epileptic Disorders, 6*(4), 241–245.

Parnia, S., Waller, D.G., Yeates, R., & Fenwick, P. (2001). A qualitative and quantitative study of the incidence, features and aetiology of near death experiences in cardiac arrest survivors. *Resuscitation, 48*(2), 149–156.

Persinger, M.A. (1983). Religious and mystical experiences as artefacts of temporal lobe function: A general hypothesis. *Perceptual and Motor Skills, 57*(3 Pt 2), 1255–1262.

Persinger, M.A., & Healey, F. (2002). Experimental facilitation of the sensed presence: Possible intercalation between the hemispheres induced by complex magnetic fields. *Journal of Nervous and Mental Diseases, 190*(8), 533–541.

Roberts, J.K., & Guberman, A. (1989). Religion and epilepsy. *Psychiatry Journal University of Ottawa, 14*(1), 282–286.

Rolls, E.T., Kringelbach, M.L., & de Araujo, I.E. (2003). Different representations of pleasant and unpleasant odours in the human brain. *European Journal of Neuroscience, 18*(3), 695–703.

Ruby, P., & Decety, J. (2003). What you believe versus what you think they believe: A neuroimaging study of conceptual perspective-taking. *European Journal of Neuroscience, 17*(11), 2475–2480.

Sabom, M. (1998). *Light and death: One doctor's fascinating account of near-death experiences*. Grand Rapids: Zondervan.

Saver, J.L., & Rabin, J. (1997). The neural substrates of religious experience. *Journal of Neuropsychiatry and Clinical Neuroscience, 9*(3), 498–510.

Spiegel, H., Aronson, M., Fleiss, J.L., & Haber, J. (1976). Psychometric analysis of the Hypnotic Induction Profile. *International Journal of Clinical and Experimental Hypnosis, 24*(3), 300–315.

Spilka, B., & McIntosh, D.N. (1995). Attribution theory and religious experience. In R.W. Hood (Ed.), *Handbook of religious experience* (pp. 421–445). Birmingham: Religious Education Press.

Spratling, W.P. (1904). *Epilepsy and its treatment*. Philadelphia: WB Saunders.
Stace, W.T. (1960). *Mysticism and philosophy*. New York: Macmillan.
Tucker, D.M., Novelly, R.A., & Walker, P.J. (1987). Hyperreligiosity in temporal lobe epilepsy: Redefining the relationship. *Journal of Nervous and Mental Diseases, 175*(3), 181–184.
van Lommel, P., van Wees, R., Meyers, V., & Elfferich, I. (2001). Near-death experience in survivors of cardiac arrest: A prospective study in the Netherlands. *Lancet, 358*(9298), 2039–2045.
Waldron, J.L. (1998). The life impact of transcendent experiences with a pronounced quality of noesis. *Journal of Transpersonal Psychology, 30*(2), 103–134.
Waxman, S.G., & Geschwind, N. (1975). The interictal behavior syndrome of temporal lobe epilepsy. *Archives of General Psychiatry, 32*(12), 1580–1586.
Willmore, L.J., Heilman, K.M., Fennell, E., & Pinnas, R.M. (1980). Effect of chronic seizures on religiosity. *Transactions of the American Neurological Association, 105*, 85–87.
Wuerfel, J., Krishnamoorthy, E.S., Brown, R.J., Lemieux, L., Koepp, M., Tebartz van Elst, L., et al. (2004). Religiosity is associated with hippocampal but not amygdala volumes in patients with refractory epilepsy. *Journal of Neurology, Neurosurgery, and Psychiatry, 75*(4), 640–642.

Consciousness: A Riddle and a Key in Neuroscience and Spirituality

Daniel Jeanmonod

Abstract The concept of consciousness is omnipresent in the fields of neuroscience, quantum physics, philosophy and spirituality. In neuroscience, it represents the highest and least understandable component of brain function, unescapable but difficult to describe, its mechanism(s) remaining until today elusive. The dominant view in this field is that consciousness is an emergent product of the brain. In quantum physics, the relevance of the observer brings it into the center of discussions about interpretations of reality. It constitutes the hallmark of some models of quantum physics. In philosophy, it is also discussed at a fundamental level, sometimes refuted, other times placed in the center of the process of reality, as in the different philosophical idealistic approaches. In spirituality, it plays a central role in different eastern, monistic-idealistic, particularly hindu traditions. The aim of this chapter is, looking at a large scope going from the oldest spiritual to the most modern scientific approaches, to highlight the surprising and promising convergence of their data and concepts. It has been conceived to serve as a field overview for the different detailed reports presented in this volume.

Consciousness, Neuroscience and Quantum Physics

It is only relatively recently that neuroscience has addressed the issue of consciousness. Its detailed mechanisms remain elusive, in spite of various efforts to elucidate them. The necessity/adequacy of the concept of conscious experience has even been questioned by some authors (Dennett 1991), surprisingly enough when one considers that the most obvious and unquestionable evidence we have at disposition

D. Jeanmonod (✉)
Center for Ultrasound Functional Neurosurgery, SoniModul Ltd, Solothurn, Switzerland
e-mail: daniel.jeanmonod@gmail.com

H. Walach et al. (eds.), *Neuroscience, Consciousness and Spirituality*,
Studies in Neuroscience, Consciousness and Spirituality 1,
DOI 10.1007/978-94-007-2079-4_5, © Springer Science+Business Media B.V. 2011

about reality is the fact that we are indeed conscious in and of it! Some approaches, close to buddhist views, have considered the possibility of multiple neurobiological consciousnesses (e.g. visual consciousness, etc.), corresponding to different functional brain domains (Crick and Koch 2003; Weiskrantz 2003). A differentiation of consciousness from other higher order brain processes, for example attention, has been examined experimentally (Koch and Tsuchiya 2006), and the interest for subconscious mechanisms was recently rekindled in the neurobiological community. Self-consciousness however, which arises in large, i.e. complex enough mammalian brains and may be considered as the crown of consciousness processes, could not be localized to any one single brain structure. There is indeed no evidence for an "ego center", and a reduction of self-consciousness can only been observed after widespread, diffuse and bilateral reductions of cortical partners, as seen in dementia.

The function of the brain hemispheres is supported by a highly complex non-linear oscillatory coherent system, comprising many billions of interaction possibilities between thalamus and cortex and between cortical areas. A correlation has been observed between the conscious state and the production of high frequencies, i.e. above 13 Hz (Llinas et al. 1998). A subtotal dominance of low frequencies (between 1 and 13 Hz) correlates with an unconscious state, i.e. sleep (Steriade et al. 1990, 1997). During wakefulness, cognitive (Michels et al. 2008) and meditative processes (Austin 1998) have been characterized by variable increases of low and high frequencies. An even stronger increase of low and high frequency production correlates with clinical states of hyperfunctionality, as seen in functional brain disorders like phantom pain, tinnitus, tremor, epilepsy and psychosis (Jeanmonod et al. 1996; Llinas et al. 1999). Thanks to the combined increase in both low and high frequencies, both hypo- and hyperstates may appear in the relevant disease domains (akinesia and tremor, cognitive reduction and hallucinations, respectively, etc.).

The central unsolved issue of consciousness remains the existence of "qualia". Who is it that experiences, moments after moments, consciousness? Who has/uses his/her brain and its consciousness(es)? Who experiences these brain productions like emotions, ideas, memories? We do indeed rarely say that we are our brain, but rather that we have a brain, like we do for the rest of the body. Is consciousness really an emergent function of the brain? Is there evidence for non-local consciousness? What about intuitions and moments where we feel more connected than it seems? The scientific field of "psi" speaks for a different relationship between consciousness and the brain, a view supported by both quantum physics and spirituality: there is indeed serious statistical evidence for sharing of information beyond space and time and for mind/matter interactions (Radin 1997). This might indeed be an instance of non-locality in our macroscopic, human experiential world that has been described in quantum physics by Einstein-Podolsky-Rosen (EPR) correlations. Different quantum physical approaches to neurobiology (Stapp 2007; Penrose 1995) are considering central nervous system substrates for quantum collapse and entanglement at the synaptic and axonal levels. The electromagnetic field, as the product of the complex nonlinear thalamocortical network dynamics, has also been proposed by Mc Fadden (2000) as a candidate.

The whole group of "double slit" quantum physical experiments has received an immense and well deserved interest. In substance, an observer looking at a beam of light changes the behaviour of its constituents by choosing a measurement set-up, in accordance with the particle/wave duality. In addition, these experiments provide support for two most relevant propositions: first, the effect of the observer on the observed elements in an experiment cannot be explained by a neuronal process, and indicates the existence of a non-material dimension to consciousness (Schwartz et al. 2005). This goes with interpretations of quantum physics centered on the observer (Goswami 2001a; Stapp 2007). Second, there is an interaction between the quantum physical microdimension and the macroscopic world, in which the observer is active, thus contradicting comments claiming that quantum physical phenomena can only be active at the particle level and do not apply at the macrolevel. There is thus support for mind/matter interactions as seen in "psi" psychokinetic (PK) experiments and for the research centered on "weak" quantum interactions (Walach 2003; Schmidt et al. 2004; Lucadou et al. 2007). These elements are compatible with the view according to which consciousness would be primary to the brain (and even to all other material objects) and not its product. Similar to the context existing between a TV set and a TV broadcast, could consciousness be considered as a non-local, non-material field basically intrinsic to reality and using brains to generate multiple local daughter-consciousnesses? This view is fully in line with the oldest spiritual teachings of the human race, found beautifully described in the non-dualistic (or monistic) and idealistic hindu tradition of vedanta.

Consciousness and Spirituality

The non-dualistic (or monistic) and idealistic vedanta tradition, based on mystical experiences collected first in the Upanishads (Zaehner 1992), describes a unique unmanifested all-encompassing and ineffable Reality, named Brahman, which manifests itself as a fully interconnected network of different living or inanimate phenomena. These are seen as separate only as the result of an illusion, the roots of which lie in our ignorance of the real, hidden and non-dual nature of Reality. There is evidence for the existence of multiple levels of reality, which exist parallel to each other, do not disturb each other, and never affect the root Reality, source of all things. If one considers the teachings of vedanta and of other non-dualistic traditional systems (shivadvaita, buddhism, taoism, sufism, neoplatonism, jewish and christian mystical teachings, shamanism), one may be struck by the impressive homogeneity in the description of these two basic dimensions of unity and multiplicity/multidimensionality. They make us aware of the fact that differentiation is not separation, an awareness of high relevance in the social domain. In spiritual terms, the Source of all things is inside all things, contains them all (the One in all, and all in the One), but remains unaffected by them. This describes the paradox, unsolvable for our mental logics, of transcendence and immanence of the Source. We find again here the quantum physical combined evidence of non locality/unity (also called entanglement, see the EPR paradox above) and of multiple states or levels (see the double slit experiments above).

A metaphysical scheme arises based on these non-dualistic teachings, and is characterized by the existence of a succession of reality levels in a progression from the Source to the multiplicity of its manifestations. A basic requirement for this manifestation process to occur is that, from its start, a counterpart to manifestation arises, which observes/experiences it. This primordial duality in unity has been described in beautiful terms in the hindu interplay between Siva the first observer and Sakti the primordial energy and manifestation.

Starting from the material end, the first level is the physical, material, local, i.e. measurable and objective, level or sphere, which comprises, for the individual, the body, the brain but also, as a subtle emanation of it, the mind with its mental sphere. It may be proposed that, in spite of its subtlety, this local mind remains part of the material dimension, as it is produced by the brain and subserves all functions necessary for life in the local, material world. The second level is the non-material, decreasingly local, indirectly measurable and subjectively describable sphere. It consists of two sublevels, the "vital" or "energy" level and the even more subtle and less local "causal" or "mental" level. The first one relates to different phenomena described in the so-called energy medicine and which have to do with "energy" centres and auras in and around the material body (Goswami 2001b; Brennan 1988). The second one relates to scientific approaches of a basic field of information/energy, which connects all manifestations with each other (Mc Taggart 2008; Laszlo 2003). It has been called the akashic field in the hindu and the alaya-vijnana in the buddhist traditions. Next in this sequential description, but in fact primordial to all levels, comes the Potentiality (or Noumen) which becomes actual/phenomenal through the three depicted levels. An attempt to define it is possible only by negations, like non-material, non-local, non measurable, basically undescribable, i.e. purely experiential. This Source has been called, from East to West, the Tao, Paramasiva, Parabrahman, Sunyata, the Ground of Being, the Ultimate Reality, Ahura Mazda, Allah, En Sof, Spirit, the Oversoul, the One (without a second), the divine Father/Mother, the Godhead. Adding the idealistic component to the monistic one underlines the basic relevance of consciousness for phenomenal experience, causing the arising of misinterpretations and illusions hiding the ultimate reality. This in the same way as a dream state prevents the discovery of its own experiential and illusory nature as well as of the dreamer, transcendent and immanent to the dream.

Mystics have all, along centuries, given us the same amazing and reassuring message from the Source. An ultimate reality from which we may not, in spite of cumulated personal local evidence, be separated. The Good One, blissful awareness, giving birth to the Divine Father and Mother and their infinitely numerous offsprings, very special parents essentially unable to abandon their children.

It is proposed here, in summary, that the presented multidimensional evidence of a convergence of science and spirituality may soundly induce the following thought: these common elements of knowledge stem from a deep, or root level of reality, which conditions the discoveries of two seemingly totally different domains.

References

Austin, J.H. (1998). *Zen and the brain*. Cambridge: The MIT Press.

Brennan, B.A. (1988). *Hands of light. A guide to healing through the human energy field*. New York: Bantam Books.

Crick, F., & Koch, C. (2003). Consciousness and neuroscience. In B.J. Baars, W.P. Banks, & J.B. Newman (Eds.), *Essential sources in the scientific study of consciousness*. Cambridge: The MIT Press.

Dennett, D.C. (1991). *Consciousness explained*. Boston: Little Brown.

Goswami, A. (2001a). *The physicist's view of nature. Part 2: The quantum revolution*. New York: Kluwer Academic/Plenum Publishers.

Goswami, A. (2001b). *Physics of the soul*. Charlottesville: Hampton Roads Publishing Company.

Jeanmonod, D., Magnin, M., & Morel, A. (1996). Low-threshold calcium spike bursts in the human thalamus: Common physiopathology for sensory, motor and limbic positive symptoms. *Brain, 119*, 363–375.

Koch, C., & Tsuchiya, N. (2006). Attention and consciousness: Two distinct brain processes. *Trends in Cognitive Sciences, 11*, 16–22.

Laszlo, E. (2003). *The connectivity hypothesis. Foundations of an integral science of quantum, cosmos, life and consciousness*. Albany: State University of New York Press.

Llinas, R., Ribary, U., Contreras, D., & Pedroarena, C. (1998). The neuronal basis for consciousness. *Philosophical Transactions of the Royal Society of London B, 353*, 1841–1849.

Llinas, R.R., Ribary, U., Jeanmonod, D., Kronberg, E., & Mitra, P. P. (1999). Thalamocortical dysrhythmia: A neurological and neuropsychiatric syndrome characterized by magnetoencephalography. *Proceedings of the National Academy of Sciences of USA, 96*, 15222–15227.

Lucadou, Wv, Römer, H., et al. (2007). Synchronistic phenomena as entanglement correlations in generalized quantum theory. *Journal of Consciousness Studies, 14*, 50–74.

Mc Taggart, L. (2008). *The field. The quest for the secret force of the universe*. New York: Harper.

McFadden, J. (2000). *Quantum evolution*. London: Harper Collins Publishers.

Michels, L., Moazami-Goudarzi, M., Jeanmonod, D., & Sarnthein, J. (2008). EEG alpha distinguishes between cuneal and precuneal activation in working memory. *NeuroImage, 40*, 1296–1310.

Penrose, R. (1995). *Shadows of the mind. A search for the missing science of consciousness*. London: Vintage Science.

Radin, D. (1997). *The conscious universe*. San Francisco: Edge/Harper Collins Publishers.

Schmidt, S., Schneider, R., Utts, J., & Walach, H. (2004). Remote intention on electrodermal activity – Two meta-analyses. *British Journal of Psychology, 95*, 235–247.

Schwartz, J.M., Stapp, H.P., & Beauregard, M. (2005). Quantum physics in neuroscience and psychology: A neurophysiological model of mind-brain interaction. *Philosophical Transactions of the Royal Society B: Biological Sciences, 1458*, 1309–1328.

Stapp, H.P. (2007). *Mindful universe. Quantum mechanics and the participating observer*. Berlin: Springer.

Steriade, M., Jones, E.G., & Llinas, R.R. (1990). *Thalamic oscillations and signaling*. New York: Wiley.

Steriade, M., Jones, E.G., & McCormick, D.A. (1997). *Thalamus volume I: Organisation and function*. Amsterdam: Elsevier Science.

Walach, H. (2003) Generalized entanglement: Possible examples, empirical evidence, experimental tests. In: R. Schneider & R. Chez (Eds.), *Proceedings of generalized entanglement from a multidisciplinary perspective*, Freiburg.

Weiskrantz, L. (2003). Disconnected awareness for detecting, processing and remembering in neurological patients. In B.J. Baars, W.P. Banks, & J.B. Newman (Eds.), *Essential sources in the scientific study of consciousness*. Cambridge: MIT Press.

Zaehner, R.C. (1992). *Hindu scriptures*. New York: Everyman's Library.

Generalized Entanglement – A Nonreductive Option for a Phenomenologically Dualist and Ontologically Monist View of Consciousness

Harald Walach and Hartmann Römer

Abstract The conundrum with current models of consciousness is that they either deny consciousness its own causal role, defying everyday experience and phenomenology, or they concede consciousness its own causal activity, without explaining a potential interaction. While the first, physicalist, option is very much in line with most current reasoning within neuroscience it faces serious theoretical problems and has to exclude a range of phenomena in order to be convincing. The second, dualist model, is phenomenologically more satisfying, but cannot explain how such an interaction might work. This problem has beset philosophy since Descartes. We propose here a model that is ontologically monist, in line with the general intuition of the natural sciences, and at the same time phenomenologically dualist, true to our subjective experience. This is possible if we follow the track laid out by Generalized or Weak Quantum Theory. Such a model predicts generalized entanglement. This can be seen as a coordinating notion aligning two systems through a generalized non-local correlation. Using this model one can easily conceive of the mind-body relationship as a form of generalized entanglement correlating two systems with each other. In an extension, the same mechanism can be used to redefine spirituality as a coordination of single individuals with one Whole.

H. Walach (✉)
Faculty of Cultural Sciences, Viadrina European University,
Frankfurt (Oder), Germany

Samueli Institute, European Office, Frankfurt (Oder), Germany
e-mail: walach@europa-uni.de

H. Römer
Institute of Physics, University of Freiburg, Freiburg, Germany
e-mail: roemer@physik.uni-freiburg.de

H. Walach et al. (eds.), *Neuroscience, Consciousness and Spirituality*,
Studies in Neuroscience, Consciousness and Spirituality 1,
DOI 10.1007/978-94-007-2079-4_6,

Introduction: The Mainstream View and Its Problem

The Standard Physicalist View and Arguments in Favour

An implicit consensus within the neuroscience research community is that consciousness is produced by the brain (Damasio 1999; Metzinger 2000). The arguments for this assumption seem to be quite strong:

1. We know, from a long history of neuropsychology and neuropathology, that damage to certain brain areas leads to circumscribed and reproducible deficits (Damasio 2000). If the whole brain is damaged or if anesthesia interrupts neural transmission consciousness breaks down. The accumulated data make it plausible to postulate that brain activity of a certain kind is a necessary condition for consciousness. Note that brain activity alone is not sufficient. We have strong brain activity each night and yet fall unconscious during sleep. Also, in comatose patients brain activity can be recorded and yet they are unconscious. However, under normal circumstances brain activity is strongly associated with consciousness (Alkire et al. 2008).
2. We know that if certain areas of the brain are damaged, the thalamus for instance or areas of the brainstem, consciousness is irreversibly lost (Tononi 2004).
3. So far, nobody has observed conscious activity without brain activity. A few anomalies are around (see the chapter by Pim van Lommel in this book), but by and large, the correlation between brain activity and consciousness is very strong.
4. If we follow the useful physiological heuristic that every organ has a specific function and that the anatomy and physiology of an organ support this function, then it makes sense to assume that the brain is the organ whose function it is, among others, to produce some kind of consciousness. This is in strict analogy to other observations, for instance that it is the function of the lungs to exchange gas between the blood and the environment (and not to produce thoughts), and that it is the function of the gut to absorb nutrients and water and excrete waste products (and not to generate mathematical reasoning). In that sense it seems reasonable to assume that one major function of the brain, among others, is to produce cognition and as a consequence consciousness.
5. If we look at the correlation between brain anatomy, brain size and the evolutionary record it seems plausible to assume that an increase in brain complexity and size relative to the rest of the body is the driving force behind the success of the human race in evolutionary terms and that this success is paired with an increase in conscious activity (see Rossano in this book).
6. We can build technical equipment that mimics some aspects of brain architecture, for instance its strong connectivity, and implement some elements of learning as physiologically seen in single neurons, such as strengthening and weakening of synaptic connections. Such neuronal networks can simulate some cognitive activity such as learning and pattern recognition, or decision between complex alternatives in expert systems.

7. If we study the elements of brain activity, neurons and their physiology, we can see that their activity gives rise to the modifications correlated at least partially with changes in cognitive performance, emotional valence and bodily activity.
8. The concept of emergent properties can make it plausible how a system, by the virtue of its structure and complexity, can give rise to completely new properties (Metzinger 2003). For instance, a system of amino-acids in a certain structure and environment can suddenly give rise to properties that the single amino-acids did not have, namely the capability to sustain their own environment, or to reproduce or to move, or other phenomena generally associated with life (Maturana 1980). In the same vein, a complex neuronal system that is strongly interconnected, might give rise to a completely new property hitherto unknown, such as consciousness. There is nothing in the elements of the system, and nothing in previous kinds of systems that would predict the occurrence or the nature of the property to emerge. Hence, such complex emerging properties, such as consciousness are completely new, completely unexpected and wholly dependent on the physical organization of the system and its substrate, the brain (Baianu and Poli in press; Kronz and Tiehen 2002; van Gulick 2001). Although once emerged the new property, consciousness, might have a strong causal influence on its substrate, it still cannot exist without it.

All these data seem to suggest with overwhelming plausibility that brain activity and conscious activity are so strongly dependent on each other that a causation of conscious activity through brain activity is a plausible, if not inescapable conclusion.

Problems with the Standard Physicalist View

However, although currently espoused in one version or another by the majority of neuroscientists and a large number of active philosophers of mind, such a standard physicalist viewpoint has also attracted criticism that has not been alleviated by arguments from the mainstream camp as yet. A few arguments that speak against the plausibility of the mainstream view are the following:

1. Although the correlation between brain activity and conscious activity are very strong they are none the less only correlations. For instance, there are empirical instances of conscious activity seemingly without accompanying brain activity (see Pim van Lommel's chapter in this book). Similar types of brain activity can be associated with quite different types of conscious states. For instance strong delta and theta waves are characteristic both for certain sleep states, epileptic states (Petsche and Brazier 1972), and states of deep meditative absorption (Aftanas and Golocheikine 2001, 2002). Thus, similar physiological patterns are associated with different brain activities and diverse phenomenological states.
2. There is no theory as yet that really transforms the correlational hypothesis of neuroscience into a truly causative theory demonstrating convincingly that brain activity must be the cause of consciousness. There is no argument that shows that

the brain is not only a necessary but also sufficient condition for consciousness. To use a blunt example: The increased usage of refrigerators is strongly correlated with the decline of births in Western countries. However, no one would claim that using refrigerators decreases birth rates. The explanation is more complex and has to be sought in the social conditions where technical progress, social change, affluence and the changing role models for women are brought into the picture.

3. While the language of the brain is comparatively monotonous, conscious experience is extremely rich. In the brain, there is a pattern of electrical discharges and conductivity between neurons, mediated by different types of neurotransmitters and a rich variety of receptors. But the result of all this complex machinery is always the same: neurons either depolarize or don't. The only difference that can be observed from the outside is a change in rhythmicity and spiking of activity of certain neurons. But the language of the brain is always electric discharges, while the language of conscious experience is that of different qualities. Qualia, the subjective feel of certain experiences, can nowhere be found in the brain itself (Chalmers 1996). To jump from a physical description into a mental description without a mediating model is to make a category mistake (Hoche 2008).
4. Although we can build machinery that mimics certain aspect of brain activity there is no evidence whatsoever that this also produces consciousness (Searle 1992).
5. A physicalist view of the world is against all odds of our phenomenal experience. Here we have the clear subjective experience that we are agents operating through our body but not that our body and our brain processes determine what we do. Although in many instances there is a clear traceable consequence from neuronal processes to experience, and although some of these experiences are so compelling that we hardly can escape them – think of hunger, thirst, sexual desire, other forms of passion – nevertheless in nearly all instances there are examples how individuals, out of their free will, decided to not act on such impulses (Libet 1999), defying such physicalist causation.
6. The worldview that underlies a physicalist view of the world is predicated on an obsolete Newtonian physics (Schwartz et al. 2005). Although most of the time and for large scale objects Newtonian physics are a valid and good approximation, when it comes to the basic theoretical understanding of the relationship between mind and matter this is not good enough. A true physicalist picture of the world has to use the best and most fundamental theory of matter. This is quantum mechanics. But quantum mechanics, at least as a fundamental theory, only works if we presuppose conscious activity that takes into account the measurement result. This is one way of reminding us of the strong self-referentiality. And many authors, starting with Gödel, have pointed to the fact that as a consequence of such strong self-referentiality an essential incompleteness arises in the following sense: There are always statements whose truth cannot be decided by means internal to the system. One always has to make use of a reference frame outside the system, whose explanation or understanding is sought. If this is true for the framework of natural numbers already, as shown by Gödel, it is even true more so for the whole framework of a physicalist theory of consciousness.

7. We all operate and experience as covert dualists. Although we might be wrong here, as we were with the idea that the sun is turning around the earth, it seems to be very difficult to produce a plausible theory that would allow us to understand how immaterial events such as thoughts, or decisions, might impact on material events (Bieri 1989). To just say that the alleged immaterial events do not really exist is rhetoric, not science.
8. The physicalist view can only be maintained at the prize of excluding many phenomena that have been well documented, are relevant to many people (Ross and Joshi 1992), yet are neglected by mainstream science, such as experiences of telepathy (Schmidt et al. 2004), precognition (Utts 1996), mystical experiences (Walach 2007b), or similar ones. These phenomena are, as one common denominator, experiences of non-locality, where conscious experience seems to have access to information that is not available through known physical channels of information transfer and interaction (see also Pim van Lommel's chapter in this book). As a consequence, there needs to be a different theoretical framework for treating these phenomena, if they are to be taken seriously. We maintain that there is good reason to do so (Walach and Schmidt 2005).

The Problem

The problem, thus, seems to be twofold: The standard physicalist view of the world and of consciousness in particular, does not really account well for consciousness as a non-material phenomenological reality. If we think of pink elephants, smell the smell of aged Pinot Noir, taste white truffles, experience the pang of being in love, then we do not have elephants, wine, truffles or love in our brains but always electrical activity. Exactly how this comparatively similar event is translated into quite a different and rich language of phenomenology no physicalist theory has made plausible as yet. On the other hand, the same problem ensues: quite how an immaterial event such as a decision to not follow the impulse to smoke or drink alcohol, for instance, that lies at the heart of breaking addictions impacts on the physiology of the brain and on its whole architecture, is difficult to understand. Bieri has aptly described the conundrum as a trilemma (Bieri 1995). This consists of three sentences that each taken for itself is plausible, but together produce a contradiction: (1) The world of material events is causally closed. (2) Mental events are not physical events. (3) Mental events are causally effective. We can subscribe to two of the three sentences and produce a contradiction with the third.

The fact that most people seem to opt for a physicalist solution, at least – and mostly only – in theory, does not make the contradiction go away.

Hence, there is scope for an alternative approach. This approach should fulfill several requirements:

1. It should be true to the strongest theory of matter we have, quantum mechanics.
2. It should not be reductionist in the sense that it should allow for conscious experience both as partially autonomous of and in some way causative for material events.

3. At the same time it should account for the strong correlation observed between brain events and conscious events.
4. It should allow for phenomenological duality. At the same time it should, if possible, conform to the basic intuition of unity and monism that inspires science.

How is this at all possible? In the chapter by Römer and Walach in this book we assumed that physical or physiological and mental or phenomenal properties pertain to one and the same system (see also Römer (2004)). Loosely speaking, they are two sides of the same medal. Moreover, we argued that physiological and phenomenal observables are complementary in the sense of a Generalized Quantum Theory. This implies a correlation between the measured values of physiological observables on the one hand and phenomenal observables on the other hand, although, due to their complementarity, it is in general impossible to attribute sharp and definite values to both of them simultaneously. In what follows, we are going to develop a somewhat different description, which keeps mind and brain somewhat further apart. This time, they are associated with different subsystems of a larger system containing both of them. This is similar to an approach by H. Primas (2003) with the important difference that Primas considers a partition of an "unus mundus" into one mental and one material domain, whereas we assume many minds and brains. This alternative framework, seems to be particularly appropriate, if mind is considered as "soul", for problems of free will, for transpersonal phenomena or near death experiences (see van Lommel in this book). In such a model, the relationship between mind and matter is given by generalized entanglement correlations to be described in the next section. We will be drawing on a generalized formalism of quantum mechanics. This allows us to derive a non-local coordinating principle, generalized entanglement. Such a postulated mechanism would be exactly the coordinating principle we are seeking: it coordinates two tightly correlated systems, allowing for a phenomenological duality.

Weak Quantum Theory and Generalized Entanglement

Predecessor Ideas: Leibniz

Leibniz was the first prominent author to query a strong physicalist hypothesis in the discussions following Descartes. Locke and Boyle had, following physicalist tendencies, developed ideas that made conscious experience secondary to physical events. Leibniz countered this argument by his well known thought experiment (Bieri 1995): if we imagine the brain as a big machine which we could enter and inspect, walking through all paths and channels, we would never encounter a thought or another conscious activity, only physical activities. Thus, he conceived of physical and mental systems as two parallel systems, like two clocks running in perfect harmony. He coined the phrase "pre-established harmony" for this. In his essay

"Betrachtungen über die Lebensprinzipien und die plastischen Naturen – Contemplation on the Principles of Life and the Plastic Natures" he wrote:

> Souls follow their own laws,..., while bodies follow theirs, namely the rules of motion. Nevertheless, these two entities of completely different kind meet and are coordinated like two clocks, which have been perfectly set in the same way, although they may be of totally different making. It is exactly this which I call pre-established harmony. (Leibniz 1966, p. 65 f.)

Leibniz' idea of pre-established harmony did not meet with a lot of enthusiasm at the time and was buried by mechanistic reasoning following Newton. It is only with the advent of quantum mechanics that Leibniz' philosophy can be seen as a pre-emptive imaginative leap of insight that is much more akin to the type of reasoning that manifests itself in quantum theory than in Newtonian mechanics. The reason for this is the implicit non-locality that is germane both to Leibniz and to quantum mechanics.

Nonlocality in Quantum Theory

The formalism of quantum physics describes quantum systems by a state function that defines the whole system with all potential measurement outcomes at once. If, for instance, the system is a multi-particle system then the state function of the system fixes the joint probabilities for the outcomes of measurements on all of its constituent particles. This is much more information than the probabilities of measured values for each of the particles separately. Since the system with all its elements is governed by one single function, the analysis of such a situation yields insights into a hitherto unrecognised phenomenon that Schrödinger dubbed "entanglement" (Schrödinger 1935). It means that all parts of a quantum system that are governed by one state function behave in a coordinated fashion, and only certain combinations of measurement outcomes are likely or possible. Exactly which combination will be discovered on measurement is unpredictable. But it is predictable that if one measurement outcome is observed at one part of the system, then other outcomes are more likely for the other part of the system. In other words: which outcomes will be seen is unclear, but what is clear is that only certain joint results are likely or possible. The interesting thing about this quantum correlation is the fact that it holds, theoretically, across space and time, i.e. no matter how far distant elements of a system are in space, nor in time, provided the system is isolated well enough against interactions with the environment. This gives rise to what Einstein had called "spooky actions at a distance" (Einstein et al. 1935). It appears as if elements belonging to one system behave in a coordinated fashion although there is no signal travelling between the elements of the system informing them of their "theoretically supposed" behaviour or about what measurement value the counterpart of the system has just taken. In fact, entanglement correlation cannot be used for causal influences or signal transfer. (See e.g. (Lucadou et al. 2007)). This global coordination or correlation thus gives rise to what has been called quantum non-locality or quantum correlation or EPR-type correlations, named after Einstein, Podolsky and Rosen who were the first to exploit this quantum feature in order to demonstrate the potential lack of coherence of quantum theory.

As it turned out, empirical tests showed that quantum theory is quite correct and that quantum systems do have this peculiar property of behaving in a correlated, coordinated fashion no matter how widely parts of the system are separated in space or time. Nowadays, intricate tests of quantum entanglement have proven that such coherent quantum states can be maintained over many kilometres, making quantum teleportation of information or quantum encryption a technical feasibility, as well as the potential application of quantum computing (Salart et al. 2008). All these potential applications build on the reality of quantum entanglement and quantum non-locality.

However, it is very important to realise that such quantum entanglement proper is highly dependent on the fact that a quantum system can be isolated and maintained in isolation for a time period relevant enough for a measurement to be taken. In fact, each interaction of a quantum system with its environment is such a measurement, and as soon as such a measurement has taken place, quantum entanglement may be destroyed, and a classical, non-local world may ensue. The fact that quantum entanglement has been proven is only because quantum systems could be technically isolated for a long enough period of time. The challenge to engineer quantum entanglement for application purposes is associated with the technical difficulty of maintaining quantum systems in separation and preventing them from interacting with their environment. Technically this requires intricate precautions like very strong magnetic fields, ion-trappings in such fields, cooling, or optical and other devices that allow for quantum coherence.

In normal systems, such as the brain or other physiological macro-scale systems, interactions are much too numerous to maintain any trace of quantum entanglement, quantum coherence or non-locality.

Generalized Non-locality

The quantum physical formalism can be generalized and extrapolated to all kinds of systems beyond quantum physical systems proper, as we have shown (Atmanspacher et al. 2002; Filk and Römer 2011; see also the chapter by Römer and Walach in this book). We use the very same formal instruments that quantum physics uses. We then drop a couple of formal requirements and definitions. But we retain the most decisive element of quantum theory, the handling of non-commuting operations. This is intimately related to the fact, fundamental in quantum physics and assumed to be valid in its generalized form, that measurement will necessarily be related to a change of the state of the system and that the order in which different quantities are measured will in general be relevant. In the formalism of Generalized Quantum Theory this is implemented by non-commuting operators associated to observables of the system.

This is the formal expression of complementarity, which is decisive for quantum mechanics (see below). Once complementarity or non-commutativity is allowed into the formalism, we discover a strange and exciting prediction: Local elements of a system pertaining to its different parts whose descriptions are complementary to the global description of the system as a whole are non-locally correlated just as elements of a quantum system are correlated by entanglement correlations. In other words: entanglement or non-locality might not be restricted to quantum systems proper, but might be a feature of our world at large, provided systems obey the respective requirements of complementarity between local and global observables. Put still differently: what has become obvious in quantum mechanics, first formally through the theoretical description and then proven through experimentation, namely entanglement, might be a universal feature of our world. It became obvious in quantum mechanics, because the tight formalism of quantum mechanics made the conclusion inescapable and the precise theoretical description made experimental predictions possible that allowed for empirical testing.

Weak or generalized quantum theory stipulates that this feature holds true for all sorts of systems, provided global and local descriptions of the system are complementary. This is, at the moment, a theoretical stipulation that has to be confirmed by empirical evidence. In favour of it speaks some theoretical intuition that salient structural features of quantum physics are realised in a wider framework. In particular, measurement will change the state of a system under quite general circumstances. For instance, the mental state of a conscious individual will change by the very "measuring" act of becoming consciously aware. Moreover, what is a sufficiently rich description for material systems might also be useful to describe more complex systems, and the principle of analogy stipulates that what is true at one systemic level of description likely also holds at a higher level of systemic description. Generalized non-locality or generalized entanglement, thus, is, at this time, a theoretical prediction. Römer (2011) has given many examples where generalized entanglement may be at work. Just to list a few cases:

- It can be used to reconstruct everyday phenomena that are widely known, described in all cultures and at all ages, yet defy a cogent and plausible reconstruction within the framework of local theories, such as Newtonian mechanics. Parapsychological phenomena, such as telepathy, telekinesis, remote viewing and precognition can be reconstructed as non-local correlations, without the requirement of special signals or violating accepted laws of physics (Lucadou et al. 2007).
- Generalized entanglement has been used to reconstruct certain highly contentious areas of medicine, such as homeopathy, spiritual healing, or the Chinese medical concept of Chi (Walach 2003, 2005).
- The model of generalized non-locality can be used to understand transference and counter-transference phenomena in psychotherapy and close human relationships which are empirically well described but extremely difficult to theoretically conceptualise (Walach 2007).

- Finally, the model of generalized entanglement would lend itself to a theoretically elegant description of a coordinating mechanism within the body, between bodily systems and also between the mental and the physical system, as is proposed in this chapter.

Complementarity

Before we go into detail here, we must pause to examine the formal requirement that the model stipulates as a precondition for anything to be non-locally correlated, namely complementarity. Non-locality or entanglement is just a special case of complementarity, namely complementarity between local and global descriptions of a system. Clearly, all further discussions hinge on an appropriate understanding of the notion of complementarity. Niels Bohr, one of the founding fathers of quantum mechanics who had introduced the term complementarity into physics, never defined it clearly. On closer scrutiny, one can see that he uses at least three different descriptions of the term (Fahrenberg 1992; Hoche 1990).

1. On one level, he used the two mutually exclusive set-ups of wave and particle detection of light to determine complementarity at the experimental level.
2. Then, complementarity referred to two descriptions of a particle that are mutually contradictory yet necessary to describe it, such as location and momentum. While in classical physics these descriptions could be measured independently, in quantum mechanics we have the strange situation that measuring one means giving up any definite knowledge of the other, and vice versa. This is where Heisenberg's uncertainty relationship comes into play. It is in fact a formalisation of the complementarity relationship of two descriptors, yielding an uncertainty. While in quantum theory proper this relationship is defined, with Planck's constant defining the amount of uncertainty or non-commutativity, in the case of the Weak Quantum Theory this relationship is unrestricted, hence could be much smaller or much larger.
3. Finally, Bohr used the term complementarity for an epistemological relationship, where two general approaches or concepts were incompatible with each other, yet both belonged to it. Thus, he thought that the perspective of natural sciences and religion or concepts such as conscious and unconscious are complementary.

While the usages of the term complementarity in (1) and (2) are well defined in quantum mechanics, it is this latter usage beyond the realm of quantum mechanics that poses some difficulties. It might be helpful if we used a definition that is more general than the one used in quantum mechanics proper, distilled from Bohr, but never verbally produced by him (Atmanspacher 1996; Meyer-Abich 1965): *We can call "complementary" two descriptions of one and the same entity, event or system that are maximally incompatible with each other, yet have to be applied conjointly to describe this entity, event, or system.*

We are much in line with this definition except for the fact that we request incompatibility but not maximal incompatibility.

Table 1 Potential and proposed examples of complementary relationships outside physics

Global	Local	Area
Freedom	Structure	Education
Holistic	Linear	Processing, thinking
Love	Hate	Relationships
Appetence	Avoidance	Motivation
Explicit, unconscious	Implicit, conscious	Memory
Mercy	Justice	Law
Unity	Separation	Development
Function	Structure	Systems description
Whole	Part	Systems theory
Community	Individual	Society, systems description
Good	Evil	Morality, ethics

Some of these pairs may not constitute truly complementary relationships, or, rather, may sometimes be complementary in the sense used here, and sometimes not. This is a consequence of the fuzziness of our everyday language and the fact that we do not really distinguish between complementarity and opposites

It is useful to pause here to ponder on the implications. For most of our cultural, philosophical and scientific heritage we have not had much dealing with such concepts. Most of our everyday world follows the bivalent logic that Aristotle famously formalised and made the cornerstone of our scientific world-view: One of two assertions which contradict each other must be true. Something either has a description or it does not. This is the principle of the excluded third. In this logic, there is no place for complementarity, where one has to use incompatible descriptions to describe something. Although not formally part of our scientific culture, complementarity or the mode of thinking derived from it has been part of our heritage nevertheless. Our everyday world ("Lebenswelt" in the sense of Husserl) is full of examples. In personal relationships we often encounter situations where we both love and hate someone. Psychology is full of conflict situations that do not obey the simple logic of bivalent relationships. The famous conflict between appetence and avoidance, that is at the base of many neurotic problems, or the relationship between implicit and explicit, holistic and algorithmic processing, emotional and rational, conscious and unconscious, extensive and lexical memory, to name but a few, are examples of how complementarity might in fact be theoretically important also for areas other than quantum mechanics proper. Table 1 gives a few examples.

The most general and perhaps most important of these pairs might be "Whole" – "Part" or "Separation" and "Unity". These are two very general descriptions that govern almost all relationships. Depending on the level of analysis every part is at some point part of a whole that unifies it into a larger whole. In this case the complementarity between individual and community comes into play. If this is the case, then we would expect, following Generalized Quantum Theory, a non-local correlation or generalized entanglement between the parts of the whole, or between all elements of the system that belong to the system.

Thus, the model would actually predict a non-local, coordinating mechanism in any system that can be separated into subsystems and has sufficient cohesion to be called a whole. The same can be said of systems that are united to form a larger system. In other words, how an ensemble is partitioned also determines, whether one observes such correlations or not. Still put differently, entanglement as a systems property might be dependent on the observer.

Generalized Entanglement

Non-local Coordinating Principle Between Body and Consciousness

We now have our elements in place to propose a different, non-reductionist viewpoint in which we can combine a phenomenological duality with an ontological monism. We may assume that the underlying reality is one, beyond the distinction between mind and matter. Atmanspacher has formally shown that it is possible to construe the two systems as derived from one underlying unity through a spontaneous breaking of time-reversal symmetries (Atmanspacher 2003). This first breaking of an underlying symmetry would yield the distinction of two phenomenologically different systems, mind and matter, or material and conscious systems. Within one human being these systems might be coordinated with each other through the "mechanism" of generalized entanglement, or in other words, these two systems might be non-locally correlated. We have put inverse commas around the term "mechanism", as we normally mean by it any local mechanism in the sense that something is affecting something else using a signal exchange process or an interaction exchanging energy. A non-local process is clearly without exchange of energy and does not use signalling; this is its very definition. We are hereby making clear that we take this process to in fact fulfil the condition of explaining the "mechanics" of something without signal exchange. As our language does not offer any term at this time, we have no other option than saying it is similar to a mechanism, yet it is not a mechanism in the classical sense of the word. The mechanism is, as it were, an anti-mechanism not functioning mechanistically through signal exchange processes or energetic interaction, but non-locally, without such interaction, yet coordinated.

In such a model, consciousness and its physical substrate, the brain, or rather the whole body, can be seen as intimately linked, as in Leibniz' example of the two clocks that are of different make yet intricately coordinated. There is, however, no coordinating "something", as this "something" is the non-local correlation between the two systems. This model would explain why we have two phenomenologically different systems that are extremely tightly correlated. Hence we have a clear phenomenological duality with an underlying unity. Observe in passing that this model is not an ad-hoc parallelism, but is formally derived from the strongest theoretical model available to us so far, from quantum theory.

Theoretically, thus, the model is feasible and plausible. There is one caveat, though: at the moment our generalized or weak quantum theory is a systems theoretical description of very general scope. It can be applied. It can be used to make some predictions, such as the entanglement between a physical and mental system. It can be used to reconstruct a situation. However it is not precise enough to make more concrete predictions. And more importantly, it is nothing but a theoretical option at the moment that awaits some direct empirical verification. We hold that it has strong face validity and explanatory strength. This might recommend it and allow us to view its consequences with some confidence. If those provisions are duly taken into account we can see that the model provides us with what we have been looking for: a plausible account of a dualist phenomenological model of how a mental and a physical system might interact without postulating dualist ontology.

Spirituality: Non-local Correlation Between Whole and Individual

By the same token, we can now extrapolate the prediction. If we concede that such a non-local correlation operates between parts of a system and the whole system, then it is only a small step to accept that there is one system that can be called the Whole, comprising everything. By definition each subsystem is a part of this whole and is at the same time unified by it. Thus, the basic complementarity between part and whole also holds true here. We can now re-define spirituality as alignment of an individual with the Whole. Spiritual practice, such as meditation, prayer, contemplation, Chi Gong, Tai Chi, or Yoga, to name but the more prominent examples, can then be conceived as actions designed to increase the alignment of the individual with the Whole.[1] Thus, in the same way as elements within our body are coordinated by the organism at large, producing health, and our mental and physical life are coordinated to produce our conscious experience, we can conceive of all individual elements being coordinated and orchestrated into one Whole. This would at the same time give a very precise meaning to the common adage that all is dependent on everything else. There might indeed be a non-local reverberation of single events on other events or the whole. Thus, what Leibniz had called pre-established harmony would find a new and potentially naturalistic description in a global non-local correlatedness of all events with each other and the Whole.

References

Aftanas, L.I., & Golocheikine, S.A. (2001). Human anterior and frontal midline theta and lower alpha reflect emotionally positive state and internalized attention: High-resolution EEG investigation of meditation. *Neuroscience Letters, 310*, 57–60.

[1] Exactly what this "alignment" means would have to bet he content of another chapter, if not book. It likely means to bring tendencies of individualisation or separation in conformity or balance with the whole.

Aftanas, L.I., & Golocheikine, S.A. (2002). Non-linear dynamic complexity of the human EEG during meditation. *Neuroscience Letters, 330*, 143–146.

Alkire, M.T., Hudetz, A.G., & Tononi, G. (2008). Consciousness and anesthesia. *Science, 322*(5903), 876–880.

Atmanspacher, H. (1996). Erkenntnistheoretische Aspekte physikalischer Vorstellungen von Ganzheit. *Zeitschrift für Parapsychologie und Grenzgebiete der Psychologie, 38*, 20–45.

Atmanspacher, H. (2003). Mind and matter as asymptotically disjoint, inequivalent representations with broken time-reversal symmetry. *Biosystems, 68*, 19–30.

Atmanspacher, H., Römer, H., & Walach, H. (2002). Weak quantum theory: Complementarity and entanglement in physics and beyond. *Foundations of Physics, 32*, 379–406.

Baianu, I.C., & Poli, R. (in press). From simple to highly-complex systems: A paradigm shift towards non-Abelian emergent system dynamics and meta-levels. In B. Iantovics (Ed.), Conference Proceedings: Understanding Intelligent and Complex System. Alba Iulia: Acta Universitatis Apulensis.

Bieri, P. (1989). Schmerz: Eine Fallstudie zum Leib-Seele-Problem. In E. Pöppel (Ed.), *Gehirn und Bewusstsein* (pp. 125–134). Weinheim: VCH.

Bieri, P. (1995). Why is consciousness puzzling? In T. Metzinger (Ed.), *Conscious experience* (pp. 45–60). Thorverton: Imprint Academic.

Chalmers, D. J. (1996). *The conscious mind. In search of a fundamental theory*. New York/Oxford: Oxford University Press.

Damasio, A.R. (1999). How the brain creates the mind. *Scientific American, 281*, 74–79.

Damasio, A. (2000). *The feeling of what happens. Body, emotion, and the making of consciousness.* London: Vintage.

Einstein, A., Podolsky, B., & Rosen, N. (1935). Can quantum-mechanical description of reality be considered complete? *Physical Review, 47*, 777–780.

Fahrenberg, J. (1992). Komplementarität in der psychophysiologischen Forschung. Grundsätze und Forschungspraxis. In E.P. Fischer, H.S. Herzka, & K.H. Reich (Eds.), *Widersprüchliche Wirklichkeit. Neues Denken in Wissenschaft und Alltag: Komplementarität und Dialogik* (pp. 43–77). München: Piper.

Filk, T., & Römer, H. (2011). Generalized quantum theory: Overview and latest developments. *Axiomathes, 21*, 211–220.

Hoche, H.-U. (1990). *Einführung in das sprachanalytische Philosophieren*. Darmstadt: Wissenschaftliche Buchgesellschaft.

Hoche, H.-U. (2008). *Anthropological complementarism. Linguistic, logical, and phenomenological studies in support of a third way beyond dualism and monism*. Paderborn: Mentis Verlag.

Kronz, F.M., & Tiehen, J.T. (2002). Emergence and quantum mechanics. *Philosophy of Science, 69*, 324–347.

Leibniz, G. W. (1966). Betrachtungen über die Lebensprinzipien und über die plastischen Naturen. In E.Cassirer (Ed.), *Hauptschriften zur Grundlegung der Philosophie* (pp. 63–73). Hamburg: Meiner.

Libet, B. (1999). Do we have free will? *Journal of Consciousness Studies, 6*, 47–57.

Lucadou, Wv, Römer, H., & Walach, H. (2007). Synchronistic phenomena as entanglement correlations in generalized quantum theory. *Journal of Consciousness Studies, 14*, 50–74.

Maturana, H.R. (1980). Autopoiesis: Reproduction, heredity, and evolution. In M. Zeleny (Ed.), *Autopoiesis, dissipative structures and spontaneous social order* (pp. 45–79). Boulder: Westview.

Metzinger, T. (2000). *Neural correlates of consciousness: Empirical and conceptual questions.* Cambridge: MIT Press.

Metzinger, T. (2003). *Being no one: The self-model theory of subjectivity*. Cambridge: MIT Press.

Meyer-Abich, K.M. (1965). *Korrespondenz, Individualität und Komplementarität*. Wiesbaden: Steiner.

Petsche, H., & Brazier, M.A.B. (Eds.). (1972). *Synchronization of EEG activity in epilepsies.* Wien: Springer.

Primas, H. (2003). Time-entanglement between mind and matter. *Mind and Matter, 1*, 81–121.
Römer, H. (2004). Weak quantum theory and the emergence of time. *Mind and Matter, 2*(2), 105–125.
Römer, H. (2011). Verschränkung. In M. Knaup, T. Müller & P. Spät (Eds.), Post-Physikalismus (pp. 87–121). Freiburg: Karl Alber.
Ross, C.A., & Joshi, S. (1992). Paranormal experiences in the general population. *The Journal of Nervous and Mental Disease, 180*, 357–361.
Salart, D., Baas, A., Branciard, C., Gisin, N., & Zbinden, H. (2008). Testing spooky actions at a distance. *Nature, 454*, 861–864.
Schmidt, S., Schneider, R., Utts, J., & Walach, H. (2004). Remote intention on electrodermal activity – Two meta-analyses. *British Journal of Psychology, 95*, 235–247.
Schrödinger, E. (1935). Discussion of probability relations between separated systems. *Proceedings of the Cambridge Philosophical Society, 31*, 555–563.
Schwartz, J.M., Stapp, H.P., & Beauregard, M. (2005). Quantum physics in neuroscience and psychology: A neurophysiological model of mind-brain interaction. *Philosophical Transactions of the Royal Society B: Biological Sciences, 1458*, 1309–1328.
Searle, J.R. (1992). *The rediscovery of the mind*. Cambridge: Massachusetts Institute of Technology Press.
Tononi, G. (2004). An information integration theory of consciousness. *BMC Neuroscience, 5*, 42.
Utts, J. (1996). An assessment of the evidence for psychic functioning. *Journal of Scientific Exploration, 10*, 3–39.
van Gulick, R. (2001). Reduction, emergence and other recent options on the mind/body problem. A philosophical overview. *Journal of Consciousness Studies, 8*, 1–34.
Walach, H. (2003). Entanglement model of homeopathy as an example of generalizsed entanglement predicted by weak quantum theory. *Forschende Komplementärmedizin und Klassische Naturheilkunde, 10*, 192–200.
Walach, H. (2005). Generalized entanglement: A new theoretical model for understanding the effects of complementary and alternative medicine. *Journal of Alternative and Complementary Medicine, 11*, 549–559.
Walach, H. (2007a). Generalisierte Verschränkung – Ein theoretisches Modell zum Verständnis von Übertragungsphänomenen. *Zeitschrift für Psychotraumatologie, Psychotherapiewissenschaft, Psychologische Medizin, 5*, 9–23.
Walach, H. (2007b). Mind – body – spirituality. *Mind and Matter, 5*, 215–240.
Walach, H., & Schmidt, S. (2005). Repairing Plato's life boat with Ockham's razor: The important function of research in anomalies for mainstream science. *Journal of Consciousness Studies, 12*(2), 52–70.

Complementarity of Phenomenal and Physiological Observables: A Primer on Generalised Quantum Theory and Its Scope for Neuroscience and Consciousness Studies

Hartmann Römer and Harald Walach

Abstract We argue in this chapter that complementarity is a feature governing the relationship between neurophysiological aspects and phenomenological aspects of our mind. Hence a formal framework that is derived from quantum theory is applicable, generalized or weak quantum theory. This is a formal axiomatic framework that relaxes some of the requirements of quantum theory proper. Thereby it becomes relevant to more diverse kinds of systems, for instance to our mind. Basic elements of quantum theory are retained, such as the notion of observables, measurement, system, state of a system, and most importantly the handling of complementary or incompatible observables, such as physical and mental aspects of a human being. Allowing for complementary observables, however, also introduces by formal necessity an aspect peculiar to quantum theory, entanglement. We introduce the framework briefly and discuss how it might be useful for consciousness studies. We first show that complementarity has to be used to describe mental and physical states of the human mind. We show that the neuroreductive credo is not consistent with the analysis resulting from generalised quantum theory and that complementarity is an irreconcilable feature of our conscious existence. Hence generalised entanglement also becomes a notion that needs to be taken into account.

H. Römer (✉)
Physikalisches Institut der, Universität Freiburg, Freiburg, Germany
e-mail: hartmann.roemer@physik.uni-freiburg.de

H. Walach
Institute of Transcultural Health Studies, Viadrina European University, Frankfurt (Oder), Germany

Samueli Institute, European Office, Frankfurt (Oder), Germany
e-mail: walach@europa-uni.de

H. Walach et al. (eds.), *Neuroscience, Consciousness and Spirituality*,
Studies in Neuroscience, Consciousness and Spirituality 1,
DOI 10.1007/978-94-007-2079-4_7,

Introduction

To all of us introspection provides a privileged access to our own state of mind, open at any time and on any occasion. Moreover, access to other people's minds is given by interpreting their communications such as behaviour, words, postures and facial expressions under the overwhelmingly plausible hypothesis that not only their bodily appearance but also their mental organisation is very similar to ours. The huge body of cultural knowledge of mankind about its own interior life is almost exclusively derived from this single source. Adopting a philosophical term we shall call all knowledge coming directly or indirectly from introspection *phenomenal data.*

Only very recently, as seen from a historical perspective, these phenomenal data on the human mind have been supplemented by *neurophysiological data* on the neuronal activity in our brain. The first device for obtaining neurophysiological data was the EEG (electroencephalogram), giving signals of the neuronal activity in various parts of the brain with low spatial and only moderate temporal resolution. More detailed signals can be gathered by fMRI (functional magnetic resonance imaging), PET (positron emission tomography), SPECT (single photon emission tomography), MEG (magnetoencephalography) with better local resolution. In addition, information on the activity of individual neurons is accessible by means of precisely placed microelectrodes, either in animals or in clinical cases of neurosurgery for epileptic patients, for instance.

The contribution of neurophysiological data to our vast corpus of knowledge and experience concerning the human mind is very small indeed. Even so, particular epistemological dignity and significance is attributed to them in the spirit of the prevailing reductionist scientific world view. This high esteem culminates in what might be called the *strong neuroreductive credo*:

> All features of the human mind can (at least in principle) be reduced to and understood in terms of neurophysiological data.

At present, most neuroscientists in the western world would probably subscribe to this credo.

In this chapter, we investigate the mutual relationship of phenomenal and neurophysiological data and argue that in many cases it will be complementary in a quantum theoretical sense (Römer 2004) (see also chapter by Walach and Römer in this volume). Such an argument requires a theoretical framework, which (a) comprises essential features of physical quantum theory and (b) allows treating self observational and neurophysiological data on an equal footing under a notion of generalised measurement.

Such a formal framework is available as "*Weak Quantum Theory*" or "*Generalised Quantum Theory*" (GQT) (Atmanspacher et al. 2002, 2006; Filk and Römer 2011).

This is a generalisation of physical quantum theory, applicable to all sorts of systems in a very general sense. In GQT quantum concepts like complementarity and entanglement are formally well defined and applicable notions. GQT will be the framework of our considerations. We shall proceed as follows: The next section contains the bare minimum of GQT for making this presentation reasonably self sustained. In section on "Statement of the problem" the problem of complementarity between phenomenal and neurophysiological observables will be described, and in section on "Arguments for complementarity" our arguments in favour of complementarity will be given.

Basics of Generalised Quantum Theory

Generalised Quantum theory arose from physical quantum theory in algebraic formulation by weakening or omitting axioms. Thus a broadening of the range of applicability was achieved. Notions taken from physical quantum theory are:

System: A *system* is everything which (at least in principle) can be separated from the rest of the world and be turned into the object of a study. It may be possible to identify *subsystems* within a system.

State: A system can reside or can be thought to reside in different *states* without losing its identity as a system. The notion of states contains an epistemic element, because it also expresses the amount of knowledge about a system. One may further distinguish between *pure states,* which correspond to maximal attainable knowledge about the system and *mixed states,* in which maximal knowledge is not given. In contradistinction to quantum physics, in GQT the set Z of all states z need not be describable in terms of an underlying vector space.

Observable: *Observables* correspond to all features of a system, which can be investigated in a (more or less) meaningful way. If a system has subsystems, one may distinguish between *global observables* pertaining to the system as a whole and *local observables* pertaining to subsystems.

Measurement: A *measurement* of an observable A is made by performing the investigation which belongs to A and arriving at a result α, which can claim the status of a fact. Exactly how this is to be done depends on the detailed description of the system. The set of all possible results α of a measurement of the observable A is called the *spectrum of A* and is denoted by specA. The result of a measurement of A will depend on the state z of the system but will in general not be completely determined by z.

GQT is defined by a set of axioms, for whose precise form we refer to the original publications (Atmanspacher et al. 2002, 2006; Filk and Römer 2011). Here we only point out the most salient features:

Observables A can be identified with functions associating to every state z a state A(z). In general we have $z \neq A(z)$. In classical systems this is only true for mixed states, because a measurement will increase our knowledge of the system, whereas pure states will remain unchanged by measurement. In quantum like systems we generically have $z \neq A(z)$ even for pure states. Observables A and B can be concatenated by applying A after B to states z: $AB(z) = A(B(z))$.

Observables A and B are called *compatible*, if AB = BA and *incompatible* or *complementary*, if AB ≠ BA. Observables A and B are compatible if and only if the corresponding measurements are interchangeable. If observables are compatible one can say that the measurements of these observables commute, otherwise they do not commute. In Filk and Römer (2011) a slightly generalised formulation of GQT was given in which the action of observables on states is only defined for proposition observables (see below). Such differences do not matter for the argument in this contribution.

Complementarity: Complementarity of observables is a genuine quantum feature and does not occur in classical systems, i.e. in systems that can be described using classical mechanics. The most distinguishing feature between classical and quantum or non-classical systems is the fact that in classical systems measurements leave the system undisturbed, hence the sequence of measurements of different observables is irrelevant, whereas in a quantum system measurements change the system and hence the sequence is relevant. This gives rise to what in quantum theory is termed non-commutativity of measurements, which is a formal expression for complementarity, which again is a defining feature of systems that need a non-classical treatment. In a more general setting beyond quantum physics, complementarity will arise whenever a change of the state z of a system by performing measurement on it is inevitable. This is true in an exemplary way for systems containing conscious individuals with the ability of self observation, such as ourselves, because self observation necessarily changes our state of mind. This is likely also relevant for all natural systems that are not only reactive but to some degree agents since every action – measurement in our terminology – will result in a change of state and hence a further measurement will yield a different result.

Entanglement: Entanglement is another defining feature of quantum physics which results immediately from the fact that complementary descriptions are necessary to treat quantum systems. It denotes the fact that subsystems of a quantum system behave in a correlated or coordinated fashion, for instance when measured, because they belong to one system and cannot, strictly speaking, be separated. A similar kind of entanglement is to be expected, for the same formal reasons, in GQT when a global observable A is complementary to a local observables B_i pertaining to subsystems. In such a situation the state z of the system is an *entangled state* z, for which

$$AB_i(z) \neq B_iA(z).$$

Propositions are special observables P with PP = P and $\mathrm{spec}P \subset \{\mathrm{yes,no}\}$. They simply correspond to yes-no questions about the system. For every proposition P there is a negation $\neg P$ compatible with P. For compatible propositions P_1 and P_2 there exists a *conjunction* $P_1 \wedge P_2 = P_1 P_2$ and an *adjunction* $P_1 \vee P_2 = \neg(\neg P_1 \wedge \neg P_2)$. Certain laws of ordinary propositional logic (excluding associativity) are assumed to hold for compatible propositions.

For the arguments to be presented in section on "Arguments for complementarity" we have to mention some axioms of GQT referring to propositions:

If z is a state and P is a proposition, and if a measurement of P in the state z gives the answer "yes" then P(z) is a state for which P is true with certainty. This emphasizes the constructive nature of measurement as preparation and verification.

The following property generalises the spectral property of observables in ordinary quantum theory. To every observable A and every element $\alpha \in$ specA there belongs a proposition A_α which is just the proposition that α is the outcome of a measurement of A. Then

$$A_\alpha A_\beta = A_\beta A_\alpha = 0 \quad \text{for} \quad \alpha \neq \beta, \alpha, \beta \in specA, \tag{1}$$

$$AA_\alpha = A_\alpha A, \quad \bigvee_{\alpha \in specA} A_\alpha = 1 \tag{2}$$

where 0 and 1 are just the trivial propositions which are never and always true, respectively. Moreover, an observable B commutes with A if and only if all B_β commute with all A_α. The sets of projectors $\{A_\alpha\}$ or $\{B_\beta\}$ are called *complete sets of commuting propositions.* If the proposition A_α yields the answer "yes" in the state z, the state $z_\alpha = A_\alpha z$ is an *eigenstate of the Observable A with eigenvalue* α, a state for which a measurement of A will give the result α with certainty.

GQT has found a considerable number of applications, for which we refer to Atmanspacher et al. (2004, 2008); beim Graben and Atmanspacher (2006); Lucadou et al. (2007); Römer (2004, 2006a, b, 2011); Walach (2003)

Statement of the Problem

Neurophysiological data pertaining to states of the brain and phenomenal data of the human mind differ so much that they almost seem to come from different worlds.

Neurophysiological data belong to the realm of biology, chemistry, and ultimately to that of physics. In all cases, they are obtained by an external observer at the end of a chain of devices and causal relationships and they are explained mainly in terms of causality by notions such as stimulus and response, or cause and effect, for instance if the magnetic moment of a haemoglobin molecule gives rise to a magnetic signal that is picked up by a magnetic resonance spectroscopy device, converted by a computer program into a voxel in a three-dimensional statistical phase space, analysed by a statistical program and converted into a false-colour map on the computer screen, and ultimately interpreted by a neuroscientist as a proxy for metabolic demand in certain brain areas.

On the other hand, phenomenal data are immediately available to an internal "observer" by introspection. Here, a subjective human mind has direct access to some of its own activities and contents. In addition, notions of denotation, referentiality to self and other, intentionality and second level self-referential evaluation such as in emotionality, alien to physical data, are vital for understanding phenomenal data. These typically refer to something else, usually outside the human mind, are often colored by and related to intentions, plans and desires, and go along with emotional validations. The inner perspective of an internal "observer" is often called the ***first person perspective*** as opposed to the *third person perspective* of a physical observer.

There is another very important difference between neurophysiological and phenomenal data. Viewed as a physical system, the human brain can almost certainly be described by classical physics, and quantum physics likely does not play an important role for understanding it. There are, however, some claims that the human brain and consciousness need to be understood in terms of quantum physics (Atmanspacher 2006; Beck 2001; Beck and Eccles 1992; Hagan et al. 2002; Hameroff and Penrose 1996). In these attempts, true quantum processes are supposed to be active either in the synapses between the neurons, or in the microtubuli inside the neurons. But quantum physics almost exclusively rules the microworld, and if these quantum approaches are to produce more than just some small random noise, mechanisms of very low plausibility, amplifying these fluctuations and making them macroscopically relevant have to be invoked (Hepp 1999).

On the other hand, as already mentioned in the previous section, the human mind **as seen from an internal first person perspective is a paradigmatic case of quantum** behaviour in the broader sense of GQT, because an introspective registration of the current state of mind will inevitably alter it. Of course, the reason for this quantum-like behaviour of the human mind is not quantum physics but a partial structural analogy with quantum physics in the sense of and as described by GQT. In fact, GQT is a general phenomenological theory for systems of all kinds incorporating both classical and quantum mechanics as special cases but mainly devised for macroscopic systems with behaviour analogous to quantum physical ones. The uncertainties of the outcome of a measurement in GQT need not be genuine quantum indeterminacies. In many cases they are of more innocent origin, such as incomplete knowledge and inevitable perturbations by measurements. Beim Graben and Atmanspacher (2006) have shown that even systems obeying classical mechanics can show quantum features of GQT like complementarity after suitable "coarse grained" partitioning of the state space. This remark will be important in the next section. In contradistinction to quantum physics, the general formalism of GQT does not allow for a derivation of "no go" theorems for the existence of underlying classical "hidden variable" systems in the way Bell's inequalities (Nielsen and Chuang 2000) rule out local hidden variables and the Kochen-Specker theorem (Kochen and Specker 1967) rules out context free hidden variable theories. A system with quantum-like behaviour according to GQT **may in some cases have a classical mechanical refinement.**

We want to clarify the relationship between neurophysiological and phenomenal data in the framework of GQT. To achieve this, a little obstacle has to be overcome: The very notion of an observable requires an observer, and observables can only be compared if they pertain to the same system and are measured by the same or at least equivalent observers.

However, neurophysiological and phenomenal data are normally taken by **different observers, an external one in the first and an internal one in the second** case. But phenomenal data can be communicated, as a rule, to the same external observer who also takes the neurophysiological data without complete loss of their salient features. In this sense, we can speak of a human being as a system of GQT with both neurophysiological observables N_1, N_2, N_3,... and phenomenal observables P_1, P_2, P_3,.... In quantum theoretical language, the external observer takes over the role of a *superobserver*, who observes a measurement of an internal observer.[1]

The main problem we have to deal with in applying GQT can be stated as follows:

The human brain is a classical system, and for such systems all observables are commuting and compatible without any chance for complementarity. Now the strong neuroreductive credo spelled out in the Introduction claims that every feature of the human mind can be described in terms of neurophysiological data. This seems to imply that every phenomenal observable is a function of neurophysiological observables, symbolically:

$$P = f(N_1, N_2, N_3, \ldots)$$

If this is true, then also all phenomenal observables have to commute with one another and with all neurophysiological observables. On the other hand, we have argued that complementarity is typical for phenomenal observables, and we want to show, that neurophysiological and phenomenal observables will often be complementary.

One way out is of course to question the strong neuroreductive credo, but, although we have strong doubts about the credo, we shall try to develop stronger arguments which work without this step. In one of the arguments presented in the following section we shall even use a weak version of the neuroreductive credo which follows from the strong version without implying it. We would like to call it the *weak neuroreductive credo* (WNC):

> Every state of the human mind has a neuronal correlate and different states have different correlates.

We see no reason to exclude the possibility that the same state of the human mind may have different neuronal correlates.

Arguments for Complementarity

To be able to argue in favour of complementarity for two observables A and B we need a convenient criterion for complementarity. The axioms quoted at the end of section on "Basics of generalised quantum theory" suggest that the existence or nonexistence of joint eigenstates of A and B should be decisive. In section

[1] In the special case that a person registers his or her own neurophysiological data, there is also the possibility to "internalise" these data. We shall not elaborate on this situation, which is largely analogous to the more important situation described previously.

"Basics of generalised quantum theory" we already saw that two observables A and B are compatible if and only if the associated complete sets of compatible propositions $\{A_\alpha\}$ and $\{B_\beta\}$ commute with one another. This implies that A and B are compatible if and only if there is a complete set $\{A_\alpha B_\beta\} = \{B_\beta A_\alpha\}$ of joint compatible propositions. Starting from these propositions we can construct states, which are simultaneously both eigenstates of A with eigenvalue α and of B with eigenvalue β. As a corollary we can state that the observables A and B are complementary if and only if there is at least one α in specA for which no common eigenstate $z_{\alpha\beta}$ exists. A forteriori A and B are certainly complementary if they do not possess any common eigenstate. In the following we give three arguments for the possibility of complementarity between phenomenal and neurophysiological observables which make use of this criterion

(A) As described in section on "Statement of the problem", a "measurement" of a **phenomenal observable A is first performed by introspection, and the result is** subsequently communicated to an external observer who may also measure a neurophysiological observable B. Both with respect to A and B the external observer is in the position of a superobserver taking measurements of a system, inside which another and different measurement is performed. Now, if a person performs a measurement of the phenomenal observable A, the very act of self observation and conscious registration of its result will inevitably change the mental state of this person. By the weak neuroreductive credo this change of the mental state will be accompanied by a change of the neurophysiological state which is registered by the external observer. Hence, a measurement of a phenomenal observable always is accompanied by a change of the neurophysiological state which is measured by the observable B. In contradistinction, a common eigenstate $z_{\alpha\beta}$ of A and B would be unaffected by a measurement of either A or B. This means that there is no common eigenstate of the phenomenal observable A and the neurophysiological observable B, and hence the relationship between A and B must be one that is complementary.

(B) The difference between substance ontology and process ontology is a recurrent subject of contemporary philosophy. Römer (2006a, b) provides a detailed discussion in terms of GQT.

Substance observables pertain to properties of stable substances, whereas process observables refer to changes and transitions. Typically, substance propositions are expressed by nominal sentences, and process propositions are expressed by verbal sentences. It was argued in Römer (2006a, b) that substance observables should be complementary to process observables. The reason is a vital difference in their relationship to a time observable T. Substance observables commute with T, and an eigenstate of a substance observable can be assumed to be an eigenstate of T, too. In sharp contrast to this, process observables do not commute with T and will change the value of T. This means that there are no common eigenstates of substance and process observables. This implies complementarity between them. A neurophysiological state is described by the states of neurons, and thus neurophysiological observables are substance observables, if we think about it, since they can be ascribed one precise state variable at a time. On the other hand, phenomenal observables most of the time are process observables. Sometimes there is no precise state-description of a phenomenal variable, e.g. when a phenomenal state cannot be fully explicated, and sometimes phenomenal variables are such that they change subsistent states of other phenomenal variables or impact on neurophysiological observables. Hence, we have at least to expect some complementarity between neurophysiological observables and a large class of phenomenal observables.

(C) The third argument employs the notion of complementary partitions as introduced by beim Graben and Atmanspacher (2006). The human language is not rich enough for a complete description of all phenomenal states of the human mind, and every characterisation in terms of phenomenal observables is coarse grained and therefore contains an inevitable element of vagueness. A complete description of the neurophysiological state at any one point in time of the human brain is practically impossible, because it would consist of a description of the states of roughly 10^{12} neurons. The state of a few neurons gives a very incomplete description of the total neurophysiological state and even the most modern imaging procedures have a spatial resolution orders of magnitudes more coarse grained than the distance of two neighbouring neurons. In addition, the temporal resolution is poor compared to typical neuronal time scales. Thus, imaging procedures yield only rough space-time averages. One should also keep in mind that neurophysiological states are frequently characterised by referring to phenomenal observables with their inherent fuzziness. Thus, phenomenal and accessible neurophysiological observables only give coarse grained partitions of the full set of states.

Moreover, the topologies of phenomenal and neurophysiological states are quite different.

Two clearly separable phenomenal states may correspond to very similar neurophysiological states, whose difference cannot be resolved by realisable neurophysiological observables. Vice versa, clearly different neurophysiological states may give rise to very similar phenomenal states. In fact, the proverbial covariation problem or lack of correlation between phenomenal psychological states and corresponding physiological states has haunted psychophysiology since its inception. In such a situation, an eigenstate of a neurophysiological observable will imply indeterminacies of phenomenal observables and an eigenstate of a phenomenal observable will be beset with indeterminacies of neurophysiological observables. In this case there will be no common eigenstates of certain phenomenal and neurophysiological observables, and the criterion for complementarity will be fulfilled. Of course, this situation of complementarity is not always realised. It is not expected to hold for sensomotoric phenomena or for dispositional states like hunger or sexual arousal, which are associated with clearly distinguishable neuronal excitation patterns. For instance, different parts of the retina are mapped onto distinguishable regions of the visual cortex, and different parts of the human body correspond to different regions of the parietal somatosensory cortex. In these cases, the complementarities described under the points (A) and (B) are irrelevant in the same sense as complementarity in quantum mechanics is negligible as long as classical mechanics is a valid approximation. Indeed, a visual excitation pattern, or a state of hunger are not essentially changed by becoming conscious.

In other situations, however, complementarity arising out of a certain partitioning of a whole will be of decisive importance. For example, this will be the case for the subtle and highly unstable stream of consciousness, which is redirected by every act of conscious registration. Two direct consequences of this complementarity between phenomenal and neurophysiological observables arise and should be noted:

- Detailed "thought reading" by means of neuronal imaging is impossible.
- Even if the strong neuroreductive credo holds true, it refers to an unrealisable situation. Because of their complementarity to realisable neurophysiological observables, phenomenal observables are indispensable for a full description of the human mind in the same sense as in quantum mechanics it is impossible to dispose of spatial observables in favour of momentum observables.

Thus, it seems, that complementarity is an irreducible notion which needs to be applied to the relationship between phenomenal, subjective states or first-person descriptions, and neurophysiological states or third-person descriptions. Arguably, at least for some cases, namely where registration of a phenomenal state through a conscious event takes place and thus changes the very state including a potential neuro-physiological correlate, such complementarity is irreducible. If that is the case, though, our generalised or weak formalism of quantum theory applies, and by the same token, we can expect generalised entanglement to play a role. This is further explored in the chapter by Walach & Römer in this book.

References

Atmanspacher, H. (2006). Quantum approaches to consciousness. In E.N. Zalta (Ed.), *Stanford encyclopedia of philosophy*. Stanford: Metaphysics Research Lab.

Atmanspacher, H., Römer, H., & Walach, H. (2002). Weak quantum theory: Complementarity and entanglement in physics and beyond. *Foundations of Physics, 32*, 379–406.

Atmanspacher, H., Filk, T., & Römer, H. (2004). Quantum Zeno features of bistable perception. *Biological Cybernetics, 90*, 33–40.

Atmanspacher, H., Filk, T., & Römer, H. (2006). Weak quantum theory: Formal framework and selected applications. In A. Khrennikov (Ed.), *Quantum theory: Reconsiderations of foundations – American Institute of Physics, conference proceedings*. New York: Melville.

Atmanspacher, H., Bach, M., Filk, T., Kornmeier, J., & Römer, H. (2008). Cognitive time scales in a Necker-Zeno-model for bistable perception. *The Open Cybernetics and Systems Journal, 2*, 234–251.

Beck, F. (2001). Quantum brain dynamics and consciousness. In P. Van Looke (Ed.), *The physical nature of consciousness* (pp. 83–116). Amsterdam: John Benjamins.

Beck, F., & Eccles, J.C. (1992). Quantum aspects of brain activity and the role of consciousness. *Proceedings of the National Academy of Science of the USA, 89*, 111357–111361.

beim Graben, P., & Atmanspacher, H. (2006). Complementarity in classical dynamical systems. *Foundations of Physics, 36*(2), 291–306.

Filk, T., & Römer, H. (2011). Generalized quantum theory: Overview and latest developments. *Axiomathes, 21*, 211–220.

Hagan, S., Hameroff, S. R., & Tuszynski, J.A. (2002). Quantum computation in brain microtubules: Decoherence and biological feasibility. *Physical Review E, 65*, 61901-1–61901-11.

Hameroff, S., & Penrose, R. (1996). Conscious events as orchestrated space-time selections. *Journal of Consciousness Studies, 2*(1), 36–53.

Hepp, K. (1999). Toward the demolition of a computational quantum brain. In P. Blanchard & A. Jadczyk (Eds.), *Quantum future from Volta and Como to the present and beyond* (pp. 92–104). Berlin: Springer.

Kochen, S., & Specker, E. (1967). The problem of hidden variables in quantum mechanics. *Journal of Mathematics and Mechanics, 17*, 59–87.

Lucadou, Wv, Römer, H., & Walach, H. (2007). Synchronistic phenomena as entanglement correlations in generalized quantum theory. *Journal of Consciousness Studies, 14*, 50–74.

Nielsen, M.A., & Chuang, I.L. (2000). *Quantum computation and quantum information*. Cambridge/New York: Cambridge University Press.

Römer, H. (2004). Weak quantum theory and the emergence of time. *Mind and Matter, 2*(2), 105–125.

Römer, H. (2006a). Complementarity of substance and process. *Mind and Matter, 4*, 69–89.

Römer, H. (2006b). Substanz, Veränderung und Komplementarität. *Philosophisches Jahrbuch, 113*, 118–136.

Römer, H. (2011). Verschränkung. In M. Knaup, T. Müller & P. Spät (Eds.), Post-Physikalismus (pp. 87–121). Freiburg: Karl Alber.

Walach, H. (2003). Entanglement model of homeopathy as an example of generalizsed entanglement predicted by weak quantum theory. *Forschende Komplementärmedizin und Klassische Naturheilkunde, 10*, 192–200.

Hard Problems in Philosophy of Mind and Physics: Do They Point to Spirituality as a Solution?

Nikolaus von Stillfried

Abstract I suggest that there exists an interesting and little known relationship between Neuroscience, Consciousness and Spirituality. To illustrate this, I first outline the paradoxical relation between the subjectivity of mind (i.e. consciousness) and its objective material correlate (i.e. neuroscience). I then give support to the notion that this paradox is rationally unsolvable by showing that it is isomorphic to the wave-particle paradox in quantum physics, where the impossibility to rationally resolve it has eventually been accepted as a fundamental property of reality, called the complementarity principle. Next, I point out that spiritual (mystical) traditions have also arrived at very similar paradoxical descriptions of reality, which lends additional plausibility to the insights from quantum physics and philosophy of mind (and vice versa!). Finally, and most importantly, I suggest that since mystical practices offer ways to individually transcend logical paradoxa by developing non-dual, transrational states of consciousness, they may provide a solution to fundamental theoretical problems such as those outlined above and should thus be regarded as an indispensible part of any advanced research methodology.

N. von Stillfried (✉)
Institute for Environmental Health Sciences, University Medical Center Freiburg, Freiburg, Germany
e-mail: n.stillfried@web.de

H. Walach et al. (eds.), *Neuroscience, Consciousness and Spirituality*,
Studies in Neuroscience, Consciousness and Spirituality 1,
DOI 10.1007/978-94-007-2079-4_8, © Springer Science+Business Media B.V. 2011

The Hard Problem of Consciousness

On the theoretical level, any discussion involving neuroscience and consciousness sooner or later has to address the question of how the two relate to each other, often called "the mind-body problem" (e.g. Young 1990) or "the hard problem of consciousness" (Chalmers 1995b).[1]

The conceptualisation of this problem, in the form in which it is familiar to us today, is usually attributed to René Descartes who in the first half of the seventeenth century most prominently introduced the distinction of "res cogitans" vs. "res extensa", i.e. of 'the thinking substance' vs. 'the substance which extends in space' (my translation). More recent characterizations of conscious experience do not focus so much on its 'thinking'-nature, but on its subjective qualitative aspects, the so called 'qualia', in other words the experience that it is 'like something' to be conscious (see e.g. Chalmers 1995a; Nagel 1974; Shear 1997).

Whatever particular words are used, at the heart of the issue is essentially the realization that subjective experience and objective material reality seem to be of categorically different quality. As soon as this distinction is made, the question arises of how these categories relate to each other. Our experience clearly tells us that they are strongly correlated. But how is this possible, given their fundamentally distinct natures?

Throughout history, several avenues have been pursued in trying to provide an answer to this question, ranging from monistic reductions of mind to matter and vice versa, to dualist constructs and the denial of the existence of subjective consciousness altogether. Here is not the place to enter into a detailed discussion of all the available proposals. It shall suffice to say that thus far none of these approaches has been able to gain anything like general acceptance among philosophers of mind. This is quite understandable given that all of them have obvious shortfalls: Monist explanations suffer from the difficulty to explain how qualia should be reducible to a material reality totally devoid of subjectivity or, vice versa, how mass and spatial extension of matter should derive from a mental reality which does not even in the most rudimentary sense display these characteristics. Dualist approaches, on the other hand, leave open the question of how such different 'substances' are coordinated or interact with each other so as to result in the intimate correlation between them. A middle way has been put forward under the heading 'neutral monism' or 'dual aspect monism', proposing that consciousness and matter are two aspects or manifestations of a third, neutral substance. Here, of course, the problem is how to conceptualize this mysterious third substance, which somehow must unite material and non-material, subjective and objective properties.

---

[1] Here, the label 'hard' was meant to accentuate the differentiation between the practically demanding but paradigmatically not so challenging questions regarding the form a particular content of consciousness takes on, in correspondence to the characteristics of the correlated neuronal processes, and the more fundamental question about why there is any conscious, i.e. subjective and qualitative experience at all, and how it relates to its physical counterpart.

While Descartes still had the cultural freedom (or duty) to attribute such impossible tasks to God, this hypothesis has not enjoyed a lot of interest in more recent times. As I will try to argue, however, there might be good reasons to reconsider a somewhat updated version of it.

My argument is that not only do all existing theoretical frameworks fail to give a coherent explanation of the relationship between consciousness and its neurobiological counterpart, but that such a framework is in fact *in principle* impossible to conceive of rationally.

Wave-Particle Duality and the Complementarity Principle

This somewhat audacious conjecture, namely that there cannot be a rational solution to the 'hard problem' even though it is a real problem, is to some extent inspired by a similar situation in the natural sciences, namely quantum physics. Here a centuries-long dispute focused on the question whether light is composed of particles or waves. It began, to my knowledge, as a disagreement between Isaac Newton (1704/1979), who speculated that light was a stream of particles (then called corpuscles) and one of his contemporaries, Christiaan Huygens, who believed that light was a wave (Huygens 1690/1912). It was seemingly decided in favour of the latter by the observation of interference patterns in the double slit experiments by Thomas Young (1807), because such interference effects can only be explained by the assumption that light is wave-like. At the beginning of the twentieth century, however, investigations of the so called photoelectric effect by Albert Einstein (1905, 1909) and Robert Andrews Millikan (1916), among others, showed again that light had properties that could only be explained if one regarded light as composed of particles, later to be called photons. The problem was aggravated by the work of Luis de Broglie who showed that the electron, which was then thought of as a particle, also required a wave-type description. In fact, De Broglie argued, all matter has to be attributed, in addition to its discrete corpuscular nature, a wave nature (De Broglie 1925, 1926).

This situation presented to the physics community a rather serious problem, because the wave and particle nature of a photon, electron or any other object mutually exclude each other: A wave is in multiple places at once and can interfere with itself while a particle is localised in a limited region of space and cannot produce interference on its own. As Werner Heisenberg put it:

> The two pictures are of course mutually exclusive, because a certain thing cannot at the same time be a particle (i.e., a substance confined to a very small volume) and a wave (i.e., a field spread out over a large space) (Heisenberg 1958, p. 49)

In short, there is nothing wave-like to a particle and nothing particle-like to a wave. So how could these opposing characteristics be reconciled?

The only 'solution' to this problem which remains contested but unsurpassed until today was eventually formulated by Niels Bohr in form of the complementarity principle (Bohr 1928). It basically states that the problem cannot be solved but instead it has to be regarded as a fundamental principle of physical reality that its components can only be comprehensively described by two mutually exclusive descriptors.

As I have shown in more detail elsewhere, this paradox shows remarkable structural similarities to the hard problem of consciousness (see e.g. von Stillfried 2010). Not only have all attempts to reduce the wave-nature to particle-nature (or vice versa) failed or at best led to formulations of new paradoxes. It is also an up to now notoriously unsolved question how the two 'aspects' of a particular quantum inform each other in such a way, that, for example, the location in which a quantum is detected (in its particle-nature) is clearly indicative of the interference it has undergone earlier in its wave-nature.

I am by no means the first to suggest that the complementarity principle might apply to consciousness and body in an analogous fashion as it does to wave and particle. A number of authors have voiced similar views (e.g. Brody and Oppenheim 1969; Edelheit 1976a, b; Fahrenberg 1979, 2007; Feigl 1972; Hoche 1990, 2007, 2008; Nakagomi 2003; Primas 2007, 2009; Tang 1996; von Stillfried and Walach 2006a, b; Walach 2005, 2007; Walach and Römer 2000; Walach et al. 2006). Max Velmans has also pointed to a similarity with "quantum complementarity" (Velmans 1991, 1993, 1995, 2002, 2009), but maintained that "psychological complementarity" differs in some important ways (Velmans 2000, 2009). Thomas Filk and Albrecht von Müller have pointed out similarities between quantum physics and consciousness (Filk and von Müller 2009) but do not understand consciousness and matter as complementary (Filk, personal communication, October 28, 2009).

More historically, Bohr himself regarded the physical and the psychological aspect of existence as complementary (Bohr 1934, p. 24), even though he never published any detailed exploration of this idea. Other founding fathers of quantum theory seem to have shared his view: Werner Heisenberg for example points out that complementarity to him is a compelling analogy (e.g. Heisenberg 1971, p. 115). Wolfgang Pauli obviously thought along the same lines, but also did not provide much detail to support the idea.

> It would be most satisfactory of all if physics and psyche could be seen as complementary aspects of the same reality. (Pauli 1955, p. 207–208)

Despite these and probably other authors drawing a connection between the hard problem in the philosophy of mind and that of complementarity in quantum physics, the view of consciousness and matter as complementary has, to my knowledge, not yet pervaded the current discourse on the 'hard problem of consciousness' in any major way. One reason for this could be that, although complementarity may in some sense provide a solution, it does not really provide an answer, at least not of the kind most philosophers probably hoped for. The kind of answer our rational minds seem to be seeking is a logically consistent, rationally understandable, *unitary* description of reality. Why this is, I can only guess. It may have to do with our experience of being *one* individual living in *one* reality. It may be the result of a cultural conditioning by monotheistic religions or binary logical systems. It may be a feature of our mind that developed due to some evolutionary advantage. It may be the inner compass of a soul remembering primordial oneness and seeking to return.

Whatever the reasons for our longing for *one* answer, for *the* understanding of *the* ultimate reality are, the complementarity principle does not satisfy it; two mutually exclusive descriptions are required for a complete description of our observations on reality, in other words: the most fundamental descriptions we can arrive at are paradoxa .

Spirituality

Not all is lost, however; on the contrary: In some peculiar way, the complementarity principle may provide us with a hint on how to continue our search. By confronting us with an inescapable paradox both at the very basis of physical reality, as well as with respect to our most intimate first person experience, it very clearly indicates an absolute limit of rationality as a means to gain ultimate understanding of ourselves and the universe. In doing so, it lends credibility to spiritual traditions which for millennia have claimed just that: the inadequacy of the rational mind when it comes to anything absolute.[2]

Instead of relying on rationality to decipher the ultimate nature of reality, these mystical traditions teach methodologies that have been developed in order to calm and transcend the mind and arrive at an immediate experience of the absolute. Interestingly, some of these techniques, such as the 'Koan' practice in 'Rinzai Zen', actually consist in the intensive engagement with a paradox (sometimes over decades) until the mind capitulates, the paradox disappears and a 'higher' state of consciousness is experienced (Oshima 1985 and personal communication, 2005). Alternatively, a plethora of diverse techniques are available, involving among other elements for example certain modes of sustained attention, breath, movement, psychoactive substances, ritual, visual stimuli, music, contemplation of sacred texts etc., sometimes in combination with extensive guidelines for ethical behaviour or precise dietary prescriptions. Apart from that, mystical experiences are also known to occur spontaneously and unintentionally.

And, according to what is reported from mystical experiences, there actually *is* an experiencable unity behind or beyond the paradoxical duality of existence in this universe. In fact, surveys of reports from mystical experiences quickly reveal that the unity of opposites is one of the most common features of mystical experiences:

> [f]undamental opposites appear as unified, laws of logic as abolished, and normal intellectual functions as replaced by a 'higher' mode. (Wulff 2000, p. 397–440, see also e.g. Stace 1960 or Daniels 2003 and others)

[2] I use the word spirituality here to denote the experiential aspect of transcendence, the mystical core of all religions.

For this reason they are also referred to as "non-dual", "acategorial" or "transrational" states of consciousness (e.g. Atmanspacher and Fach 2005; Gebser 1986 and Taylor 1984).

Accordingly, both Western and Eastern mystical traditions almost ubiquitously provide a wealth of teachings about opposites, paradoxa and their relationship to ultimate reality. One of the most well known examples is probably the description of the universe as composed of the opposite forces "Yīn" and "Yáng" in Daoism and Confucianism (Mou 2001) and the according graphic representation ☯, also called the "Tàijítú", literally: "diagram of the supreme ultimate" (Chunqiu 2003). Interestingly, when in 1947 Niels Bohr was ennobled with the Danish Order of the Elephant for his achievements, he chose precisely this symbol for his coat of arms.[3]

It also comes as small surprise, then, that the logical analysis known as 'fourfold negation', which seems to most accurately describe situations which we might characterise by complementarity, was also adopted in the introspective tradition of Buddhism with regard to questions concerning the ultimate nature of reality: For example Nagarjuna, arguably one of the most influential Buddhist teachers after Gautama Buddha himself (in approx. the second or third century CE), analyzed different questions regarding reality (such as "Do things exist out of themselves or are they caused by others?"). Using the so-called "tetralemma", he showed that all four possible answers (alternative A is true, alternative B is true, both are true, neither is true) are flawed[4] (Napper 2003). (Comparing this analysis to complementarity, we can see that in some way the question "what is a quantum?" can neither be answered by "a particle" nor "a wave" nor "both" nor "neither".)

Obviously, again, such a (non-)answer is not really satisfying to the rational mind, in fact, plainly speaking, it hardly makes any rational sense at all. The problem seems to be that while the truth behind these statements may indeed be experienced, it is impossible to convey it using words alone. This may be one of the reason why mysticism is called just that, etymologically rooted in *gr.* mystos = keeping silent (Daniels 2003).[5]

---

[3]A photograph of this coat of arms is available at http://www.nbi.dk/hehi/logo/crest.html (last accessed April 20th 2010)

[4]It may be reassuring to note that Nagarjuna also made clear, that this applies primarily to so called ultimate truths (paramartha satya), whereas on the level of so called conventional, instrumental or relative truth (samvriti satya) definite answers and binary logic can be considered adequate (e.g. Scott 1995).

[5]Of course, if we analyze it precisely, no experience can ever be conveyed *exactly* using only words or any other means of communication. Firstly, language is coarse and simplistic compared to experience which is subtle, fluid and highly complex. Secondly, any communication can only serve to call forth in the 'receiver' a new or remembered experience of his or her own, which will therefore never be exactly the same as the 'sender's'. The difference between normal experiences (such as seeing the colour red) and a mystical experience is that for the former it is more likely that two people have both had it. A mystical experience is in this sense only mystical for someone who has not had it. And for someone who has never seen colours, seeing red is in this sense mystical.

A common reaction then of rational people to such seemingly absurd claims about reality is to doubt the sincerity, sanity and/or sobriety of the person making them. A more subtle critical position is held by those who maintain that we can never be sure to experience reality directly because it always has to be mediated by the senses and the respective perceptual neural structures. Therefore, it is often claimed, non-dual experiences resemble hallucinations in that they are merely a product of some neuronal processes rather than telling us anything about reality. With the knowledge, however, that not even the sharpest thinkers in physics have been able to come up with a satisfying or even just a different solution to the paradox presented by physical reality (as studied by the most objective methodologies science can muster) we might now be somewhat more inclined to suspend judgement for just a moment and reconsider: What has complementarity taught us about reality?

Interestingly, we can observe that even when we use pure logic to interpret complementarity, say of wave and particle or mind and matter, it is almost inevitable to arrive at similar nonsensically meaningful statements as Nagarjuna. Let us, as a playful exercise, think through such a rational interpretation e.g. by asking: Why is it that reality is complementary? What does the existence of the complementarity principle tell us about reality? Since we know that the complementary categories are mutually incompatible, meaning that there is nothing that they have in common, it becomes logically inevitable that any reality uniting them must have nothing in common with either. From all we know, however, they (being either in one place or in many, being either material or mental) are the only possibilities of existing. All that is left, then, is nothing(ness); leading to yet another all-time favourite of mystical paradoxes, the equivalence of all and nothing. The only escape from this conclusion is that the complementary categories are in fact two different realities, which are not contained in a common reality. In this case, however, they cannot both have come into existence out of nothing (because that would again unite them) so they have to have existed forever without beginning. Infinity, however, is again a concept that is rather common in descriptions of mystical experiences, but that our rational mind tends to be rather uncomfortable with.

If I follow the argument unfolding here, there is limited use in producing more non-sense words about the ultimate nature of reality. Let us therefore just quickly sum up by returning to the starting point: What does it all mean with respect to the hard problem of consciousness?

In my opinion, mystical experiences have to be considered an invaluable tool for gaining understanding of the relationship between consciousness and matter. As Ken Wilber put it:

> The "hard problem" – the jump to qualia (i.e. how can exterior quantities give rise to interior qualities?) – is finally solved, not by seeing that every exterior has an interior, since that merely says that they are correlative (and leaves the hard problem still pretty hard) – but by *developing* to the nondual realm, whereupon the problem is radically (dis)solved. The solution is what is seen in satori,[6] not anything that can be stated in rational terms (unless one has

[6] Satori denotes a mystical peak or enlightenment experience in Zen Buddhist terminology, literally (*jap.*) "understanding" (author's note).

> had a satori, and then rational terms will work fine). The reason the hard problem cannot be solved – and has not yet been solved – in rational and empirical terms is that the solution does not exist at those levels. Philosophical geniuses trying to solve the mind-body problem at that level have failed (by their own accounts) not because they are stupid, but because it can't be solved at that level, period. (Wilber 2000, Chapter 14, note 15, p. 282)

For this reason, practices facilitating trans-rational states of consciousness should be studied on a broad base in academia not only from a third person objective but also from a first person experiential point of view by individual scientists.[7]

In which specific ways the integration of such practices into the scientific endeavour will transform science with regard to both its content and its culture, I could at best speculate at this moment in time.

Possibly it will engender and necessitate a whole new science of inner epistemology yet to be invented (as suggested e.g. by Walach and Runehov 2010).

Probably, the effects are going to be quite varied among individuals, disciplines and cultural contexts.

What I am certain, however, is that to the extent it will allow us to experience aspects of reality which otherwise are not accessible, it will make our understanding of the universe and ourselves more comprehensive and render more adequate our according interactions.

Acknowledgment The author gratefully acknowledges helpful comments from Harald Walach and Stefan Schmidt and funding from the Fetzer-Franklin Fund.

References

Atmanspacher, H., & Fach, W. (2005). Acategoriality as mental instability. *The Journal of Mind and Behavior, 26*(3), 181–206.

Bohr, N. (1928). The quantum postulate and the recent development of atomic theory. *Nature, 121*(3050), 580–591.

Bohr, N. (1934). *Atomic theory and the description of nature*. New York: Cambridge University Press (Republished 1961).

Braud, W., & Anderson, R. (1998). *Transpersonal research methods for the social sciences: Honoring human experience*. Sage Publications, Inc.

Brody, N., & Oppenheim, P. (1969). Application of Bohr's principle of complementarity to the mind-body problem. *The Journal of Philosophy, 66*, 97–113.

Chalmers, D.J. (1995a). *The conscious mind: In search of a fundamental theory*. New York: Oxford University Press.

Chalmers, D.J. (1995b). Facing up to the problem of consciousness. *Journal of Consciousness Studies, 2*(3), 200–219.

Chunqiu, L. (2003). The taiji diagram: A meta-sign in chinese thought. *Journal of Chinese Philosophy, 30(2)*, 195–218.

Daniels, M. (2003). Making sense of mysticism. *The Transpersonal Psychology Review, 7*(1), 39–55.

[7]Such a statement is capitalizing on quite similar and earlier ones claiming e.g. "state dependent knowledge" (Tart 1986) or "transpersonal research methods" (Braud and Anderson 1998).

De Broglie, L. (1925). Recherches sur la théorie des quanta. *Annales de physique,* 3, 22–128 (Republished (1992) in Ann. Found. Louis de Broglie (17) p. 22).
De Broglie, L. (1926). *Ondes et mouvements.* Paris: Gauthier-Villars.
Edelheit, H. (1976a). Complementarity as a rule in psychological research – Jackson, Freud and the mind/body problem. *The International Journal of Psycho-Analysis, 57*, 23–29.
Edelheit, H. (1976b). Complementarity as a rule in psychological research. Jackson, Freud and the mind/body problem. *The International Journal of Psychoanalysis, 57*(1–2), 23–36.
Einstein, A. (1905). Über einen die Erzeugung und Verwandlung des Lichtes betreffenden heuristischen Gesichtspunkt. *Annalen der Physik,*(17), 132–148 (Republished (2005) Annalen der Physik 14(11): 164–181).
Einstein, A. (1909). Über die Entwicklung unserer Anschauungen über das Wesen und die Konstitution der Strahlung. *Physikalische Zeitschrift, 10*, 817–825.
Fahrenberg, J. (1979). Das Komplementaritätsprinzip in der psychophysiologischen Forschung und psychosomatischen Medizin. *Zeitschrift für Klinische Psychologie und Psychotherapie, 27*, 151–167.
Fahrenberg, J. (2007). *Gehirn und Bewusstsein: Neuro-Philosophische Kontroversen*. Göttingen: Hogrefe.
Feigl, H. (1972). *The "mental" and the "physical"*. Minneapolis: University of Minnesota Press.
Filk, T., & von Müller, A. (2009). Quantum physics and consciousness: The quest for a common conceptual foundation. *Mind and Matter, 7*(1), 59–79.
Gebser, J. (1986). *The ever-present origin*. Athens: Ohio University Press.
Heisenberg, W. (1958). *Physics and philosophy; The revolution in modern science*. New York: Harper.
Heisenberg, W. (1971). *Physics and beyond*. New York: Harper & Row.
Hoche, H.U. (1990). Anthropologische Komplementarität und die 'Einheit der Sache'. Versuch einer skeptischen Lösung eines skeptischen Zweifels. In H. U. Hoche (Ed.), *Einführung in das sprachanalytische Philosophieren* (pp. 107–129). Darmstadt: Wissenschaftliche Buchgesellschaft.
Hoche, H.U. (2007). Reflexive monism versus complementarism: An analysis and criticism of the conceptual groundwork of Max Velmans's reflexive model of consciousness. *Phenomenology and the Cognitive Sciences, 6*(3), 389–409.
Hoche, H.U. (2008). *Anthropological complementarism. Linguistic, logical, and phenomenological studies in Support of a third way beyond dualism and monism*. Paderborn: Mentis.
Huygens, C. (1690). *Traité de la Lumière* (S. P Thompson, Trans). Leyden: University of Leyden (Republished (1912) as "Treatise on Light" , London: McMillan).
Millikan, R.A. (1916). A direct photoelectric determination of Planck's "h". *Physical Review, 7*(3), 355–388.
Mou, B. (2001). Becoming-being complementarity. *An account of the Yin-Yang metaphysical vision of the Yijing*. Retrieved May 2, 2008, from http://them.polylog.org/3/amb-en.htm.
Nagel, T. (1974). What is it like to be a bat? *Philosophical Review, 83*(4), 435–450.
Nakagomi, T. (2003). Quantum monadology and consciousness. *Biosystems, 69*(1), 27–38.
Napper, E. (2003). *Dependent arising and emptiness: A Tibetan Buddhist interpretation of Madhyamika philosophy*. Somerville: Wisdom Publications.
Newton, I. (1704). *Opticks: Or, a treatise of the reflections, refractions, inflections and colours of light*. London: Printed for Sam. Smith and Benj. Walford, Printers to the Royal Society (Republished (1979) Mineola, NY: Courier Dover Publications).
Oshima, Y. (1985). *Zen – anders denken?* Heidelberg: Verlag Lambert Schneider.
Pauli, W. (1955). The influence of archetypal ideas on the scientific theories of Kepler. In C. G. Jung & W. Pauli (Eds.), *The interpretation of nature and the psyche* (Bollingen series). New York: Pantheon Books. Chapter 2.
Primas, H. (2007). Non-boolean descriptions for mind-matter problems. *Mind and Matter, 5*, 7–44.
Primas, H. (2009). Complementarity of mind and matter. In H. Atmanspacher & H. Primas (Eds.), *Recasting reality: Wolfgang Pauli's philosophical ideas and contemporary science* (pp. 171–209). Berlin/New York: Springer.

Scott, D. (1995). Buddhist responses to Manichaeism: Mahayana reaffirmation of the" Middle Path"? *History of Religions, 35*(2), 148–162.
Shear, J. (Ed.). (1997). *Explaining consciousness: The hard problem*. Cambridge: MIT Press.
Stace, W.T. (1960). *Mysticism and philosophy*. Philadelphia: Lippincott.
Tang, P.C. L. (1996). Alternative representations and the complementarity model of mind-brain. In D. Peterson & D. Peterson (Eds.), *Forms of representation: An interdisciplinary theme for cognitive science* (p. 198). Exeter: Intellect Books.
Tart, C.T. (1986). Consciousness, altered states, and worlds of experience. *Journal of Transpersonal Psychology, 18*(2), 159–170.
Taylor, E. (1984). *William James on exceptional mental states*. Amherst: University of Massachusetts Press.
Velmans, M. (1991). Is human information processing conscious? *The Behavioral and Brain Sciences, 14*(4), 651–726.
Velmans, M. (1993). Consciousness, causality and complementarity. *The Behavioral and Brain Sciences, 16*(2), 409–416.
Velmans, M. (1995). The relation of consciousness to the material world. *Journal of Consciousness Studies, 2*(3), 255–265.
Velmans, M. (2000). *Understanding consciousness*. New York: Routledge.
Velmans, M. (2002). How could conscious experiences affect brains? *Journal of Consciousness Studies, 9*(11), 3–29.
Velmans, M. (2009). Psychophysical nature. In H. Atmanspacher & H. Primas (Eds.), *Recasting reality: Wolfgang Pauli's philosophical ideas and contemporary science* (pp. 115–134). Berlin/New York: Springer.
von Stillfried, N. (2010). Theoretical and empirical explorations of "Generalized Quantum Theory", Doctoral thesis, Kulturwissenschftliche Fakultät, Europa Universität Viadrina, Frankfurt a. d. Oder.
von Stillfried, N., & Walach, H. (2006a). Taking pre-established harmony beyond determinism: the complementarity principle applied to the mind-body problem. *Proceedings of: 'VIII. International Leibniz Congress'*, Hannover, Gottfried-Wilhelm-Leibnitz-Gesellschaft e.V.
von Stillfried, N., & Walach, H. (2006b). The whole and its parts: Are complementarity and non-locality intrinsic to closed systems? *International Journal of Computing Anticipatory Systems, 17*, 137–146.
Walach, H. (2005). The complementarity model of brain-body relationship. *Medical Hypotheses, 65*(2), 380–388.
Walach, H. (2007). Mind – body – spirituality. *Mind and Matter, 5*(2), 215–240.
Walach, H., & Römer, H. (2000). Complementarity is a useful concept for consciousness studies. A reminder. *Neuroendocrinology Letters, 21*, 221–232.
Walach, H., & Runehov, A. L. C. (2010). The epistemological status of transpersonal psychology: The data-base argument revisited. *Journal of Consciousness Studies, 17*(1-2), 145–165.
Walach, H., von Stillfried, N., & Römer, H. (2006). Pre-established harmony revisited: Generalised entanglement is a modern version of pre-established harmony. *Proceedings of: 'VIII. International Leibnitz Congress'*, Hannover, Gottfried-Willhelm-Leibniz-Gesellschaft e.V.
Wilber, K. (2000). *Integral psychology: Consciousness, spirit, psychology, therapy*. Boston: Shambhala Publications.
Wulff, D.M. (2000). Mystical experience. In E. Cardena, S. J. Lynn, & S. C. Krippner (Eds.), *Varieties of anomalous experience* (pp. 397–440). Washington, DC: American Psychological Association.
Young, T. (1807). *Course of lectures on natural philosophy and the mechanical arts*. London: J. Johnson.
Young, R.M. (1990). The mind-body problem. In R. C. Olby, G. N. Cantor, J. Christie, et al. (Eds.), *Companion to the history of modern science* (pp. 702–711). New York: Routledge.

Brain Structure and Meditation: How Spiritual Practice Shapes the Brain

Ulrich Ott, Britta K. Hölzel, and Dieter Vaitl

Abstract Meditation practices can be conceived as specific types of mental training with measureable effects on the function and structure of the human brain. This contribution narratively reviews recent morphometric studies that compared experienced meditators with matched controls. While meditation types and measures differed between studies, results were remarkably consistent. Differences in gray matter (GM) volume and density were found in circumscribed brain regions which are involved in interoception and in the regulation of arousal and emotions, namely insula, hippocampus, prefrontal cortex, and brainstem. The normal age-related decline in GM volume and in attentional performance was present in controls but not in meditators. These findings need to be replicated in longitudinal studies in order to confirm the causal role of meditation training. Future research has to elucidate effects of these structural changes on neural activity and mental functioning during behavioral tasks.

Introduction

For many centuries, meditation has been practiced by mystical branches of major religions for promoting spiritual development, for gaining insight into reality, and for attaining transcendental states of consciousness. From a scientific perspective, the effects of these traditional exercises are based on the plasticity of the brain.

U. Ott (✉) • D. Vaitl
Bender Institute of Neuroimaging, University of Giessen, Giessen, Germany
e-mail: ott@bion.de

B.K. Hölzel
Bender Institute of Neuroimaging, University of Giessen, Giessen, Germany

Massachusetts General Hospital, Harvard Medical School, Boston, MA, USA

H. Walach et al. (eds.), *Neuroscience, Consciousness and Spirituality*,
Studies in Neuroscience, Consciousness and Spirituality 1,
DOI 10.1007/978-94-007-2079-4_9,

Sustained efforts to focus attention and to cultivate emotional balance leave traces in the underlying neural substrate and circuitry. Over time, these changes in brain structure in turn support the intended changes in mental faculties and personality.

The current contribution reviews findings of structural differences in the brains of advanced meditation practitioners when compared to non-meditating controls. Increases in GM density and cortical thickness of specific brain regions may provide objective indicators for the enhancement of particular self-regulation skills. Meditation techniques involve the training and development of certain mental abilities or qualities, e.g. awareness of bodily sensations, focusing of attention, emotion regulation etc. Often such heightened skills and improved cognitive abilities are referred to as "expansion of consciousness". Significant improvements detectable at the cognitive-behavioral level, such as one's ability to control attention, regulate emotion, and bring awareness to bodily sensations, should also be mirrored in morphological changes at the neural level. The popular idea of "consciousness expansion through meditation" can thus be understood more scientifically by understanding how the underlying neural structures are modified by meditation practices.

Morphological Differences in Meditation Practitioners

Up to now, five studies on structural differences between meditation practitioners and controls have been conducted and will be reviewed here (for a summary of findings, see Fig. 1 and Tables 1 and 2).

The first study by Lazar et al. (2005) compared cortical thickness of 20 Buddhist insight meditation practitioners and 15 matched controls. Insight meditation practice aims at cultivating a nonjudgmental awareness of the internal and external stimuli present in each moment ("mindfulness"). On average, participants meditated for 9.1 years (SD = 7.1 years), practicing about 40 min/day.

Statistical analyses revealed differences in cortical thickness between groups in the right anterior insula and the right middle and superior frontal sulci. The cortex of meditation practitioners was significantly thicker in both regions. In the prefrontal cortex, the effect was most likely caused by an age-related decrease of cortical thickness in the control group which was absent in the meditation group. The authors argue that the strong effect in the right anterior insula could be due to the extensive training in breath awareness and in maintaining attention to visceral sensations.

Slowing of the breathing rate between a baseline condition and the first 6 min of meditation showed a strong correlation with the amount of practice and was taken as a physiological indicator of meditation experience. Within the meditation group this measure was correlated with cortical thickness in a region in the inferior occipito-temporal visual cortex and, when controlling for age, also with cortical thickness in the right anterior insula. The latter finding was taken as further evidence that training in interoceptive awareness during meditation could be responsible for increased cortical thickness in the right anterior insula, since this structure is

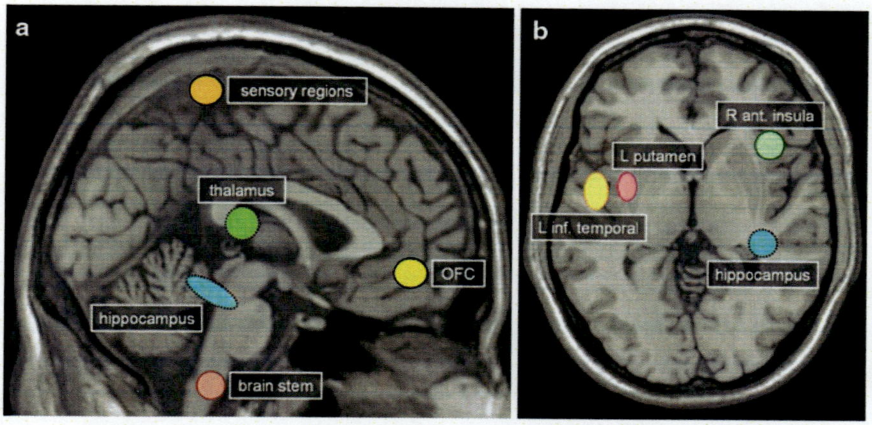

Fig. 1 Regions, in which differences between meditators and non-meditators were found: Thalamus (Luders et al. 2009), right hippocampus and left inferior temporal gyrus (Hölzel et al. 2008; Luders et al. 2009), orbito-frontal cortex (OFC; Luders et al. 2009), brain stem (Vestergaard-Poulsen et al. 2009), right anterior insula (Lazar et al. 2005; Hölzel et al. 2008), and sensory cortex (Lazar et al. 2005) (**a**) sagital view; (**b**) axial view. Regions that are not located in this plane are depicted with *dotted lines*

Table 1 Overview of morphometric studies on meditation

Authors, year	Meditation type (practice)	N Med/Con[a]	Measures	Main results
Lazar et al. (2005)	Insight meditation (9.1 years)	20/15	Cortical thickness	Med > Con: right anterior insula & prefrontal cortex Med: no decrease with age
Pagnoni and Cekic (2007)	Zen (>3 years)	13/13	GM[a] volume, attention task	Med: no age-related decline in left putamen; no decrease in response speed and accuracy
Hölzel et al. (2008)	Vipassana (8.6 years)	20/20	GM concentration	Med > Con: left inferior temporal gyrus, right anterior insula, right hippocampus
Vestergaard-Poulsen et al. (2009)	Tibetan Buddhist Meditation (16.5 years)	10/10	GM concentration and volume	Med > Con: solitary tract nucleus, left prefrontal cortex, cerebellum
Luders et al. (2009)	Zazen, Vipassana, Samatha and others (24.2 years)	22/22	GM volume	Med > Con: right orbito-frontal cortex, right thalamus, left inferior temporal gyrus, right hippocampus

[a]Note: *Med* Meditators, *Con* Controls, *GM* gray matter

Table 2 Summary of findings and interpretations

Structure	Studies	Meditation training	Mental faculties
Right anterior insula	Lazar et al. (2005) Hölzel et al. (2008)	Awareness of breathing sensations, body scan	Interoception, awareness of bodily feelings
Orbito-frontal cortex	Hölzel et al. (2008) Luders et al. (2009)	Equanimity, inhibition of automatic responding	Emotion regulation, modifying reactions to aversive stimuli
Right hippocampus	Hölzel et al. (2008) Luders et al. (2009)	Bodily relaxation while staying vigilant, distanced observing of thoughts and emotions	Regulation of arousal
Left inferior temporal gyrus	Hölzel et al. (2008) Luders et al. (2009)	Awareness of present moment, state of being	Mindful state, pleasure, connectedness
Right thalamus	Luders et al. (2009)	Attend to a chosen meditation object	Focusing of attention
Left putamen	Pagnoni and Cekic (2007)	Awareness of present moment, keeping static body position	Sustained attention
Brain stem, solitary tract nucleus	Vestergaard-Poulsen et al. (2009)	Observing a deep and regular breathing pattern	Respiratory and cardiovascular control

involved in the meta-representation of the body scheme, homeostasis, and associated visceral sensations.

Effects of meditation on GM volume and on cognitive performance were investigated in a subsequent study by Pagnoni and Cekic (2007). Here, 13 Zen meditators with more than 3 years of daily practice were compared to a same-size group of matched controls. Zen meditation was characterized as a state of openness towards the flow of mental events while maintaining a straight sitting posture and a natural breathing pattern. Analyses were performed with the voxel-based morphometry (VBM) toolbox (http://dbm.neuro.uni-jena.de/vbm) running under SPM5 (http://fil.ion.ucl.ac.uk/spm/software/spm5).

In controls, total GM volume was negatively correlated with age ($r = -0.54$, $p = 0.056$) whereas in the meditation group virtually no correlation was present ($r = 0.006$, $p = 0.83$). The Age × Group interaction for total GM volume failed to reach significance (ANCOVA: $t(19) = 1.82$, $p = 0.08$). However, a significant cluster for this interaction was found in the left putamen (combined threshold of $p = 0.001$, uncorrected, and cluster size $k > 1{,}000$ voxels), where GM volume even showed a trend to increase with age in the meditation group (controls: $r = -0.80$, $p = 0.0011$; meditators: $r = 0.55$, $p = 0.063$).

The authors also assessed cognitive performance of participants with a computerized attention task, which required monitoring of a series of digits and responding to target stimuli by pressing a button as fast as possible. Accuracy of responses and reaction times were used as performances measures.

Target sensitivity and speed of responses decreased significantly with age in the control group but not in the meditation group. According to the authors, this effect could be directly related to the differences in the left putamen, a region involved not only in motor control but also in attention processing and cognitive flexibility. Conscious regulation of attention and control of the body posture during meditation training could possibly counteract an age-related decline in this region and explain why elderly Zen practitioners retained a high level of cognitive performance.

Hölzel et al. (2008) studied 20 advanced practitioners of Vipassana meditation in the tradition of S. N. Goenka and 20 controls, matched for sex, age, education, and handedness. This meditation training is focused on awareness of breathing and attending to bodily sensations ("body scan"). On average, meditators had practiced 8.6 years (SD = 5.0 years) daily for 1 h in the morning and 1 h in the evening. Analysis of structural images was done with the VBM toolbox under SPM2. Results were reported for differences in GM concentration, i.e. the statistical probability that a voxel contains GM.

Meditators had a significantly higher concentration of GM in three regions: left inferior temporal gyrus, right anterior insula, and right hippocampus. The left inferior temporal gyrus was also found to be activated during meditation in a functional study with the same participants (Hölzel et al. 2007). Moreover, GM concentration in this region was correlated with the amount of meditation practice. Increased GM concentration in the right anterior insula replicated the finding by Lazar et al. (2005) and was presumably likewise related to the strong focus on interoceptive awareness in this meditation tradition. The third finding of increased GM concentration in the right hippocampus was attributed to training in arousal regulation. High levels of stress are known to impair neuronal growth in this brain region. As part of the limbic system, the hippocampus plays an important role in the appraisal of situations and emotional reactivity. The increase in GM in this region could reflect an enhanced ability to reduce autonomic arousal level and to maintain a state of inner peace and serenity in stressful circumstances.

Furthermore, GM concentration in the orbitofrontal cortex was positively correlated with meditation practice (whole-brain regression analysis for the meditation group, where the amount of practice was entered as a regressor). This region has been associated with the modification of responses to aversive stimuli, which is an integral part of emotion regulation training during meditation, namely the maintenance of equanimity when confronted with painful sensations.

A Danish research group (Vestergaard-Poulsen et al. 2009) investigated ten practitioners of Tibetan meditation involving attention of breathing, the cultivation of positive attitudes (loving-kindness, compassion), and a state of open awareness towards any content appearing in the mind. The experienced meditators (practice: M = 16.5 years; SD = 5.1 years; 2.2 h per day) were compared to an age-matched control group of equal size. High-resolution structural scans were analyzed with the VBM toolbox under SPM5.

A significant higher concentration of GM in meditators was found in circumscribed parts of the medulla oblongata, namely the solitary tract nucleus. This region of the brain stem is involved in the control of respiration and the vagal modulation of cardiac function. Increased GM concentration was also found in the prefrontal cortex (left superior and inferior frontal gyrus) and in the anterior lobe of the cerebellum. No correlation with the amount of practice was found. The authors argue that a ceiling effect in their group of highly experienced meditators could be responsible for the absence of the correlation.

The most recent study by Luders et al. (2009) compared 22 long-term practitioners (M = 24.2 years, SD = 12.4 years) of different traditions (Zen, Samatha, Vipassana and "others") with 22 control datasets matched for gender and age, taken from a database of normal adults. Data processing was performed with SPM5 and the VBM toolbox. Global analysis of GM volume was supplemented by a region-of-interest analysis based on a review of the findings of Lazar et al. (2005) and Hölzel et al. (2008). Therefore, regions-of-interest included the left inferior temporal gyrus, the right insula, the right hippocampus, and the right superior and middle frontal gyri.

Results were consistent with the findings by Hölzel et al. (2008). Meditators showed significantly more GM volume in the left inferior temporal gyrus, the right hippocampus, and the right orbito-frontal gyrus. In addition, meditators had more GM volume in the right thalamus. However, no differences in the right insula were detected and no correlation was found with the duration of practice. The authors suggest that morphological changes are likely to occur primarily within the first years of practice. Their sample contained only longstanding practitioners (at least 5 years, mostly above 10 years); hence a significant correlation could not be expected. The authors explain the lack of differences in the right insula with the heterogeneity of practices of the meditators in their study. The finding of higher GM volume in the thalamus was related to its function to gate sensory information and to focus attention.

Discussion

The reviewed findings suggest that the sustained efforts of meditation practitioners to modulate attention, arousal, and emotional responses could change the underlying neural circuitry in the thalamus, hippocampus, orbitofrontal cortex, and brainstem. Furthermore, the regular engagement in **introspection** is likely to improve the ability to discern subtle visceral sensations and to increase the awareness of the momentary bodily and emotional states. On the neural level, it has been shown that a meta-representation of bodily sensations is actually generated in the right anterior insula (Craig 2009), which is enlarged in those meditators practicing the body scan.

However, the authors of all reviewed studies stress the need of longitudinal studies to investigate the causal role of meditation regarding the observed differences and to rule out the alternative explanation, namely self-selection. Perhaps people who decide to begin meditation have certain pre-existing differences in brain structure compared to those who don't, or perhaps those with a certain neural constitution are more likely to maintain a long-term meditation practice. In particular, the lack of an age-related decline in gray matter has to be interpreted with great caution since people with cognitive impairments are likely to discontinue meditation practice. Thus, participants of control groups have to be matched also regarding such kinds of selection pressure, e.g., by recruiting them from a population of chess players participating regularly in tournaments. In a similar way, longitudinal studies will have to employ active control groups. Meditation training needs to be compared with other sorts of mental training to identify specific effects of the respective meditation technique. The following are a few of the key questions that have to be addressed by future studies within this emerging field of contemplative neuroscience:

1. Future research should compare different meditation traditions and techniques, in order to differentiate between common and specific effects. For example, studies should directly compare meditations with different kinds of attention regulation (guided vs. volitional; cf. Newberg and Iversen 2003), different ways of focusing attention (focused meditation vs. open awareness), and different intentional goals (e.g., cultivating compassion vs. attention training vs. relaxation) – all of which will likely rely on different neural mechanisms and produce different neural and behavioral effects.
2. In order to grant a better understanding of the relevance of morphological changes, it will be indispensible to investigate how structural changes are related to brain function and behavior. For example, are morphological differences associated with functional brain activation patterns (detectable by functional magnetic resonance imaging and electroencephalography) during the performance of relevant tasks? Is the interplay between different brain regions (functional connectivity) impacted by meditation practice? And most importantly, are morphological differences reflected in subjective measures of well-being and objective measures of behavior and performance?
3. Future studies should also investigate how neural connections between brain regions change as a result of practice. New imaging modalities, (e.g. fractional anisotropy in diffusion tensor imaging) have to be applied in order to apply quantifiable analyses to such complex processes.
4. Future research should also track the changes in morphological measures across short periods of time, in order to figure out the time frame within which such modifications occur. Gray matter changes detectable in anatomical magnetic resonance imaging have been reported after a period of as few as 7 days (Driemeyer et al. 2008). Also, the amount of training should be actively manipulated, to detect how much training is required to obtain a measurable effect.

A Glimpse into the Future

Studies addressing these questions are constantly emerging. In other domains, it has been shown that differences in regional gray matter are directly related to functional abilities (Gaser and Schlaug 2003; Ilg et al. 2008; Maguire et al. 2000; Mechelli et al. 2004; Milad et al. 2005).

In the field of meditation research, a recently published study investigates the relationship between pain sensitivity and cortical thickness in Zen meditators, linking morphological findings in meditation to changes in behavioral measures (Grant et al. 2008). Zen meditators showed lower thermal pain sensitivity (defined as the temperature required for producing a subjective experience of moderate pain) compared to non-meditators (Grant and Rainville 2009). When these findings on pain sensitivity measures were related to regional cortical thickness, Zen meditators showed greater cortical thickness in the right mid-anterior cingulate cortex and secondary somatosensory cortex bilaterally when compared to non-meditating control subjects (Grant et al. 2008). These brain regions are known to be involved in pain processing. A correlation analysis confirmed that individual pain sensitivity was associated with cortical thickness across the two subsamples. Pain sensitivity was reduced in participants with greater cortical thickness.

This study illustrates how relationships between morphological findings and behavioral measures should be tested in order to shed light on the neural mechanisms underlying abilities attributed to meditation training. However, it has to be kept in mind that cross-sectional studies do not allow the causal attribution of differences to the meditation training. In the above study, it is possible that both individual motivation to engage in (and maintain) meditation practice and a person's specific pain sensitivity might have a common neural basis. In order to rule out such an alternative explanation, longitudinal studies are indispensible.

The first longitudinal study to test the effect of mindfulness meditation training on brain structure has recently been presented by Hölzel et al. (2011; cf. Lazar et al. 2009). Sixteen participants underwent an 8-week Mindfulness-Based Stress Reduction (MBSR; Kabat-Zinn 1990) course. MBSR is a group program that aims at the cultivation of mindfulness by employing different meditation practices, such as the body scan, yoga, awareness of breathing, and open awareness meditation. Anatomical magnetic resonance images were acquired before and after the training and analyzed for changes in gray matter concentration. Changes were hypothesized in those structures previously identified in the study by Hölzel et al. (2008), namely the hippocampus, right anterior insula and left inferior temporal lobe. These regions had been identified to show differences between meditators and non-meditators in at least two out of the five published studies reviewed above (see Table 2).

Data analysis confirmed longitudinal increases in regions of interest for the left hippocampus and left inferior temporal lobe. Changes in the right anterior insula could not be confirmed. Additionally, exploratory whole-brain analyses identified significant increases in gray matter concentration in other parts of the brain that are involved in introspective processes, as well as emotion and arousal regulation. This is the first longitudinal evidence that supports some of the cross-sectional differences found in earlier studies. However, the generalizability of the study by Hölzel et al. (2011) is limited, as the sample size was very small and no control group was included. Particularly, MBSR is a complex group program, and its positive effects are likely in part attributable to meditation-unspecific effects, such as social interactions in the group. Future studies should control for such effects, e.g., by including active control conditions, such as the Health-enhancement program

(HEP; MacCoon et al. 2009), which was specifically designed to control for non-specific intervention effects associated with MBSR.

In a further analysis, Hölzel et al. (2010) investigated the morphological correlate of longitudinal changes in perceived stress following MBSR. Changes in scores on the perceived stress scale (PSS; Cohen and Williamson 1988) from before and after the 8 week program were significantly correlated with changes in gray matter concentration in the right basolateral amygdala. The more participants' subjective stress scores were reduced, the more decrease in gray matter concentration was found within this region. The data illuminate a change in neural architecture underlying modifications in one aspect of subjective well-being that resulted from mindfulness meditation training.

Summary

Morphometric studies have found differences between mediation practitioners and controls in a number of brain regions. While the assumption is plausible that these differences result from meditation practice, longitudinal studies are required to elucidate causal connections between the practice of different meditation techniques and structural changes in circumscribed brain structures.

Clearly, morphometric analyses have to be supplemented with functional and behavioral data acquired during relevant tasks. The recent studies exemplify approaches that are able to reveal the mechanisms that facilitate the benefits ascribed to meditation practice. In addition to shedding light on the mechanisms underlying the cultivation of beneficial qualities in meditators, the findings of contemplative research have the potential to inform larger inquiries into the basic mechanisms of the human nervous system, such as attentional and emotional self-regulation.

References

Cohen, S., & Williamson, G.M. (1988). Perceived stress in a probability sample of the United States. In S. Spacapan & S. Oskamp (Eds.), *The social psychology of health* (pp. 31–67). Newbury Park: Sage.

Craig, A.D. (2009). How do you feel – Now? The anterior insula and human awareness. *Nature Reviews Neuroscience, 10*, 59–70.

Driemeyer, J., Boyke, J., Gaser, C., Buchel, C., & May, A. (2008). Changes in gray matter induced by learning – Revisited. *PloS One, 3*(7), e2669.

Gaser, C., & Schlaug, G. (2003). Brain structures differ between musicians and non-musicians. *Journal of Neuroscience, 23*(27), 9240–9245.

Grant, J.A., & Rainville, P. (2009). Pain sensitivity and analgesic effects of mindful states in Zen meditators: A cross-sectional study. *Psychosomatic Medicine, 71*, 106–114.

Grant, J., Duerden, E., Duncan, G., & Rainville, P. (2008, August 17–22). *Cortical thickness and pain sensitivity in advanced Zen meditators*. Poster presented at the 12th World Congress on Pain, Glasgow.

Hölzel, B.K., Ott, U., Hempel, H., Hackl, A., Wolf, K., Stark, R., et al. (2007). Differential engagement of anterior cingulate and adjacent medial frontal cortex in adept meditators and non-meditators. *Neuroscience Letters, 421*, 16–21.

Hölzel, B.K., Ott, U., Gard, T., Hempel, H., Weygandt, M., Morgen, K., et al. (2008). Investigation of mindfulness meditation practitioners with voxel-based morphometry. *Social Cognitive and Affective Neuroscience, 3*, 55–61.

Hölzel, B.K., Carmody, J., Evans, K.C., Hoge, E.A., Dusek, J.A., Morgan, L., Pitman, R.K., & Lazar, S.W. (2010). Stress reduction correlates with structural changes in the amygdala. *Social Cognitive and Affective Neuroscience, 5*, 11–17.

Hölzel, B.K., Carmody, J., Vangel, M., Congleton, C., Yerramsetti, S.M., Gard, T., & Lazar, S.W. (2011). Mindfulness practice leads to increases in regional brain gray matter density. *Psychiatry Research: Neuroimaging, 191*, 36–42.

Ilg, R., Wohlschlager, A.M., Gaser, C., Liebau, Y., Dauner, R., Woller, A., et al. (2008). Gray matter increase induced by practice correlates with task-specific activation: A combined functional and morphometric magnetic resonance imaging study. *Journal of Neuroscience, 28*(16), 4210–4215.

Kabat-Zinn, J. (1990). *Full catastrophe living*. New York: Delta Publishing.

Lazar, S.W., Kerr, C.E., Wasserman, R.H., Gray, J.R., Greve, D.N., Treadway, M.T., et al. (2005). Meditation experience is associated with increased cortical thickness. *NeuroReport, 16*, 1893–1897.

Lazar, S.W., Hölzel, B.K., & Evans, K.C. (2009, March 18–22). *Neurobiological underpinnings of mindfulness and meditation*. Paper presented at the 7th Annual International Scientific Conference of the Center for Mindfulness in Medicine, Health Care, and Society, Worcester.

Luders, E., Toga, A.W., Lepore, N., & Gaser, C. (2009). The underlying anatomical correlates of long-term meditation: Larger hippocampal and frontal volumes of gray matter. *NeuroImage, 45*, 672–678.

MacCoon, D.G., Sullivan, J.C., Davidson, R.J., Stoney, C.M., Christmas, P.D., Thurlow, J.P., & Lutz. A. (2009, September 1). *Health-enhancement program (HEP) guidelines*. Permanent URL: http://digital.library.wisc.edu/1793/28198.

Maguire, E.A., Gadian, D.G., Johnsrude, I.S., Good, C.D., Ashburner, J., Frackowiak, R.S.J., et al. (2000). Navigation-related structural change in the hippocampi of taxi drivers. *Proceedings of the National Academy of Sciences USA, 97*(8), 4398–4403.

Mechelli, A., Crinion, J.T., Noppeney, U., O'Doherty, J., Ashburner, J., Frackowiak, R.S., et al. (2004). Structural plasticity in the bilingual brain. Proficiency in a second language and age at acquisition affect grey-matter density. *Nature, 431*, 757.

Milad, M.R., Quinn, B.T., Pitman, R.K., Orr, S.P., Fischl, B., & Rauch, S.L. (2005). Thickness of ventromedial prefrontal cortex in humans is correlated with extinction memory. *Proceedings of the National Academy of Sciences USA, 102*(30), 10706–10711.

Newberg, A.B., & Iversen, J. (2003). The neural basis of the complex mental task of meditation: Neurotransmitter and neurochemical considerations. *Medical Hypotheses, 61*(2), 282–291.

Pagnoni, G., & Cekic, M. (2007). Age effects on gray matter volume and attentional performance in Zen meditation. *Neurobiology of Aging, 28*, 1623–1627.

Vestergaard-Poulsen, P., van Beek, M., Skewes, J., Bjarkam, C.R., Stubberup, M., Bertelsen, J., et al. (2009). Long-term meditation is associated with increased gray matter density in the brain stem. *NeuroReport, 20*, 170–174.

Neurophysiological Correlates to Psychological Trait Variables in Experienced Meditative Practitioners

Thilo Hinterberger, Niko Kohls, Tsutomu Kamei, Amanda Feilding, and Harald Walach

Abstract "Meditation" has frequently been used as an umbrella term for diverse consciousness practices. Although neuropsychological state and trait measures in persons experienced in meditation practice have been reported during the last years, there is no consensus about their phenomenological meaning and correlation with experiences. In this study we aimed to investigate the neuronal, psychological and phenomenological commonalities of various meditation styles by correlating 64

T. Hinterberger (✉)
Section of Applied Consciousness Sciences, Department of Psychosomatic Medicine, University Hospital Regensburg, Regensburg, Germany

Department of Environmental Health Sciences, University Medical Center Freiburg, Freiburg, Germany

Samueli Institute of Information Biology, VA, USA
e-mail: thilo.hinterberger@klinik.uni-regensburg.de

N. Kohls
GRP – Generation Research Program, Human Science Center, Ludwig Maximilian University of Munich, Munich, Germany

Samueli Institute of Information Biology, VA, USA

T. Kamei
Shimane Institute of Health Science, Izumo, Japan

A. Feilding
Beckley Foundation, Oxford, UK

H. Walach
Institute for Transcultural Health Studies, European University Viadrina, Frankfurt (Oder), Germany

H. Walach et al. (eds.), *Neuroscience, Consciousness and Spirituality*, Studies in Neuroscience, Consciousness and Spirituality 1,
DOI 10.1007/978-94-007-2079-4_10,

channel of EEG (electroencephalogram) data with questionnaire measures tapping into mindfulness (FMI) and exceptional and spiritual experiences (EEQ). Significant correlations between EEG measures and the mindfulness score, amount of meditation experience, and exceptional experiences such as visionary dreams were found. The heuristic approach of classifying spiritual and meditative techniques on three different dimensions – neuronal, phenomenological and psychological trait – seems to be a promising way for developing a taxonomy of meditative states that is not only based on a superficial, technological surface level description of a particular mind-body practice.

Introduction

Meditative practices and their accompanying altered states of consciousness have become a focus of attention in neuroscience and health research recently (Cahn and Polich 2006; Vaitl et al. 2005). "Meditation" has thereby frequently been used as an umbrella term for diverse practices. Such practices aim at facilitating altered states of consciousness associated with meditative and contemplative mind-body practices stemming from different cultural traditions. If the respective practices are embedded in a certain spiritual tradition or a religious background framework, they may also be called spiritual or religious practices. When these techniques have been stripped of their religious and spiritual connotations, they may also be understood as secular techniques geared towards changing states and ultimately traits of consciousness. The mindfulness based stress reduction program (MBSR) – a standardized 8 week program developed by John Kabat-Zinn that aims at improving health by reducing stress – is probably the most prominent and best investigated example of a secularised form of meditation (Kabat-Zinn 1994; Shapiro et al. 2006).

However, the classification of meditation and states produced by different techniques is not as easy as it may seem at first. A recently conducted systematic review that was commissioned by the National Center for Complementary and Alternative Medicine (NCCAM) analyzed over four hundred clinical trials on meditation identified the following seven clusters of meditation practices (Ospina et al. 2007, 2008): *Mantra Meditation* (key component: repeating a word, sound or symbol), *Mindfulness Meditation* (key component: cultivating awareness, acceptance, nonjudgment, and attention to the present moment), *Qigong* (key component: different breathing techniques combined with various physical exercises in order to increase the flow of the "life energy" that is known as "Qi" in the Chinese tradition), *T'ai Chi* (key component: moving meditation that utilizes soft and slow and flowing bodily postures in order to obtain and foster flexibility, relaxation well-being, and mental concentration, as well as balancing of "Qi"), *Yoga* (key component: combining breathing techniques with bodily postures), *Miscellaneous Meditation Practices* (techniques that combine different approaches to meditation, without giving prominence to one) and *Undefined Meditation Practices* (practices that were not properly or only vaguely described in the papers). However, the authors of the study have explicitly addressed some reservations concerning this taxonomy as each subgroup was found to be quite heterogeneous. As a natural consequence, according to the authors a variety of different techniques has been labelled "meditation" or "meditative practice" in clinical trials. Correspondingly, the authors have concluded that meditation practices do not appear to have a common theoretical perspective, and that there is a need to develop a consensus on a working definition of meditation applicable to a heterogeneous group of practices (Ospina et al. 2007).

Thus, the most pressing conceptual problem within meditation research is lack of consensus concerning a clear operational definition. Nevertheless, we suggest that the impossibility of finding a both comprehensive and clear operational definition might be inherently associated with the term meditation and will have to await a fresh attempt. Pragmatically, the majority of meditation techniques – secularly or spiritually oriented – may be regarded as belonging to a subfamily of self-regulation strategies and may correspondingly also be considered being a subset of mind-body-techniques (Walach et al. 2010). Thirty years ago, West has proposed to define meditation as "*an exercise, which usually involves training the individual to focus the attention or consciousness in a single object, sound, concept or experience.*" (West 1979). A recent definition has conceptualized meditation in a similar manner "*as a family of complex emotional and attentional regulatory training regimes developed for various ends, including the cultivation of well-being and emotional balance*" (Lutz et al. 2008), thereby highlighting the functional relationship between meditation and well-being. Nevertheless, one should recall that meditative techniques were not developed as health improvement strategies in the first instance. Rather, health benefits are normally seen to be side effects of meditation practice. Recently, Cardoso et colleagues (Cardoso et al. 2004) have proposed an operational definition employing five criteria in order to characterize a certain procedure as meditation: (1) the use of a specific technique (clearly defined), (2) muscle relaxation at some moment during the process and (3) "logic relaxation" (i.e. no intention to analyzing or judging psychophysiological effects as well as creating expectations) (4) being a self-induced state, and (5) use of self-focus skill. However, this approach of operationally defining meditation may be disputed as well, as it gives a lot of leeway concerning the type of technique as well as the definition of "self-focus skill". Apart from that, questions concerning the paradoxical nature of "logic relaxation" and specifically concerning the ability to intentionally withhold expectations also spring to mind, as it is well known within social, clinical and experimental psychology that expectancies are supposed to shape experiences in the course of time (Kirsch 1999). Meditation may be defined both as a state of consciousness as well as an extended process of mental exercising. However, it is probably the latter, broader definition that reflects the important process character of meditation in real life. Thus, meditators are eventually supposed to produce new expectations or alter existing ones, and mental representations associated with the practice of meditation, which in turn will also impact upon the immediate experiential quality during meditation. Correspondingly, the differentiation between state and trait effects of meditation should be taken into account, as the process of reframing experiences on the basis of culturally and experientially shaped expectancies is inevitable from a contextualist perspective. To give an example: paying undivided attention to something is a skill most humans have, if they need it, but do not normally employ as a matter of fact. In this sense it is a capacity, and when realized, a state. However, meditators cultivate such states and thus, gradually improve this capacity making it increasingly a trait. Another example is being present without a judgmental attitude. While cultivating present moment awareness without judgment during meditation, as a series of states, acceptance may arise as a stable trait (Kohls et al. 2009), as well as an enhanced capacity to cultivate present moment awareness as a trait.

Modulation of attention can be defined as a pivotal common denominator of most types of meditative practice. A frequently employed, albeit somewhat coarse-grained classification system for meditative techniques addresses the distinct quality of the proposed attentional shift by differentiating between *concentration* or *focused attention* (FA) meditation and *mindfulness* or *open monitoring* meditation (OM) (Cahn and Polich 2006; Goleman 1977; Lutz et al. 2008). Whereas the FA techniques, such as Buddhist Samatha meditation aim at focusing on distinct mental or sensory content or objects, such as an image, or a sound, the open monitoring techniques such as Mindfulness practices as they are found in Soto-Zen or Vipassana aim at obtaining a conscious stance that can be defined as attentive but non-judgemental observation. However, some techniques such as Rinzai-Zen, Vedic or Transcendental Meditation (TM) show actually an overlap between the two categories and are difficult to classify by means of this binary classification system. Some researchers have suggested that most meditation techniques can actually be positioned somewhere along a continuum with the two poles mindfulness and concentration (Andresen 2000; Ivanovski and Malhi 2007). However, this is also likely to be too unidimensional. More likely the two categories concentration and mindfulness are orthogonal and techniques can be ordered according to the emphasis they place on either dimension or even according to the dynamic interplay of the dimensions during one meditation session or in different types of meditation. Also, it is doubtful whether such a bivalent classification into attention focusing and mindfulness techniques is really useful. For in order to be mindful to everything, it is necessary to train attention, and once attention is well trained, mindfulness to the present moment arises. Thus, it seems, that these two apparent opposite types of meditation are really two aspects of how focused attention really is and onto what the focus is directed. Perhaps a more useful image is a circle, where the two supposed opposites are at some point united to give a whole, where different aspects are in the foreground at different moments in time.

Moreover, the differentiation between state and trait effects of meditation and the level of proficiency should be taken into account: It is conceivable that some forms of meditation place initially more emphasis on FA and later focus on OM (or vice versa), once a distinct level of proficiency is accomplished. In other words a novice and a proficient meditator practicing the same form of meditation may actually exercise different techniques or utilize aspects of concentration and unfocused attention to various degrees while seemingly practicing the same form of meditation. Such differences of emphasis might even pertain to single meditation sessions, where focusing and mindfulness aspects may change as the session progresses.

Thus, one has to distinguish between the "objective" description of a meditation technique in terms of a theoretical framework and the phenomenological first hand descriptions of meditative experiences as they are experienced by the respective practitioners and in a respective session. These first hand reports are also dependent on the cultural context and the theoretical and practical framework, in which meditative practices are embedded. To give an example, a mindfulness meditation breathing technique may lead to completely different descriptions of first hand experiences as well as exhibit different impact upon health variables if it is practiced by a Buddhist monk in Dharamsala in order to achieve spiritual insights or by executives in New York in order to improve their coping with job-related stress.

On the other hand, there are also commonalities amongst the various meditation techniques across traditions. Different forms of meditation as practiced in various Buddhist traditions, such as Zen and Tibetan Buddhism, quiet QiGong practice, as well as Christian contemplation share some commonalities during meditation sessions as well as long term changes in psychosocial traits: (1) All those techniques are characterized by the meditator usually sitting in silence in a state of wakeful awareness, relaxed, yet attentive. This specific state is an act of being present without cognitively evaluating stimuli and situations, being aware of each moment in time without prejudice. This can be achieved by different techniques, for example by being attentive to the space which the meditator is in, as in open mindfulness, or keeping attention on the breath by counting breaths or just observing the act of breathing, like in Soto-Zen, or attending to the process of thinking without getting caught up by this process, or on energetic flow processes in the body, or focusing attention on any other object and letting it rest there. (2) All these techniques teach the meditator to reach a non-judging observing state. Thus, meditation is an effortless but highly attentive set of states aiming at inducing a distinct shift in the observational perspective. (3) As a result of the distinct changes associated with meditative states described above one can additionally expect significant changes in some general psychological traits. Usually, people who meditate on a regular basis share a common set of values and ideals that are associated with a distinct shift of the self model towards a less ego-centered direction (Legrand and Ruby 2009). One of them is an aspiration for increased mindfulness in daily life. This raised level of mindfulness, possibly but not necessarily associated with a spiritual belief system might also open up the meditator's mind to encounter exceptional experiences such as visionary dreams or spiritual experiences.

In the present study we assessed neuronal correlates of meditative states in meditators with varying experience from various traditions of Western and Eastern origin by measuring EEG during their meditation session. In order to additionally investigate psychological properties as well as their correlations with the physiological brain states, the EEG measurement in this study was accompanied by questionnaires assessing exceptional and spiritual experiences as well as self-attributed degree of

mindfulness. For the sake of clarification let us shortly introduce the two concepts here (see also methodological section):

(A) **Exceptional and spiritual experiences**: Exceptional experiences touch on areas outside the common sense reality of our everyday world, e.g., a sense of enlightenment or certainty, a feeling of unity, presentiment or telepathic experiences (Kohls et al. 2008; Kohls and Walach 2006, 2007). Spiritual experiences can be regarded as a particular subcategory of exceptional experiences and can be considered as experientially touching upon a universal, comprehensive or transcendental reality that need not necessarily be interpreted in a formal or traditional religious framework. Frequently, existing frameworks are used for interpreting such experiences. They are then termed religious experiences. Spiritual practices like prayer, or different forms of contemplation as well as meditation may be seen as designed to elicit exceptional or spiritual experiences (Meraviglia 1999). We have developed a multidimensional scale, the Exceptional Experiences Questionnaire (EEQ), which differentiates such exceptional experiences into positive, negative, psychopathological and visionary experiences (Kohls 2004; Kohls et al. 2008; Kohls and Walach 2006). Our research has shown that individuals practicing different – both secular and spiritual – forms of meditation report a greater amount of exceptional experiences, and that they evaluate these experiences more positively than individuals with a lack of meditative practice (Kohls 2004). We have also seen a stable relationship of exceptional experiences and indicators of physical wellbeing (Kohls et al. 2009). Thus, we believe that exceptional experiences might be a good parameter for gauging and comparing different forms of meditation.

(B) **Mindfulness**: Mindfulness may be understood as a distinct psychological function associated with meditative techniques. Despite the fact that the concept of mindfulness was originally derived from Buddhist psychology, mindfulness can be understood in secular terms as the mental ability to focus on the direct and immediate perception of the present moment with a state of non-judgemental awareness, voluntarily suspending evaluative cognitive feedback (Hayes and Feldman 2004; Hayes and Shenk 2004). The ability to be mindful can systematically be trained (Davidson et al. 2003), and, correspondingly, practicing mindfulness or other forms of meditation may be regarded as a systematic venue for developing mindfulness (Kabat-Zinn 2005). Recent studies have shown that enhancing mindfulness through systematic training is associated with positive effects in a variety of health measures (Baer 2003; Grossman et al. 2004). Different measurement instruments for assessing self attributed mindfulness such as for example the Mindfulness and Attention Awareness Scale (MAAS) (Brown and Ryan 2003), the Kentucky Inventory of Mindfulness Scale (KIMS) (Baer et al. 2004), the Five Facets Mindfulness Questionnaire (Baer et al. 2006) or the Freiburg Mindfulness Inventory (Walach et al. 2006) are available. A relationship between the ability to be mindful and regular spiritual and meditative practices has been empirically corroborated for a variety of mind-body practices. We therefore believe that the ability to be mindful develops generically as a consequence of meditative practice, regardless of the distinct technique. Although one needs to be sceptical as to how valid such self-report measures really are, at the moment they are still the best available and most economic ways of assessing mindfulness (Grossman 2008).

Ever since the early days of Lange and James psychophysiology has been plagued by the lack of correlation between physiological indicators and phenomenology of first-person, subjective experiences (Hellhammer and Hellhammer 2008). Thus, it has become mandatory to use multilevel descriptions to elucidate experiences. While brain imaging methods such as PET, sPECT or fMRI scans (see the chapters by Beauregard and Ott in this volume) have become popular to document psychobiological changes during or as a result of meditation, EEG research also has a long tradition in meditation research, dating back to the 1950 and 1960 (Das and Gastaut 1957; Kasamatsu and Hirai 1969; West 1980). While the benefit of modern imaging techniques are the comparatively precise location of activation in deep brain structures and description of isolated functional networks, their drawback lies in the massive costs and stationarity, slow temporal resolution, noisy set-up and comparative invasiveness of the procedures. EEG measures can be used to document swift changes in micro- and macro states of large neuronal ensembles, as well as global coherence. They also lend themselves to topographical analyses as well as sophisticated coherence analyses using low resolution tomography (LORETA) (Lehmann et al. 2001, 2006). Apart from this, due to the miniaturization of equipment, EEG measures can be taken with portable devices and hence leave meditators comparatively undisturbed in their customary environment and body postures. We therefore decided to use EEG to document objective changes associated with meditative states.

EEG data lend themselves to a multitude of analyses. We decided to use approaches successfully documented by many preceding studies. We used Fourier transformed data series to analyse customary power spectra of the EEG. These are associated with overall states of brain activation. Brain activity is frequently lateralized, i.e. hemispheric activation is different dependent on tasks and activities. For instance, it is well known that in language perception and explicit analytical tasks, in right handed individuals, the left hemisphere is more active, while the right hemisphere is more engaged in pattern recognition and implicit strategies of holistic recognition. Recently, it has been suggested that increased frontal left-hemispheric activity in meditators is associated with plasticity in dealing with emotional stress (Davidson et al. 2003). Hence, differential activation of brain hemispheres during meditation might be an interesting study target and can be easily investigated using EEG. Also, earlier studies (Orme-Johnson 1977; Aftanas and Golocheikine 2001) have found stronger EEG coherence across several electrodes, suggesting that in meditative states there are coherent activities in the brain. While under normal circumstances brain activities tend to be scattered, due to many parallel processes and analyses of different features of stimuli in distant brain areas, it seems to be the case that at least under some meditative conditions cohesion of brain activation as reflected in EEG coherence is enhanced. Finally, global field power as the strength of the average electric current measured can give us some indication as to the activation status of the brain.

Therefore, in order to empirically investigate the relationship between the **type of meditative practice, level of proficiency, sociodemographic parameters,** exceptional experiences and mindfulness and EEG patterns, we have collected these data from 26 spiritual practitioners practicing different meditative techniques both from Eastern and Western origins. Such a study design allows for testing the hypothesis whether there are correlations between EEG power, lateralization, and coherence of various EEG frequency bands during meditation or resting conditions and the psychological and behavioral data assessed in the questionnaire such as meditation experience, degree of mindfulness, frequency and evaluation of exceptional experiences.

Materials and Methods

Participants

Twenty six spiritual practitioners aged 26–65 years (mean 46 years, 7 female, 19 **male) from various spiritual backgrounds and with different levels of proficiency** were measured with EEG and peripheral measures. The participants were associated with different kinds of spiritual traditions such as Zen-Buddhism (10), Qi-Gong (4), Western contemplative methods (7), or were spiritistic or mediumistic practitioners (5). Some of them were also practicing spiritual and/or shamanistic healing rituals. Six participants were ordained Buddhist monks in Japan. The inclusion criteria were that they carry out a meditative spiritual practice on a regular basis and/or be used to the practice of meditation. Nine of them were meditating every day, 11 of them more than once a week and 7 of them only once a week or less. The participants reported that they spend between 15 and 120 min for each meditation session. They had between 2 and 35 years of meditation experience (mean 15 years). With this information we could calculate the total experience in meditation which was between 12 and 13,697 h (mean 3,357 h).[1] An overview over the distribution of those measures is given in Fig. 1. All graphs show a wide range of variability which allows us to calculate a reliable correlation analysis between the experience measures and the physiology.

While the Qi-Gong practitioners were Chinese, the Buddhist practitioners had their roots in the Japanese and Tibetan culture. Possible neurophysiological differences in brain functions especially with respect to lateralisation effects in the Western and Eastern populations suggest a division of the sample into a

[1] Three spiritual practitioners had only minimal meditative experience but were engaged in other spiritual activities (e.g. by living in a monastery for certain time and the like). This large variation meditative experience is a advantage for conducting correlational analyses as we have done here.

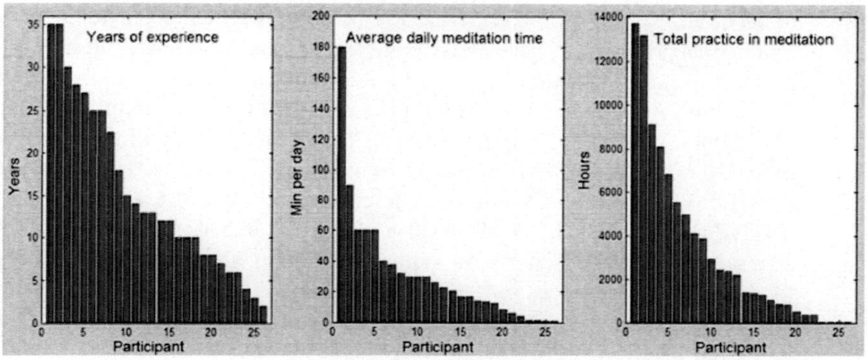

Fig. 1 Sorted distribution of meditative practice over all 26 participants. The number of years of meditative or spiritual practice is illustrated on the *left* and the average daily time spent for meditation in the *middle*. The *right* graph shows the total time participants have spent on meditation in their life as extrapolated by us. The wide variability enables us to calculate valid regression analysis between the experience and psychological trait variables

Western (15 participants) and an Eastern group (11 participants) in addition to the pooled analysis.

All meditators participated voluntarily and gave informed consent. The study was approved by the School Ethics Committee of the University of Northampton/UK and the Ethics Committee of the University Medical Center Freiburg i.Br./Germany.

Experimental Design

The measurements were carried out at various locations, predominantly in rooms which are normally used for meditation or the participants' homes. All physiological data were recorded with a 72 channels QuickAmp amplifier system (BrainProducts GmbH, Munich, Germany). EEG was measured using a 64 channels ANT electrode cap with active shielding and Ag/AgCl electrodes which were arranged according to the international 10/10 system. The system was grounded at the participant's shoulder. Data were recorded with a common average reference and filtered in a range from DC to 70 Hz at a sampling rate of 500 Hz and 22 bit resolution. For correction of eye movement and blink artefacts, the vertical electrooculogram (EOG) was measured by placing two electrodes above and below one eye. Respiration was measured with a respiration belt and the skin conductance at the second and third finger of the non-dominant hand. Additionally, for measuring heart rate variability the electrocardiogram (ECG) was measured with another two electrodes.

Before the measurement the participants had to answer a short initial questionnaire asking for some details regarding their meditation practice. Besides the frequency of meditation they should describe the posture and method of their meditative practice as precisely as possible. The measurements started with an initial 15 min baseline session in which they were asked to sit in their meditation posture for 5 min with eyes open, 5 min with eyes closed, and spend 5 min on reading a text from a book or a computer screen. After a short break a meditation session of 20–30 min duration was carried out in which participants were asked to meditate in the way they were accustomed. The meditators were offered to press a button or give a signal whenever they had a subjective experience of special interest. After the meditation a report was written mentioning all events, feelings, emotions, thoughts and experiences of the session. Finally, a 10 min guided meditation was conducted and the respective data will be reported elsewhere. After the electrodes had been removed participants were asked to answer a second questionnaire that included demographic data, the Freiburg Mindfulness Inventory (FMI), and the Exceptional Experiences Questionnaire (EEQ). The total session lasted between 2½ and 3 h.

Questionnaire Data

The following questionnaire instruments were administered to the participants before/after the meditation session.

Exceptional Experiences Questionnaire (EEQ): A four-dimensional scale developed for measuring positive and negative spiritual experiences, psychopathological experiences and visionary dream experiences (Kohls 2004; Kohls et al. 2008; Kohls and Walach 2006). A 57 item long and a 25 item short form exist. In this study, the 25-item short form of the EEQ was used, which shows good overall psychometric properties (Cronbach's alpha: $\alpha=.89$, test – retest reliability after 6 months $r=.85$) as well as acceptable properties for each factor: The four factors of the EEQ scale capture positive (7 items; $\alpha=0.88$; test–retest=0.87) and negative spiritual experiences (7 items; $\alpha=0.81$; test–retest=0.75), as well as psychopathological experiences (7 items; $\alpha=0.67$; test–retest=0.66) and visionary dream experiences (4 items; $\alpha=0.89$; test–retest=0.85). The questionnaire asks about the frequency of exceptional experiences as well as their current evaluation: individuals are not only asked to report about how often they have had an experience, but also to what degree they evaluate it as positive or negative. High scores mean that experiences have been reported frequently and evaluated more negatively. The EEQ shows adequate discriminant validity with sense of coherence, social support and mental distress and convergent validity with transpersonal trust. The four scales that were empirically corroborated by means of factor analyses can be described as follows:

1. Positive spiritual experiences: This factor embraces positive spiritual experiences of transcending the self as well as sensations of connectedness and unity with a transcendental entity or realm. Example items are "I am illumined by divine light and divine strength" and "A higher being protects or helps me".

2. Negative spiritual experiences: The second factor describes experiences of deconstruction and ego loss as well as fearful sensations of isolation and loneliness that are frequently described in the mystical literature as a consequence of following a spiritual path. Example items are "My world-view is falling apart" and "A feeling of ignorance or not knowing overwhelms me".
3. Psychopathological experiences: The third factor contains psychopathological experiences that fit into the psychotic and paranoid sphere. Example items are "I clearly hear voices, which scold me and make fun of me, without any physical causation" and "I am controlled by strange and alien forces".
4. Visionary dream experiences: The fourth factor relates to intensive dream type experiences. Two examples items are "I dream so vividly that my dreams reverberate while I am awake" and "I have meaningful dreams".

Freiburg Mindfulness Inventory (FMI) assesses awareness and nonjudgment of present moment experiences (Buchheld et al. 2001; Buchheld and Walach 2002; Heidenreich et al. 2006; Kohls et al. 2009; Walach et al. 2006). Sample items are "I am open to the experience of the present moment" and "I accept unpleasant experiences". A 30 item long and a 14 item short form do exist. In this study the 30 item long version (Cronbach's alpha=.86) was employed. High scores represent high self-ascribed mindfulness.

In the following sections, the subsequent abbreviations will be used:

EE_p	Frequency of the total EEQ score
EE1_p	Frequency of positive spiritual experiences
EE2_p	Frequency of negative spiritual experiences
EE3_p	Frequency of psychopathological experiences
EE4_p	Frequency of visionary dream experiences
EE_e	Evaluation of the total EEQ score
EE1_e	Evaluation of the positive spiritual experiences
EE2_e	Evaluation of the negative spiritual experiences
EE3_e	Evaluation of the psychopathological experiences
EE4_e	Evaluation of the visionary dream experiences
FMI	Total score of the Freiburg Mindfulness Inventory

In total we report here 15 index scores, namely the 11 questionnaire scores listed above and additionally the 3 experience related scores (years of meditation experience, daily meditation time, and total meditation time) as shown in Fig. 1, and age.

Data Pre-processing

The whole data analysis was done using Matlab version 7.3. All EEG data were visually inspected for high amplitude artefacts. After detrending the DC recorded EEG data sets all EEG channels were corrected for eye movements using a linear correction algorithm correcting each channel by a fixed correction factor. This algorithm detects eye blinks and movement events and uses those periods for determining a correction factor for each channel. The EOG was multiplied with this factor and then subtracted from the EEG. This algorithm was tested to work sufficiently in normal non-moving EEG and can also be applied in real-time online analysis as we intend to do in the future. For further analysis of the data reported here artefact-free epochs of three conditions were selected: about 5 min of eyes open, 5 min with eyes closed, and 20–30 min of meditation in a style individually selected by each participant.

Power Spectral Density

A power spectrum time series was calculated using the Fast Fourier Transform (FFT). This analysis starts from the assumption that a raw EEG time series can be represented as linear combination of ideal-typical sinusoidal curves of different frequency. Hence the EEG raw data series can be decomposed into these original sinusoidal vibratory patterns, yielding the familiar frequency bands. FFT was calculated every second in a window of 2 s resulting in a frequency resolution of 0.5 Hz. Their squared value results in the power spectral density. The following 6 frequency bands were calculated by merging the FFT coefficients: Delta (1–3.5 Hz), Theta (4–7.5 Hz), Alpha (8–11.5 Hz), Beta1 (12–16 Hz), Beta2 (16.5–25 Hz), Gamma (25.5–47 Hz). Gamma was limited to 47 Hz because of possible 50 Hz contamination caused by the electricity supply.

To obtain an overall measure for a certain condition (eyes open, closed, or meditation), all 6 band power measures which were calculated for each half second were averaged over the whole time period of the corresponding condition. Finally, to limit the number of coefficients in the statistical analysis the 64 channels were merged in 13 areas according to Fig. 2.

The global field power was calculated by averaging the band power activity in the range of 4–45 Hz of all areas. The global field power was computed separately for each of the resting conditions and each participant.

Lateralisation

Hemispheric asymmetries in band power activity were calculated from the seven band power values as described above. Instead of merging the values into 13 areas, 8 interhemispheric areas were defined as shown in Fig. 2b. The mean power of a right area was subtracted from the corresponding left area band power value. For further statistical analysis the lateralization was expressed in a relative change by normalizing the difference to the mean power in both areas. Thus, positive lateralization indices denote higher left-hemispheric lateralisation, while negative scores indicate right-hemispheric activation.

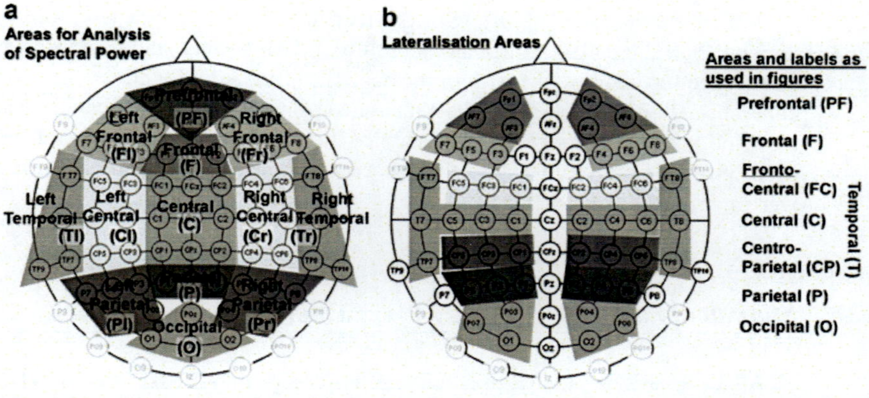

Fig. 2 The reduction scheme into 13 major areas for the analysis of the power spectral density is illustrated on the *left* while on the *right* graph the areas used for the hemispheric lateralization are defined, together with their abbreviations

Coherence Measures

The coherence of amplitude changes between areas was calculated by correlating the spectral power time series data of areas as depicted in Fig. 2a as described below. To achieve a higher time resolution in the spectral time series it was not possible to use the FFT band power values. Instead, the band power amplitudes were calculated using band pass filters resulting in a 10 samples/second time series. Depending on the frequency range of each frequency band for the first 4 frequency bands Butterworth filters of order 2 and 3 were used while for the higher frequencies filter orders from 4 to 6 were applied. This provided a stop-band attenuation between 12 and 36 dB for all bands except for the Delta band which could only be filtered with 6 dB. The stop band was defined at 0.8–0.9 times the low frequency cut-off and 1.1–1.2 times the high frequency cut-off. Before down sampling to 10 Hz, a 2nd order Savitzky-Golay filter was applied to the squared band signal values using window sizes that sufficiently smoothed the ripples in the signal.

The coherence of the signal amplitudes between channels was obtained by calculating the cross-correlation coefficients and their probability values for each frequency band across the 64 electrodes resulting in 64 × 64 matrices. In moving windows of 15 s window size and no overlap these correlations were calculated and averaged across the whole time period of a condition. To reduce the number of correlation coefficients so-called regions of interest (RoI) were defined. First, correlation coefficients were merged in the areas as shown in Fig. 2b. Then, we decided to focus on 10 different combinations of areas such as (1) frontal left-right (F_lr), (2) temporal left-right (T_lr), (3) central left-right (C_lr), (4) parietal left-right (P_lr), (5) prefrontal-occipital (Pf_O), (6) central frontal-parietal (Fz_Oz),

(7) left frontal-parietal (FP_li), (8) right frontal-parietal (FP_re), (9) all frontal-all parietal (F_P), and (10) all frontal and prefrontal-all parietal and occipital which is namely the forehead vs. the back of the head (Fh_Ba). Additionally to the amplitude coherence, the phase coherence was calculated from each frequency band. Therefore, the band pass filtered signals from each of the 64 electrodes were correlated with each other directly. Similarly to the procedure described above, those correlations were averaged in the described areas.

Statistical Analysis

Correlation coefficients were calculated by computing crosscorrelation between each of the 15 index scores with each of the EEG measures for all 26 participants. For each of the 6 EEG frequency bands 41 EEG measures were calculated, consisting of the spectral power in 13 areas according to Fig. 2 (left), the lateralization in 8 areas according Fig. 2 (right), the amplitude coherence in 10 area combinations, and the phase coherence again in 10 area combinations as described above. Using this data space, one can calculate correlation values as a function of index score, frequency band, EEG measure, and recording condition. In order to view the data in an appropriate way we produced color coded maps for each recording condition (trait condition: eyes closed, trait condition: eyes open, state condition: referenced meditation data) and each questionnaire variable. Such a map then contained a field of 6 frequency bands by 41 EEG measures as shown in Fig. 5.

In such a procedure, a rigorous correction for the multiplicity of tests is not possible, since EEG data are interdependent and a strict Bonferroni correction underestimates effects. Also, since this study is to our knowledge the first of its kind, we had no prior hypotheses to go for. Hence, all analyses are exploratory, and one needs to employ general wisdom in interpreting the correlations, taking them as effect size measures rather than judging them by their significance alone, or using conservative cutoff-points such as p values lower than 0.01.

Results

Mean Scores FMI and EEQ

Figure 3 shows the mean scores and standard deviations of the EEQ for three groups: data from the meditators in our sample (stars) in comparison with a group of spiritually practicing individuals (n = 350; triangles) and persons without such a practice (n = 299; squares) from our validation study. The mean score of the FMI was 85 (±14) which is comparatively high in relation to the reported mean of normal subjects

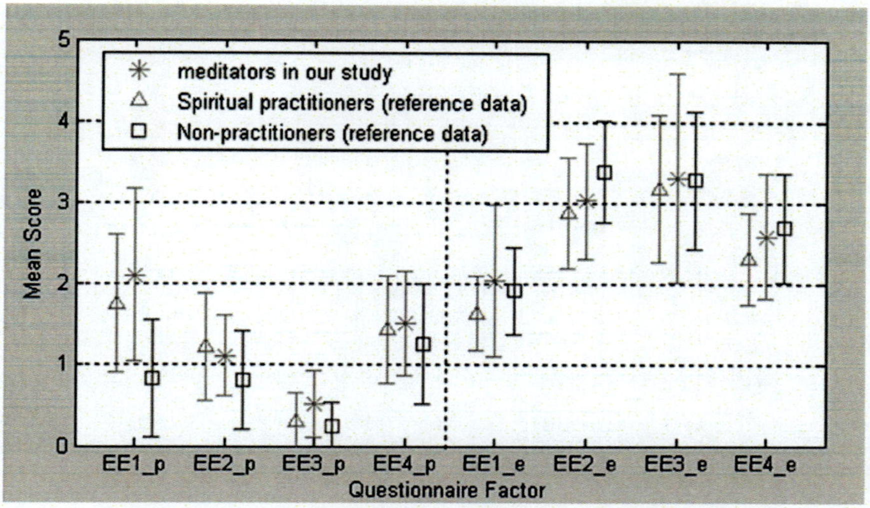

Fig. 3 Shows the mean scores of the frequency (EEx_p) and the evaluation (EEx_e) of the four factors of the EEQ. The means and their standard deviations are displayed for the 26 meditators of the present study (*stars*), as well as for a large group of spiritual practitioners (*triangles*) and non-practitioners (*squares*)

of 75 (±11) but in the range of a group of meditators after a Vipassana retreat who reported a score of 89 (±11) (Walach et al. 2006). Thus, participants of this study were more mindful than the average person.

Correlation Analysis Between EEG Measures and Questionnaire Data

All Variables Versus Global Field Power

Figure 4 depicts the correlation coefficients between global field power for each of the three conditions with the psychological variables as they were assessed with the FMI and the EEQ.

As can be derived from Fig. 4, significant negative correlations between global field power and age were found for the eyes open ($p<0.01$) and the meditation condition ($p<0.05$) as well as total years of meditation ($p<0.05$) in the eyes open condition. Additionally, a significant negative correlation between global field power and self reported mindfulness was found for the eyes open and eyes closed condition (both at $p<0.01$). With regard to the prevalence dimension of the EEQ, significant negative correlations were found for the prevalence sum score for the eyes closed condition ($p<0.05$) as well as for the positive spiritual

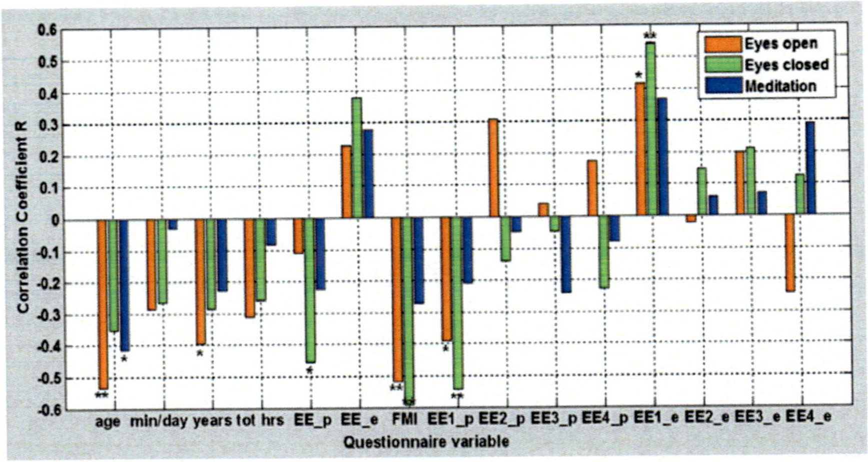

Fig. 4 Correlations between various questionnaire variables with the global field power during each of the three conditions. The eyes open and eyes closed conditions reflect the train correlations whereas the meditation correlations should be regarded as a state effect. The *asterisks* indicate significant values where * is $p<0.05$ and ** is $p<0.01$

experiences factor for the eyes open ($p<0.05$) and eyes closed condition ($p<0.1$). Concerning the evaluation dimension of the EEQ, a positive correlation between global field power and a more negative evaluation of positive spiritual experiences was found both for the eyes open ($p<0.05$) and the eyes closed condition ($p<0.01$).

State Correlations with Meditation Practice

In Fig. 5, the correlations between the years of meditative practice (a, b) and total experience of meditation and phase and amplitude coherence, lateralisation and global field power is summarized for the meditation condition for all lateralisation areas and frequency bands:

An increased coherence positively correlated with the years of meditation practice can be seen as a state variable which was predominantly visible during meditation. This correlation is also visible when correlating the total meditation experience, however not as strongly. As this Beta band coherence increase is not correlated with the age of the participants, this effect can be clearly attributed to the practice of meditative techniques.

Also, for the 'eyes open' condition, there is a negative correlation ($r=-0.25$, $p<0.01$) between the global field power in the range of 4–40 Hz and the years of experience. A tendency towards this effect can be seen in Fig. 5b) for the meditation task as well. This reflects a quieting of the brain waves in long term meditators.

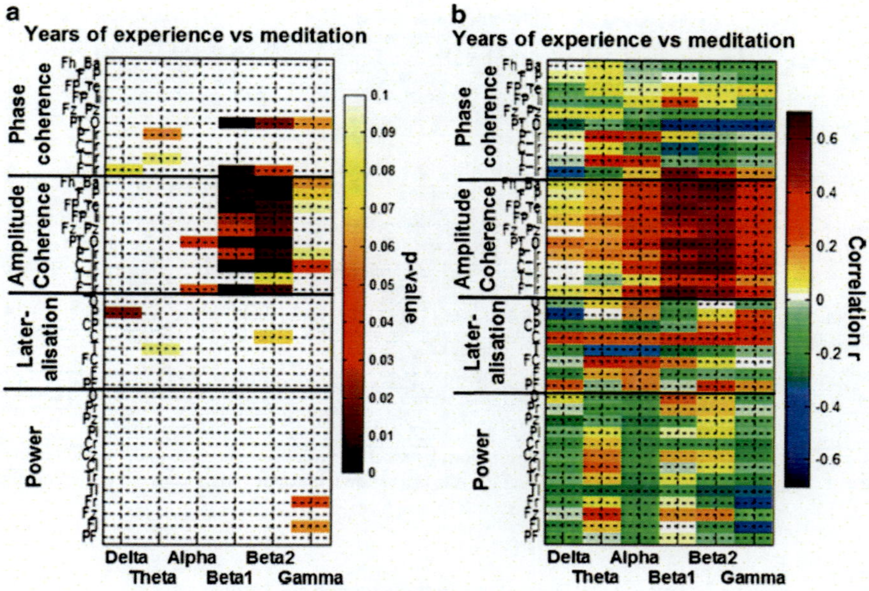

Fig. 5 The correlation between the years of meditation practice and the spectrum of the EEG (for description of frequency bands see Power Spectral Density) is depicted in graph (**b**) while the corresponding p-values are shown in (**a**)

Correlations with the Freiburg Mindfulness Inventory (FMI)

Figure 6 depicts the correlations between FMI and total experience of meditation and phase and amplitude coherence, lateralisation and global field power during meditation (a) and eyes closed (b) conditions for all lateralisation areas and frequency bands:

A globally significantly lowered Delta power during the eyes closed condition could be observed in those practitioners who reported higher mindfulness scores. This was essentially also the case when correlating the FMI scores with the data of the meditation session. Additionally, a significant correlation was found between the FMI and the Delta band lateralization both for the meditation and closed eyes condition. In depth inspection revealed that individuals higher in mindfulness tended to have higher right frontal Delta activity and left parietal Delta activity.

In the temporal lobe, a right hemispheric dominance in practitioners with higher self-reported mindfulness could be seen over a broad range from 1 up to about 16 Hz. A heightened left prefrontal high Beta and Gamma shift towards the left could also be identified as a pattern trait variable in meditators with high self-reported mindfulness, especially when looking at the other regions which also tend to a left hemispheric high Beta and Gamma dominance. However, this effect did not reach significance during meditation.

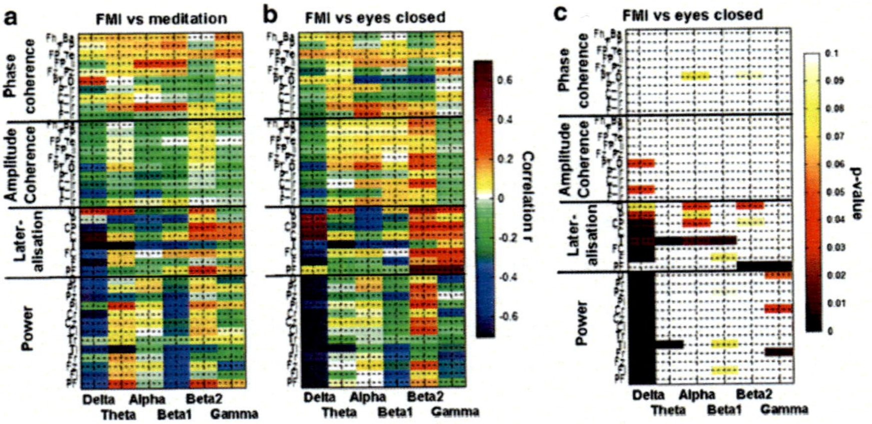

Fig. 6 The most significant correlations between the mindfulness scale and the EEG could be seen in the eyes closed and meditation condition as shown in figures (**a**) and (**b**). The statistical p-values corresponding to figure (**b**) are shown in graph (**c**)

Correlations with the Exceptional Experiences Questionnaire (EEQ)

Positive evaluation of exceptional experiences in general was correlated with a right-shifted prefrontal Gamma power during the resting state with eyes closed ($r=0.51$, $p<0.01$). The frequency of positive spiritual experiences (prevalence factor 1) was associated with an increased Delta power in resting state activity in frontal, central and parietal areas. In contrast, the prevalence of negative spiritual experiences showed a parietal right shift in Beta power. Frequency of psychopathological experiences was correlated with a decrease in Alpha amplitude coherence, between several areas, especially the interhemispheric parietal Alpha. This is shown in Fig. 7. The evaluation of visionary dream factors showed several highly significant correlations which are displayed in Fig. 8. During the resting state with eyes open, meditators who reported more positive visionary dreams showed decreased frontal-parietal phase correlations in almost all frequency bands. A strong increase of Beta power in central and parietal areas can also be seen in the resting state data as well as in the referenced meditation data (Fig. 9).

Summary of Results

A correction for multiple comparisons was not possible because the interdependencies between variables are unknown. It was also not possible to calculate an ANOVA with the results as this would require a larger sample size. Therefore, the results presented here, have to be considered as preliminary results stemming from a pilot study. Thus, it should be taken into account that some significant findings might also

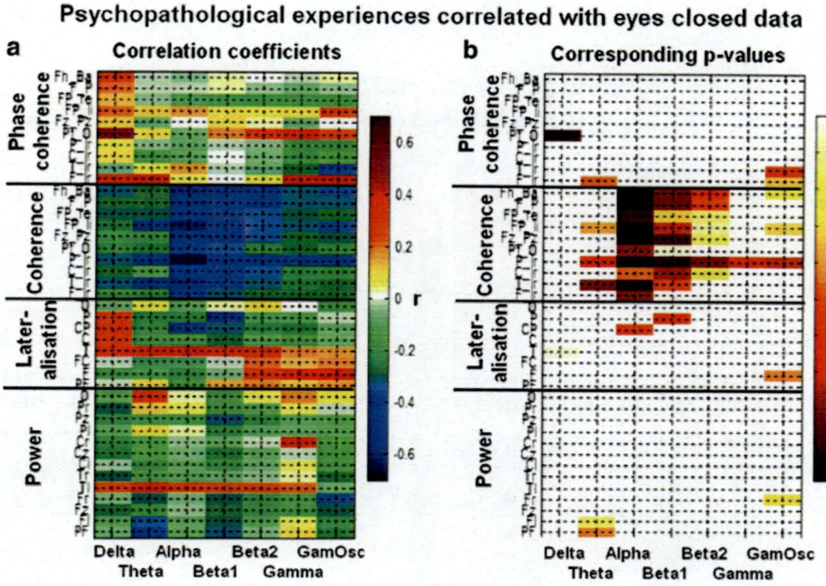

Fig. 7 The score for the second experiences factor of the EEQ, which taps into psychopathological experiences is correlated with a significant decrease in the Alpha amplitude coherence in many areas

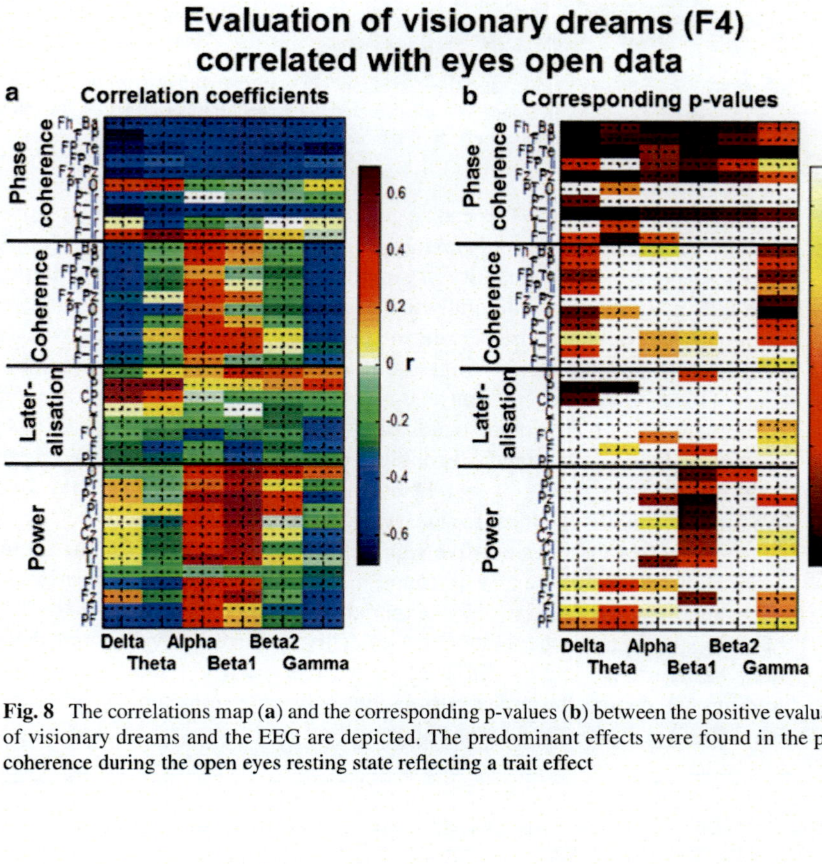

Fig. 8 The correlations map (**a**) and the corresponding p-values (**b**) between the positive evaluation of visionary dreams and the EEG are depicted. The predominant effects were found in the phase coherence during the open eyes resting state reflecting a trait effect

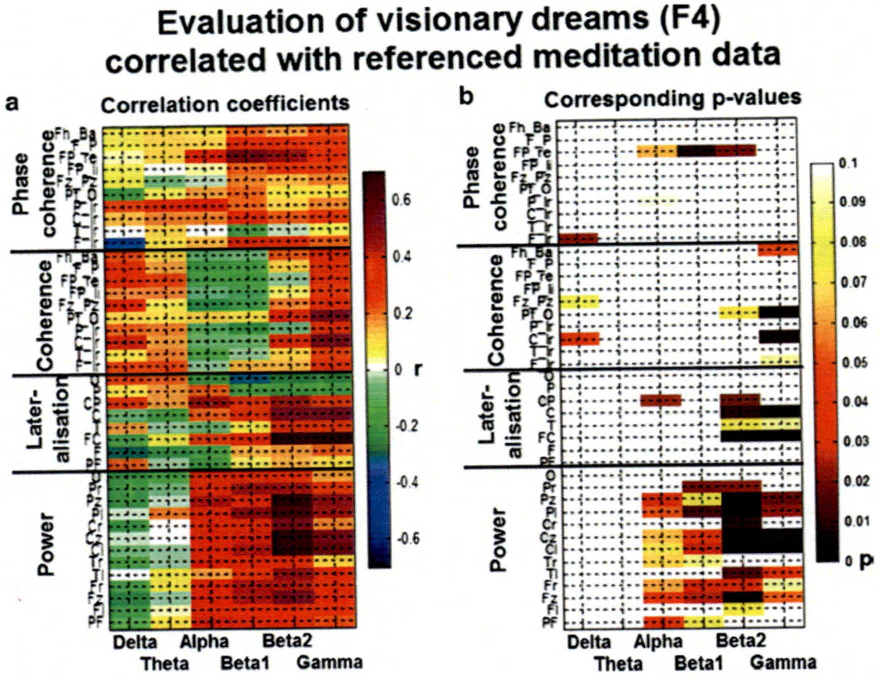

Fig. 9 The correlations map (**a**) and the corresponding p-values (**b**) between the positive evaluation of visionary dreams and the EEG are depicted. The predominant effects were found in the Beta and Gamma power during an effect of the meditation state

represent chance findings. This can be guarded against by taking a conservative stance in interpreting p-values and only taking into account strong correlations. Nevertheless, in order to compare the quantity of significant correlations identified in this study with a distribution of a structurally identical random variable matrix, an estimation of the amount of such false significant findings can be made as follows. The data matrix consisted of 18.450 single values, calculated as 15 questionnaire sores times 41 areas in all different measures (power, lateralisation, amplitude and phase coherence) times 6 frequency bands times 5 potential conditions (eyes open, eyes closed, meditation, reference, all conditions merged). When assuming a significance level of 0.001 we would expect 18 correlations below this threshold by chance. We found 75 highly significant values which is more than 4 times chance expectation. Thirty-eight of those seventy-five significances were provided by the mindfulness score alone. Seven of them were found in the variable 'years of experience', 10 in the evaluation of positive mystical experiences (EE1_e and also 10 in the evaluation of visionary dream experience (EE4_e). This supports the validity of the correlations between FMI and the EEG measures. The following Table 1 summarizes the findings and depicts the largest correlations between parameters for areas, frequency band, EEG parameter and recording condition.

Table 1 Summary of the most predominant correlation results

Questionnaire variable	Area	Frequency band	EEG parameter	EEG recording condition	Correlation r
Age	Fl	Delta	Power	All conditions	–0.63
Practice per day	Tr	Beta1	Power	Eyes closed, meditation, ref condition	0.66, 0.61, 0.63
Years of experience	Fh_Ba, C_lr	Beta1, Beta2	Ampl. Coh.	Eyes closed	0.62
Years of experience	F_lr	Beta1	Ampl. Coh.	Meditation	0.61
Years of experience	C_lr	Beta2	Ampl. Coh.	Meditation, all	0.62, 0.68
Years of experience	Fh_Ba	Beta2	Ampl. Coh.	Meditation, all	0.62, 0.61
FMI	Fz Cl Cr Pz Pr	Delta	Power	Eyes open	–0.61 to –0.65
FMI	Fl Fr Tr Cl Cr Pz	Delta	Power	Eyes closed	–0.61 to –0.73
FMI	T, C, T	Delta, Delta, Theta	Lateral	Meditation	–0.65, 0.61, –0.62
FMI	Cl Cz Cr Pz	Delta	Power	All condis	–0.62 to –0.66
FMI	Fz Tr Cl Cr Pz	Delta	Power	Ref condi	–0.61 to –0.70
EE_e*	PF	Gamma	Lateral	Eyes closed	–0.64
EE2_p**	P	Beta1	Lateral	Eyes open	–0.62
EE3_p***	P_lr	Alpha	Ampl. Coh.	Eyes closed	–0.63
EE3_p	P_lr, FP_li, F_P, Fh_Ba	Alpha	Ampl. Coh.	Ref condi	–0.61 to –0.68
EE1_e****	Fl, Fz, Cz, Pz	Delta	Power	Eyes closed	0.62 to 0.64
EE1_e	Fz, Tr, Cl, Cz, Cr, Pz	Delta	Power	Ref condi	0.62 to 0.68
EE4_e*****	C_lr, Fz_Pz, FP_re, F_P	Delta	Phase Coh.	Eyes open	–0.61 to –0.67
EE4_e	FP_re, FP_re	Gamma	Phase Coh.	Eyes open, ref condi	–0.63, –0.61
EE4_e	P, PF, PF	Delta, Beta2, Gamma	Lateral	Eyes closed	–0.61 to –0.62
EE4_e	C_lr	Delta	Phase Coh.	All condis	–0.63

Significances were found in 9 of the 15 psychological and demographic variables. The areas are labelled according to Fig. 2. The recording conditions eyes open, closed, meditation, and a reference condition (either eyes open or closed) were considered. Only correlations with $p<0.001$ are listed

*EE_e Evaluation of the total EEQ score

**EE2_p Frequency of the negative spiritual experiences

***EE3_p The frequency of the psychopathological experiences

****EE1_e Evaluation of the positive spiritual experiences

*****EE4_e Evaluation of the visionary dream experiences

In sum, the most prominent findings were:

- The amplitude coherence in the Beta range between frontal/prefrontal and parietal/occipital areas during meditation was positively correlated with years of experience of the meditators ($r = 0.61–0.68$, $p < 0.001$). A tendency towards significance for increased amplitude coherence in long-term meditators could also be seen in the Gamma band. The total time of meditation practice also shows those correlations, albeit weaker.
- Meditators scoring high on the mindfulness scale (FMI) showed a highly significant lower left temporal Theta activity – a pattern associated with internalized attention – during meditation than less mindful individuals ($r = -0.7$, $p < 0.005$).
- In meditators scoring high in the FMI Delta power was significantly reduced globally but especially in central and parietal areas ($r = -0.6$ to -0.7, $p < 0.001$).
- In contrast, Delta power in resting conditions was globally and significantly increased in meditators with highly positive evaluation of spiritual experiences.
- In resting conditions, the Alpha amplitudes' coherence was significantly reduced in people scoring high in the existence of psychopathological experiences predominantly between left and right parietal Alpha and between frontal/prefrontal and parietal/occipital Alpha.
- A positive evaluation of visionary dream experiences was correlated ($r = 0.73$, $p < 0.01$) with a right hemispheric dominance in Theta activity in predominantly frontal, central and temporal areas.
- The positive evaluation of visionary experiences and dreams is also highly significantly correlated with a right prefrontal dominance in the high Beta and Gamma band.
- The positive evaluation of visionary experiences and dreams was also correlated with decreased phase coherence in Delta band activity.
- As an additional significance in the questionnaire data we found that unmarried meditators showed significantly lower left frontal Theta power than the married ones.

Discussion

In this study we have investigated the relationship between neurophysiological state variables and psychological trait variables. As the questionnaire variables such as meditation practice, FMI, and EEQ scores exhibited a large enough variance between meditators the approach of a correlation analysis seems to be justifiable. In the literature one can find various study and analysis designs for studying the effect of meditation. When trying to compare the results we first have to look for the comparability conditions. For example, some studies have investigated the effect of meditation experience on neurophysiology during meditation by comparing EEG parameters of two groups, namely novice and experienced meditators (e.g., Aftanas and Golosheikin 2003). Their results can in some respect be compared with correlation results between meditation experience and EEG values during meditation referenced to a resting condition. Our EEG resting condition with eyes closed when correlated with psychological variables reflects a trait, since brain activity in resting conditions without specific attempts to cultivate a mindful state is measured. When we correlated psychological data with EEG parameters during meditation condition we are tapping into states.

The literature frequently reports an increased frontal theta activity or coherence effects in the alpha band. However, when looking at the meditation data in Fig. 5 there are no significant Theta and Alpha power changes which correlate with the years of meditation practice. In contrast to Faber et al. (2004), who reported an increased Theta and decreased Gamma coherence in experienced Zen meditators our data show a highly significant correlation between meditation experience and the far distant Beta band amplitude coherence. This might be an indication for cognitive efficiency due to meditation.

An additional measure of spiritual practice could be derived directly from the mindfulness scale because a higher degree of awareness and mindfulness is a central goal for most meditators. Here, meditators more mindful according to their self-report showed a highly significant lower Delta activity during conditions meditation and resting state. This could indicate heightened alertness in more mindful individuals, adding an objective marker to the growing literature about mindfulness self-reports and an indirect hint that self-reports of mindfulness can be valid proxies for the measurement of the construct. A wide consensus that can be derived from several studies is in consonance with our findings: meditation itself is not just a state of lowered cognitive activity, but more a state of heightened attention – not only during the meditation, but also in resting states. In order to avoid misunderstandings one should take a closer look at the highly significant negative correlation between the global field power and the mindfulness scale as shown in Fig. 4. This effect was mainly produced by the amount of Delta activity during the resting states. This means, practitioners scoring high in the mindfulness scale produce a lower amount of Delta activity over the whole scalp and in all resting and meditation conditions. Delta activity is normally associated with states of lowered awareness, such as in sleep or drowsiness. Hence, this finding points to heightened awareness associated with meditation not only during meditation practice, but also generalised to other activities vindicating the concept of mindfulness as a trait engendered by mindfulness and meditation practice.

The significant negative correlation between the global field power and the amount of positive mystical experiences is mainly dominated by the high Delta activity of only a few Japanese participants namely the Qi-Gong practitioners and three less experienced Soto-Zen and Za-Zen practitioners. The origin of such high Delta power remains unclear; however, it is most likely not caused by artefacts as it was measured in both, the eyes open, eyes closed, and meditation session. The positive correlation between the evaluation of positive mystical experiences with the global field power across all conditions is also produced by the Delta power. Here, meditators with very low Delta activity evaluated the mystical experiences as less positive than those with high Delta power. After realizing the strong influence of Delta activity on the global field power correlations we recalculated the global field power within the range between 4 and 40 Hz only. As a consequence, the significant correlations reported above vanish.

Those individuals reporting more positive spiritual experiences exhibited a higher global field power, especially during resting conditions which was dominated by fronto-temporal Beta activity.

The resting conditions reflect a trait in contrast to the meditation which we see more as a temporary state. From this we can assume a heightened general awareness as a trait in meditators who evaluate their experiences positively.

We found a surprisingly high number of significant correlations for the positive evaluation of visionary dream experiences. Meditators who experience visionary dreams exhibit heightened Alpha, Beta, and Gamma power with a significant shift towards the left frontal and central areas. Assuming that their meditations would also be influenced by a positive attitude towards exceptional experiences this supports the hypothesis of left frontal processing of positive emotions as reported by Davidson et al. (2003). The findings that meditators with positively evaluated visionary dreams show decreased phase synchronisations as well as increased Beta power can be associated with a generally increased mental processing of these subjects.

However, this finding seems to contradict the strong right hemispheric lateralisation of individuals higher in mindfulness, who exhibited stronger Delta-power in the right frontal areas and left parietal areas. While the lateralisation hypothesis contends that left-frontal activity is associated with more positive emotion regulation (Davidson et al. 2000) a critical test has shown that things are far more complicated. It is more likely that right hemispheric activity is associated with stronger emotions of any kind (Hagemann et al. 2005). This would be compatible with the idea that persons experienced in meditation would experience stronger emotions, likely of a positive valence, which could be visible in the right parietal asymmetry. An additional, not exclusive explanation would be that, while Davidson et al. (2003) measured lateralisation as differences in individuals beginning to learn meditation and documented effects after an initial training, we measured highly experienced individuals. The long-term practice might lead to the deployment of different resources. While emotion regulation in initial stages might entail explicit strategies, visible in stronger left-hemispheric activation, it may become an altogether more implicit and holistic strategy, reflecting in stronger right hemispheric activation in experienced individuals. This is in line with the findings of Farb et al. (2007).

To sum up, although only some of our findings could be regarded to be in line with the theory and findings from the literature, the heuristic approach of classifying spiritual and meditative techniques on three different dimensions – neuronal, phenomenological and psychological – seems to be a promising way for developing a taxonomy of meditative states that is not only based on a superficial, technological surface level description of a particular mind-body practice. Our findings suggest that mindfulness, as a trait, can develop in meditators with a long-term practice, and is visible in neurosignatures such as decreased Delta power. Distinguishing between traits and states seems to be important for further research. Also, the frequency of spiritual experiences, conceived as a stable psychological trait, is reflected in neuronal activation patterns.

Acknowledgements We are grateful to all spiritual practitioners who were involved in this study and who devoted their time and effort without reimbursement. This work has been supported by the BIAL Foundation (Portugal), the Samueli Institute (Alexandria, USA), the Beckley Foundation, UK (www.beckleyfoundation.org), and through the Beckley the Institute of Noetic Sciences (IONS, CA, USA). The IONS primary program areas are consciousness and healing, extended human capacities, and emerging worldviews (phone: 707.775.3500, www.noetic.org). Further support was given by the US Army Medical Research and Materiel Command under Award No. W81XWH-06-1-0279 P00002. The views, opinions and/or findings contained in this report are those of the authors and should not be construed as an official DoD position, policy or decision unless so designated by other documentation. In the conduct of research where humans are the subjects, the investigators adhered to the policies regarding the protection of human subjects as prescribed by the United States Code of Federal Regulations (CFR) Title 45, Volume 1, Part 46; Title 32, Chap. 1, Part 219; and Title 21, Chap. 1, Part 50 (Protection of Human Subjects).

References

Aftanas, L. I., & Golocheikine, S. A. (2001). Human anterior and frontal midline theta and lower alpha reflect emotionally positive state and internalized attention: High-resolution EEG investigation of meditation. *Neuroscience Letters, 310*(1), 57–60.

Aftanas, L. I. & Golosheikin, S. A. (2003). Changes in cortical activity in altered states of consciousness: The study of meditation by high-resolution EEG. *Human Physiology 29*(2), 143–151.

Andresen, J. (2000). Meditation meets behavioural medicine: The story of experimental research on meditation. *Journal of Consciousness Studies, 7*(11–12), 17–74.

Baer, R. A. (2003). Mindfulness training as a clinical intervention: A conceptual and empirical review. *Clinical Psychology: Science and Practice, 10*(2), 125–143.

Baer, R. A., Smith, G. T., & Allen, K. B. (2004). Assessment of mindfulness by self-report: The Kentucky inventory of mindfulness skills. *Assessment, 11*(3), 191–206.

Baer, R. A., Smith, G. T., Hopkins, J., Krietemeyer, J., & Toney, L. (2006). Using self-report assessment methods to explore facets of mindfulness. *Assessment, 13*(1), 27–45.

Brown, K. W., & Ryan, R. M. (2003). The benefits of being present: Mindfulness and its role in psychological well-being. *Journal of Personality and Social Psychology, 84*(4), 822–848.

Buchheld, N., & Walach, H. (2002). Achtsamkeit in Vipassana-Meditation und Psychotherapie. Die Entwicklung des "Freiburger Fragebogens zur Achtsamkeit". *Zeitschrift für Klinische Psychologie, Psychiatrie und Psychotherapie, 50*, 153–172.

Buchheld, N., Grossman, P., & Walach, H. (2001). Measuring mindfulness in insight meditation (vipassana) and meditation-based psychotherapy: The development of the Freiburg Mindfulness Inventory (FMI). *Journal for Meditation and Meditation Research, 1*, 11–34.

Cahn, B. R., & Polich, J. (2006). Meditation states and traits: EEG, ERP, and neuroimaging studies. *Psychological Bulletin, 132*(2), 180–211.

Cardoso, R., de Souza, E., Camano, L., & Leite, J. R. (2004). Meditation in health: An operational definition. *Brain Research Protocols, 14*(1), 58–60.

Das, N. N., & Gastaut, H. (1957). Variations de l'activité électrique du cerveau, du coeur et des muscles squelettiques au cours de la méditation et de l'extase yogique. *Electroencephalography and Clinical Neurophysiology,* (Suppl. 6), 211–219.

Davidson, R. J., Jackson, D. C., & Kalin, N. H. (2000). Emotion, plasticity, context and regulation: Perspectives from affective neuroscience. *Psychological Bulletin, 126*(6), 890–909.

Davidson, R. J., Kabat-Zinn, J., Schumacher, J., Rosenkranz, M., Muller, D., Santorelli, S. F., et al. (2003). Alterations in brain and immune function produced by mindfulness meditation. *Psychosomatic Medicine, 65*, 564–570.

Faber, P. L., Lehmann, D., Gianotti, L. R. R., Kaelin, M., & Pascual-Marqui, R. D. (2004, April). *Scalp and intracerebral (LORETA) theta and gamma EEG coherence in meditation.* Paper presented at the International Society for Neuronal Regulation, Winterthur.

Farb, N. A. S., Segal, Z. V., Mayberg, H., Bean, J., Mckeon, D., Fatima, Z., et al. (2007). Attending to the present: Mindfulness meditation reveals distinct neural modes of self-reference. *Social Cognitive and Affective Neuroscience, 2*, 313–322.

Goleman, D. (1977). *The varieties of the meditative experience*. New York: Dutton.

Grossman, P. (2008). On measuring mindfulness in psychosomatics and psychological research (invited commentary). *Journal of Psychosomatic Research, 64*, 405–408.

Grossman, P., Schmidt, S., Niemann, L., & Walach, H. (2004). Mindfulness based stress reduction and health: A meta-analysis. *Journal of Psychosomatic Research, 57*(1), 35–43.

Hagemann, D., Hewig, J., Naumann, E., Seifert, J., & Bartussek, D. (2005). Resting brain asymmetry and affective reactivity: Aggregated data support the right-hemisphere hypothesis. *Journal of Individual Differences, 26*(3), 139–154.

Hayes, A. M., & Feldman, G. (2004). Clarifying the construct of mindfulness in the context of emotion regulation and the process of change in therapy. *Clinical Psychology: Science and Practice, 11*(3), 255–262.

Hayes, S. C., & Shenk, C. (2004). Operationalizing mindfulness without unnecessary attachments. *Clinical Psychology: Science and Practice, 11*(3), 249–254.

Heidenreich, T., Ströhle, G., & Michalak, J. (2006). Achtsamkeit: Konzeptuelle Aspekte und Ergebnisse zum Freiburger Achtsamkeitsfragebogen. *Verhaltenstherapie, 16*(1), 33–40.

Hellhammer, D. H., & Hellhammer, J. (Eds.). (2008). *Stress: The brain-body connection.* Basel: Karger.

Ivanovski, B., & Malhi, G. S. (2007). The psychological and neurophysiological concomitants of mindfulness forms of meditation. *Acta Neuropsychiatrica, 19*(2), 76–91.

Kabat-Zinn, J. (1994). *Wherever you go, there you are*. New York: Hyperion Books.

Kabat-Zinn, J. (2005). *Coming to our senses: Healing ourselves and the world through mindfulness.* New York: Hyperion.

Kasamatsu, A., & Hirai, T. (1969). An electroencephalographic study of Zen meditation (Zazen). *Psychologia, 12*, 205–225.

Kirsch, I. (1999). *How expectancies shape experience*. Washington, DC: American Psychological Association.

Kohls, N. (2004). *Aussergewöhnliche Erfahrungen – Blinder Fleck der Psychologie? Eine Auseinandersetzung mit aussergewöhnlichen Erfahrungen und ihrem Zusammenhang mit geistiger Gesundheit.* Münster: Lit-Verlag.

Kohls, N., & Walach, H. (2006). Exceptional experiences and spiritual practice: A new measurement approach. *Spirituality and Health International, 7*(3), 125–150.

Kohls, N., & Walach, H. (2007). Psychological distress, experiences of ego loss and spirituality: Exploring the effects of spiritual practice. *Social Behavior and Personality, 35*(10), 1301–1316.

Kohls, N., Hack, A., & Walach, H. (2008). Measuring the unmeasurable by ticking boxes and actually opening Pandoras box? Mixed methods research as a useful tool for thinking out of the box while investigating Exceptional Human Experiences. *The Archive for the Psychology of Religion, 30*(1), 155–187.

Kohls, N., Sauer, S., & Walach, H. (2009). Facets of mindfulness: Results of an online study investigating the Freiburg mindfulness inventory. *Personality and Individual Differences, 46*(2), 224–230.

Legrand, D., & Ruby, P. (2009). What is self-specific? Theoretical investigation and critical review of neuroimaging results. *Psychological Review, 116*, 252–282.

Lehmann, D., Faber, P. L., Achermann, P., Jeanmonod, D., Gianotti, L. R. R., & Pizzagalli, D. (2001). Brain sources of EEG gamma frequency during volitionally meditation-induced, altered states of consciousness, and experience of the self. *Psychiatry Research: Neuroimaging Section, 108*(2), 111–121.

Lehmann, D., Faber, P. L., Gianotti, L. R. R., Kochi, K., & Pascual-Marqui, R. D. (2006). Coherence and phase locking in the scalp EEG and between LORETA model sources, and microstrates as putative mechanisms of brain temporo-spatial functional organization. *Journal of Physiology, Paris, 99*(1), 29–36.

Lutz, A., Slagter, H. A., Dunne, J. D., & Davidson, R. J. (2008). Attention regulation and monitoring in meditation. *Trends in Cognitive Sciences, 12*(4), 163–169.

Meraviglia, M. G. (1999). Critical analysis of spirituality and its empirical indicators - prayer and meaning in life. *Journal of Holistic Nursing, 17*(1), 18–33.

Orme-Johnson, D. W. (1977). EEG coherence during transcendental concsiousness. *Electroencephalography and Clinical Neurophysiology, 4*, 581–582.

Ospina, M. B., Bond, T. K., Karkhaneh, M., Tjosvold, L., Vandermeer, B., Liang, Y., et al. (2007, June). *Meditation practices for health: State of the research. Evidence Report/Technology Assessment No. 155* (AHRQ Publication No.07-E010). Rockville: Agency for Healthcare Research and Quality.

Ospina, M. B., Bond, K., Karkhaneh, M., Buscemi, N., Dryden, D. M., Barnes, V., et al. (2008). Clinical trials of meditation practices in health care: Characteristics and quality. *Journal of Alternative and Complementary Medicine, 14*(10), 1199–1213.

Shapiro, S. L., Carlson, L. E., Astin, J. A., & Freedman, B. (2006). Mechanisms of mindfulness. *Journal of Clinical Psychology, 62*(3), 373–386.

Vaitl, D., Birbaumer, N., Gruzelier, J., Jamieson, G. A., Kotchoubey, B., Kübler, A., et al. (2005). Psychobiology of altered states of consciousness. *Psychological Bulletin, 131*(1), 98–127.

Walach, H., Buchheld, N., Buttenmüller, V., Kleinknecht, N., & Schmidt, S. (2006). Measuring mindfulness – the Freiburg Mindfulness Inventory (FMI). *Personality and Individual Differences, 40*(8), 1543–1555.

Walach, H., Gander, M., & Kohls, N. (2010, in press). Mind-body-practices in integrative medicine. In J. Giordano (Ed.), *Complementary and integrative approaches to chronic pain.* Boston: Jones & Bartlett.

West, M. A. (1979). Meditation. *The British Journal of Psychiatry, 135*(5), 457–467.

West, M. A. (1980). Meditation and the EEG. *Psychological Medicine, 10*(2), 369–375.

Reconsidering the Metaphysics of Science from the Inside Out

Jonathan W. Schooler, Tam Hunt, and Joel N. Schooler

The only thing you need to know to understand the deepest metaphysical secrets is this: that for every outside there is an inside and for every inside there is an outside, and although they are different, they go together. (Alan Watts, *Man, Nature, and the Nature of Man*, collected public lectures, 1991)

Abstract Material reductionism – the prevailing metaphysical view that reality can be understood entirely in terms of non-conscious physical stuff – is at odds with the existence of experience, the flow of time, and the privileged present. We propose an alternative scientifically-grounded metaphysical perspective that posits: (1) Consciousness represents a fundamental aspect of reality such that all material things enjoy some varying degree of consciousness (panpsychism); (2) nervous systems entail a nested hierarchy of distinct conscious observers; (3) both experience and the flow of time suggest the reality of a subjective realm of existence; (4) the flow of time suggests a process by which all observers collectively sample segments of continuous space/time at different rates, creating a composite of experienced moments of varying thickness; (5) the possibility that consciousness can influence the duration and selection of experienced moments affords a possible opportunity for genuine free will. Although speculative, these conjectures illustrate the type of alternative metaphysics that may be able to accommodate scientific observations without abandoning the self-evident facts that experience exists and time flows.

J.W. Schooler (✉) • T. Hunt
Department of Psychological and Brain Sciences, University of California, Santa Barbara, CA, USA
e-mail: schooler@psych.ucsb.edu

J.N. Schooler
Lewis and Clark College, Portland, OR, USA

H. Walach et al. (eds.), *Neuroscience, Consciousness and Spirituality*, Studies in Neuroscience, Consciousness and Spirituality 1,
DOI 10.1007/978-94-007-2079-4_11,

Introduction

Humanity will be forever indebted to the participants of the Enlightenment who overcame the dominance of dogmatic religious beliefs, and ushered in a more rational pursuit of understanding. This commitment to rationalism led to modern science and all of the remarkable advances that it affords. However, today a new dogmatism has taken reign disguised as rationalism – the very movement that helped to overthrow the dogma of religious oppression. This set of beliefs, commonly referred to as material reductionism, asserts that the universe and all of its constituents (including us) can be entirely reduced to and understood in terms of the interactions of physical stuff that is itself lifeless and completely devoid of consciousness. As Francis Crick put it in 1994:

> You, your joys and your sorrows, your memories and your ambitions, your sense of personal identity and free will, are in fact no more than the behavior of a vast assembly of nerve cells and their associated molecules. Who you are is nothing but a pack of neurons. (p. 3)

Material reductionism, it is said, has no room for antiquated concepts of free will, spirit, or any intelligence in the universe greater than our own (Dawkins 2006; Dennett 2003). From the vantage point of this doctrine, the misguided mass of humanity who still subscribe to these obsolete notions are merely responding to the built-in tendencies of their material brains (Bloom 2004). Indeed, those few scientists and philosophers who challenge the reigning material reductionist dogma are often ridiculed as "supernaturalists" lacking in rigor, and engaging in "panicky metaphysics" (Strawson 1974).

Underlying material reductionism's rejection of spiritual views of any sort is a resolute confidence that the extant scientific concepts are sufficient to illuminate all remaining outstanding scientific (and perhaps even non-scientific) questions. It is assumed that mysteries such as the nature of consciousness will in the end be understood with exactly the same set of principles as has revealed former mysteries (Dennett 1991). The prevailing material reductionist metaphysics asserts that just as nineteenth century vitalists were misguided in their view that something special is required to understand the emergence of life, so too the nature of consciousness will ultimately be understood using exactly the same physical principles that we currently use to understand rocks and toaster ovens.

Although there is little justification for religious reactionaries who reject well-established scientific principles, this does not mean that rigorous scientists must necessarily rally around the inviolability of material reductionism. We don't have to trade one dogmatic set of beliefs for another. Dedication to the scientific method does not require that one necessarily take on faith that deeply held intuitions regarding the fundamental nature of subjectivity are entirely illusory. To the contrary, the scientific method requires that one maintain an open mind, and be wary of dogmatic views wherever they exist. As the physicist Richard Feynman (1999) observed, "Learn from science that you must doubt the experts" (p. 186).

Central to the confidence that current scientists and philosophers have in material reductionism is the notion that it has successfully worked in the past and so is likely to continue to work in the future. Science has certainly made great advances on innumerable topics. Nevertheless, there are at least two fundamental aspects of reality that have resisted satisfactory explanation within the standard material reductionist world-view: the nature of consciousness and the flow of time. As we will argue, with these fundamental mysteries still unsolved, we should be cautious in assuming that their resolution will be solved within the context of the prevailing metaphysical assumptions of today's science. This is not to say that these issues cannot be advanced with the scientific method. We fully subscribe to the scientific method. Our point is, rather, that scientists should be cautious in adopting a metaphysical stance that requires abandoning certain phenomenologically self-evident facts that are, in our view, more certain than the ostensible ***terra firma*** of modern science's current metaphysical assumptions.

In the following discussion we first consider the limitations of material reductionism with respect to experience and time, and then consider several alternative metaphysical options for integrating human consciousness and the flow of time.

This paper itself is a good example of the kind of collaboration we would like to see more of, in that it is written by three authors with disparate positions, each of whom enjoys the debate and flow of ideas (See Shariff et al. 2009 for a similar example). Hunt parts ways with the Schoolers on some key issues and we have made that clear in the text, as well as including an Afterword explaining why Hunt does so.

The Nature of Consciousness

It is a peculiar testament to the myopic vision of the prevailing material reductionist view that the psychologist and philosopher William James, though widely acknowledged as providing some of history's most insightful analyses of consciousness and psychology more generally, is often ignored when it comes to his discussions of metaphysics (although see Wallace 2010). James (1917) recognized the stronghold of material reductionism that was similarly prevalent in his day, noting:

> [P]sychologists will tell you that only a few belated scholastics, or possibly some crack-brained theosophist or psychical researcher, can be found holding back, and still talking as if mental phenomena might exist as independent variables in the world. (p. 9–10)

While acknowledging the evidence that thought is produced by the brain, James pointed out that there are alternative ways in which it might be considered. The brain might, as material reductionists assert, be the *producer* of thought. Alternatively, the brain might merely *transmit* thought, like a prism refracts but does not actually produce light. James observed the fundamental challenge to the production view of consciousness: namely, while it is relatively straightforward to postulate a productive mechanism for mechanistic things, such as a tea kettle producing steam, it is far less evident how material brains produce something as ontologically distinct as consciousness. As James (1898) noted:

> Into the mode of production of steam in a tea-kettle we have conjectural insight, for the terms that change are physically homogeneous one with another, and we can easily imagine the case to consist of nothing but alterations of molecular motion. But in the production of consciousness by the brain, the terms are heterogeneous natures altogether; and as far as our understanding goes, it is as great a miracle as if we said, Thought is 'spontaneously generated,' or 'created out of nothing.' ... All that one need do, therefore, if the ordinary materialist should challenge one to explain how the brain can be an organ for limiting and determining to a certain form a consciousness elsewhere produced, is to [ask him] to explain how it can be an organ for producing consciousness out of whole cloth. (p. 21–22)

We are not necessarily endorsing James' view of the transmission of consciousness, but we have to agree with him that it is not precluded by the facts that are generally assumed to weigh unequivocally in favor of the brain as a producer of thought.

Since the time of William James, we have made notable progress in understanding consciousness, advancing such important topics as the neurocognitive correlate of conscious states (NCC, Koch 2004), differences between conscious and unconscious thought (Baumeister et al. 2011), and the relationship between consciousness and self-reports (Schooler 2002). While such findings represent important advancements, they do nothing to address James' fundamental question of how the brain produces consciousness "out of whole cloth". This issue, the eons-old "mind/body problem," and recently renamed "the hard problem" of consciousness (Chalmers 1995a), persists because consciousness seems to differ from all other scientific topics of inquiry in its apparent lack of any material properties.

Although differing in their conclusions regarding how to deal with the problem, many philosophers acknowledge that conscious experience fundamentally challenges material reductionist explanations. The philosopher Colin McGinn (1991) goes so far as to argue that the human mind is inherently incapable of coming up with an adequate account of consciousness, a view shared by Stephen Pinker (personal communication) who in other respects strictly allies himself with the material reductionist camp (Pinker 1997, 2007). While others vary on the difficulty that they see the problem of consciousness as posing, with few notable exceptions (e.g. Dennett 1991) there is widespread agreement that this is a nut that we have yet to crack. Moreover, not only do we currently lack an explanation for the arising of subjectivity, we do not even know what such an explanation could possibly look like. There simply is no evident place for subjectivity within in the prevailing material reductionism metaphysics.

The Flow of Time

Next to the experience of consciousness, few phenomena are as self-evident as the passage of time. From our first breath to our final gasp, time inexorably marches forward. Remarkably, however, there is at present no consensus for why time flows in only one direction, or even why it seems to moves at all. As the physicist Brian

Greene observes: "[E]ven though experience reveals over and over again that there is an arrow of how events unfold in time, this arrow seems not to be found in the fundamental laws of physics" (Greene 2004, p. 144–145). There have been various speculations about what underlies the apparently forward movement of time. Chief among them is the notion of entropy (Greene 2004). Entropy, the tendency for all things in the universe to become more disorganized with time, unquestionably provides a metric for the arrow of time. However, it is far from clear how entropy resolves the underlying question of what exactly is moving forward from one moment to the next. In other words, entropy provides a description of the direction of the flow of time, but does not provide an explanation for why it flows in that direction (i.e. why shouldn't things become more ordered with time?) or why it flows at all.

In considering the nature of time, physicists often "spatialize" it. In other words, they attempt to place it on a similar footing to the traditional three dimensions of space. Though differing from spatial dimensions in important respects (Einstein 1920/2001) the notion of time as similar to a spatial dimension is a key feature of the prevailing Einstein/Minkowski interpretation of special relativity theory. Space and time are combined into one concept: space-time. The spatialization of time allows the depiction of a "block universe" in which the traditional spatial dimensions are reduced (for purposes of visual illustration) to two dimensions, and time is added as a third dimension. Louis de Broglie (1959), a French physicist who played a key role in the development of quantum theory in the beginning of the twentieth century, clarified Einstein's view (what is more accurately labeled the Einstein/Minkowski view):

> In space-time, everything which for us constitutes the past, the present, and the future *is given in block*, and the entire collection of events, successive for us, which form the existence of a material particle is represented by a line, the world-line of the particle. Each observer, as his time passes, discovers, so to speak, new slices of space-time which appear to him as successive aspects of the material world, *though in reality the ensemble of events constituting space-time exist prior to his knowledge of them.* (p. 133)

Such a depiction can be thought of as a space-time "loaf of bread," where each narrow cross-section of the loaf ("slice") constitutes a moment in time of the entire universe. According to the block universe view (widely held by today's physicists), all slices – past, present and future – already exist. It is simply that the observer is privy to only one moment (slice) at a time. Critically, as will be further described shortly, this view offers no account for the privileged quality of the present, cannot adequately explain the subjective flow of time, and leaves the source of subjective movement through the posited block universe unexplained.

An Inside-Out Ontology

A central assumption of the current scientific ontology is that personal experience is inherently flawed as a basis for rigorous knowledge and that everything that we can claim to know about the universe should be derived from scientific (empirical) investigation. Seemingly fundamental aspects of our universe, such as time (Einstein quoted in Hoffmann and Dukas 1972) and even consciousness itself (Hofstadter 2007), are characterized as mere illusions resulting from inadequately informed intuitions.

Although subjective experience is often erroneously dismissed as a fundamental source of knowledge, it is manifestly *all* that we can know about the universe and ourselves. As William James observed: "Introspective observation is what we have to rely on first and foremost and always." (James 1890/1918, p. 185). All scientific facts, literally, are communicated to us through our personal experience – there is no other way to receive information. As Descartes famously concluded, experience has more ontological certainty than external reality itself. You could be dreaming, you could be a brain in a vat, but there is simply no question but that you are an experiencing observer. Although now often overlooked, the status of subjective experience as the foundational core of any meaningful ontology was a critical element of phenomenological philosophy (Husserl 1980) and the essential insight of Descartes' (1637/1956) famous phrase "I think, therefore I am." Kant (1781/1896), Berkeley (1734/1971), Schopenhauer (1819/1995) and many other philosophers continued to build upon Descartes' insights about the necessary subjective starting point for all human knowledge.

To illustrate the unique ontological status of subjective experience, consider the following thought experiment. Imagine that an elite group of scientists, who have collectively received every imaginable recognition and accolade, told you that they had solved the hard problem of consciousness and had developed a technique that definitively discerned what possesses subjective experience and what does not. These scientists use their pioneering innovation on you and conclude that you do not in fact have consciousness ... you just think you do. Would you accept their conclusion? You surely would accept any other conclusion such a group of scientists might offer. But in this case, it seems likely, you would be absolutely certain they had come to the wrong conclusion. Ultimately, when it comes to the existence of subjectivity, one's first-person experience trumps even the most authoritative scientific evidence.

Having established that subjective experience must serve as the foundation for building one's ontology, we can now pose the following question: are there any other ontologically necessary truths that follow from subjective experience alone? That is, are there other aspects of reality that we can derive entirely from our personal experience? In this regard, the passage of time is an appropriate candidate. Experience is always and invariably extended in time. Indeed, without duration there would be no experience. By acknowledging the existence of experience, we must also acknowledge the existence of time in which experience "extends".

Closely related to the experience of subjective time is the privileged nature of the present. We can remember the past and anticipate the future but we only exist in the present, now, now, now, and now... The privileged status of the present is, however, contradicted by prevailing physical theories. The privileged present is another aspect of our experience that is ultimately more self-evident than any contradictory claims by physics or any other area of knowledge.

In sum, we see that when we inspect what are the most fundamental aspects of our experienced existence, so fundamental that they do not require any empirical support for us to be absolutely certain of them, we find that they are the very same issues (subjectivity and the flow of time) that prevailing physics is at present incapable of explaining satisfactorily. So what are we to make of this impasse between the conclusions of prevailing scientific theories and those of subjective experience?

Acknowledging the Fundamental Nature of Subjective Experience and the Flow of Time

As noted, the current inability of science to satisfactorily account for subjective experience has led to a variety of reactions. Some have suggested that this is a problem that is simply beyond the ken of mankind to solve (McGinn 1999), others have argued that it will be resolved in some yet unimaginable way, and yet others have argued that it implies that consciousness itself is an illusion (e.g. Hofstadter 2007). A few thinkers, however, have recognized that subjective experience warrants consideration as an essential aspect of the universe. As the philosopher David Chalmers (1995b) observed:

> I propose that conscious experience be considered a fundamental feature, irreducible to anything more basic... In the 19th century it turned out that electromagnetic phenomena could not be explained in terms of previously known principles. As a consequence, scientists introduced electromagnetic charge, as a new fundamental observer. Similar reasoning should be applied to consciousness. If existing fundamental theories cannot encompass it, then something new is required. (pp. 96)

Suggesting that consciousness may represent a fundamental aspect of the universe is not limited to philosophers. Although a notable minority, some physicists have made very similar suggestions. For example, physicist Andrei Linde (1990) observed:

> Note, that the gravitational waves usually are so small and interact with matter so weakly that we did not find any of them as yet. However, their existence is absolutely crucial for the consistency of our theory, as well as for our understanding of certain astronomical data. Could it be that consciousness is an equally important part of the consistent picture of our world, despite the fact that so far one could safely ignore it in the description of the well studied physical processes? Will it not turn out, with the further development of science, that the study of the universe and the study of consciousness are inseparably linked, and that ultimate progress in the one will be impossible without progress in the other (p. 27)

We will return shortly to discuss how consciousness might potentially be treated as a fundamental aspect of reality, but first let us consider the second feature of existence that is as central to subjective experience as it is elusive to prevailing physical theories – the flow of time.

As noted, standard models of physics deny the subjective flow of time any objective reality. The physicist Paul Davies observes: "From the fixed past to the tangible present to the undecided future, it feels as though time flows inexorably on. But that is an illusion." (Davies 2002, p. 32). Importantly, physicists acknowledge that the only place where the passage of time seems to have any meaning is in the context of consciousness. For example, Davies argues:

> Nothing other than a conscious observer registers the flow of time. A clock measures durations between events much as a measuring tape measures distance between places; it does not measure the "speed" with which one moment succeeds another. Therefore it appears that the flow is subjective, not objective. (p. 36)

The standard reaction to the observation that the privileged present and the flow of time only have meaning from the vantage point of a conscious observer is to dismiss these constructs as illusory. Just as we have come to terms with the facts that the earth is not flat and that species change over time, we shall, many physicists and philosophers assert, eventually come to terms with the idea of the flow of time being ultimately illusory. As Einstein observed "The past, present and future are only illusions even if stubborn ones" (quoted in Hoffmann and Dukas 1972, p. 258). However, others have recognized the inherent problem with this move: once we acknowledge the validity of subjective experience as the only source of knowledge about the universe, the privileged present and the flow of time become more certain than the physical theories that have been proposed to replace them. The philosopher David Ray Griffin (2007) observes: "The reality of time is a more fundamental and stubborn fact than the alleged facts on which its denial is based" (p. 119). In short, to abandon the certainty of time in favor of the theories of physics is like reading a map in the comfort of one's home, yet concluding on the basis of the interpretation of the map that one is lost. When ostensibly rational claims contradict a self-evident truth it is time to revisit the robustness of those claims.

Towards an Alternative Metaphysics

The current inability of prevailing physical theories to satisfactorily account for the two aspects of existence of which we can be most certain raises the genuine possibility that something very fundamental is missing from the prevailing scientific story. At a minimum, such core omissions undermine the view that prevailing theories have proven too successful to be challenged. In the following analysis, we consider several metaphysical alternatives that should be given due consideration. Importantly, we recognize the highly speculative nature of what follows. Indeed the authors themselves differ in important respects. In particular, Hunt (in press) has recently proposed an alternative perspective on a number of the following issues, including the nature of time, the meaning of information and the manner in which objectivity and subjectivity are distinguished. **The reader is also directed to the Afterword for further consideration of these issues of contention between the authors**.

Cognizant of its highly speculative nature, we present the following discussion not with the aim of persuading readers that these many contentious claims are necessarily correct, but rather to spur consideration of the types of alternative metaphysical views that might be capable of respecting *both* our intuitions drawn from subjective experience and yet be potentially consistent with what empirical science has so far revealed about the world.

Consciousness Is Inherent in Everything

If we are to take seriously the notion that consciousness represents a fundamental aspect of the universe, a natural question arises: what would this mean? This question can be traced at least as far back as the ancient Greek philosopher Democritus, who in anticipating the notion that the universe is composed of atoms, further speculated that each atom holds some primitive dim consciousness. This is a type of *panpsychism,* the view that all things participate in two realms of existence: a private realm of subjectivity, and a public realm of physical reality. Although differing in their precise construal, many brilliant thinkers throughout history have endorsed some version of the panpsychist perspective including Spinoza (1677/1985), Leibniz (1714/1989), Schopenhauer (1819/1995), James (1909), Bergson (1896/1912) and Whitehead (1929). More recent advocates of this view include Hameroff (2001), Chalmers (1995b), Griffin (2007) (though he prefers the term "panexperientialism"), Skrbina (2005) and Strawson (2008).

Although some readers not previously familiar with the notion might be tempted to reject panpsychism out of hand, careful reflection on the theory reveals it to be at a minimum worthy of consideration, and quite possibly the most parsimonious account currently available for conceptualizing the relationship between consciousness and reality. Importantly, there are both logical and observational reasons to postulate that all matter may enjoy at least a rudimentary consciousness. Before explicating these arguments, however, it is important to clarify one essential aspect of panpsychism. Panpsychism assumes that all matter partakes in at least some rudimentary form of experience, however, this is not to say that *all* objects are themselves sentient beings. To use Nagel's (1974) terminology there need be nothing "that it is like to be" a rock, for example. Rather, the panpsychist/panexperiential claim is that at some level, the constituent elements of a rock (and all other material objects) partake in at least some very rudimentary kind of experience. The fundamental units of both matter and mind are what the physicist/philosopher Alfred North Whitehead (1929), referred to as "actual entities" or "occasions of experience". In some circumstances, and in particular when present in certain organic structures, actual entities may combine to form higher-order actual entities. (For further discussion of this issue see Whitehead (1929), Rosenberg (2005), Hunt (in press), and the present discussion of the combination problem). However, for the most part, when actual entities combine they generally form "mere aggregates" that do not entail a higher-order experience. Only the constituents of mere aggregates are actual entities – the combination is not conscious. In short, most modern versions of panpsychism assume that much of nature enjoys only the most primitive type of experience.. With those preliminaries behind us, let us now consider the arguments for panpsychsm, starting with logical argument and then moving on to observational ones.

Logical Arguments for Panpsychism

The Challenge of Emergentism

The first advantage of panpsychism is that it skirts one of the fundamental limitations of the materialist alternative – the problem of emergence. According to emergentism, consciousness somehow emerges from entirely non-conscious matter in a manner similar to the way in which a liquid emerges from the unwet atoms of which it is composed. The problem with this view, as alluded to by William James, is that it is entirely unclear how something as distinct as consciousness could come "out of whole cloth". No matter where or how in the phylogenetic order (or ontogenetic, for that matter) this fantastic leap occurred it is entirely inexplicable. It is as if gravity suddenly appeared on the universal scene where it had not existed before. Ultimately, the hurdle from inert matter to consciousness is reminiscent of the classic New Yorker cartoon below (Harris 2006) in which a scientist interjects "and then a miracle occurs" between two sets of formulae on a black board. The panpsychist perspective has the notable advantage of not requiring the explanation of how something so ontologically distinct as experience suddenly emerged where none existed before. Instead, where there is matter there is mind – the two go together always.

The standard reply to such a criticism is that there are other cases of emergence, e.g. liquidity, so why shouldn't consciousness be just another example like this? However, the problem with this argument, as articulately noted by the philosopher Galen Strawson (2008) is that emergence requires the existence of a necessary dependence between the pre-existing conditions and the emergent property. As Strawson puts it:

> It seems plain that there must be a fundamental sense in which any emergent phenomenon, say Y, is wholly dependent on that which it emerges from, say X. It seems, in fact, that this must be true by definition of 'emergent'; for if there is not this total dependence then it will not be true after all, not true without qualification, to say that Y is emergent from X. For in this case at least some part or aspect of Y will have to hail from somewhere else and will therefore not be emergent from X. Plainly this is not how it is with liquidity. (p. 64)

In the case of liquidity, there are a set of comprehensible properties of the pre-existing constituents (i.e. they possess a molecular structure that slides rather than binds) thereby leading in a principled and entirely explicable way to the emergence of liquidity. No such relationship has ever been established between ostensibly non-conscious matter and consciousness. To argue that experience/consciousness can emerge from matter is more analogous to suggesting the far less tractable notion that spatial extent can emerge from non-spatial extent. Again as Strawson observes:

> If one is being metaphysically straight, the intuition that nothing (concrete, spatiotemporal) can exist at a mathematical point, because there just isn't any room, is rock solid… So if the idea of unextended-to-extended emergence is offered as an analogy for non-experiential-to-experiential emergence, I don't think it can help. (p. 64)

"I THINK YOU SHOULD BE MORE EXPLICIT HERE IN STEP TWO."

In short, the suggestion that consciousness could emerge from entirely non-conscious matter requires the existence of a bridging principle that must on the one hand explain how consciousness could emerge from a medium that lacks any trace of it, while at the same time avoiding the introduction of any new elements that are not inherent in the matter from which consciousness is said to emerge. Otherwise, one is again back to positing "then a miracle occurs".

The Advantage of Parsimony

The second core logical advantage of panpsychism is parsimony. Each human being, being the possessor of a material yet experiencing brain, knows for a fact that matter can be conscious. However, we have no evidence and absolutely no way of knowing whether there is any matter that lacks consciousness. Given that we know that matter can be conscious, and have no evidence that there exists any matter that is not conscious, parsimony strongly favors the conclusion that all matter enjoys some (albeit generally highly rudimentary) degree of sentience. Again as Strawson puts it:

> Returning to the case of experience, Occam cuts in again, with truly devastating effect. Given the undeniable reality of experience, he says, … why insist that physical stuff in itself, in its basic nature, is essentially non-experiential, thereby taking on [a] a commitment to something—wholly and essentially non-experiential stuff—for which there is absolutely no evidence whatever along with [b] the wholly unnecessary (and incoherent) burden of brute emergence otherwise known as magic? (p. 66)

Thus it seems that from a parsimony perspective it is far simpler to assume that consciousness is a fundamental property of matter/energy that extends and develops according to potentially identifiable principles, than to assume that consciousness somehow pops into existence out of material that prior to this remarkable transformation is entirely and utterly devoid of this property.

Observational Evidence for Panpsychism

As previously noted, there is no way to directly observe whether anything other than ourselves is conscious, so the notion of empirical evidence for panpsychism is admittedly a bit of an oxymoron. However, pragmatically speaking there are observations that inform our intuitions about what is conscious and what is not. That is, circumstantial evidence may be marshaled in our favor. If we observe an object's behavior that suggests the presence of intention we are inclined to believe that it is conscious. Indeed, a major reason why most of us assume that dogs are conscious but rocks are not is that dogs show behavioral evidence of intention while rocks (happily) do not.

It turns out, however, that when we look more carefully at the behavior of things to which we typically do not attribute experience our intuitions can change. The typical materialist view is that consciousness is a product of nervous systems that have reached some critical magnitude of complexity (though importantly there is no principled basis for determining how much neural complexity is required for consciousness to take place). However, many of the behaviors that lead us to attribute consciousness to organisms are also evidenced by eukaryotes that have no nervous system whatsoever (Fitch 2008). Paramecia can learn, avoid predators, find food, select mates and have sex, all without a nervous system (Margulis and Sagan 1995). If a larger creature with a nervous system evidenced these behaviors we might reasonably attribute sentience to that creature, so how can we be so sure that a single cell organism lacks it? And if some cells are conscious, on what grounds should we assume that others do not? And why should life be the defining precursor for consciousness? Isn't this a form of vitalism that is so often ridiculed by materialists? If there is nothing necessarily special about living things that enable consciousness to arise, then the premise that non-living matter might also enjoy some type of experience should not be considered *a priori* infeasible. Indeed the principled yet unpredictable behavior of particles has led a number of physicists to speculate that even they may enjoy an iota of experience. The eminent American physicist Freeman Dyson (1979) was explicit on this point, noting:

> ...mind is already inherent in every electron, and the processes of human consciousness differ only in degree but not in kind from the processes of choice between quantum states which we call "chance" when made by electrons. (p. 245)

The physicist David Bohm made a similar observation, observing "The ability of form to be active is the most characteristic feature of mind, and we have something that is mind-like already with the electron" (in Talbot 1991, p. 50). In describing the behaviors of electrons in plasma Bohm further (1987) observed:

> [T]he whole system is undergoing a co-ordinated movement more like a ballet dance than like a crowd of unorganized people... closer to the organized unity of the parts of a living being than it is to the kind of unity that is obtained by putting together the parts of a machine. (p. 38)

Many readers are likely to balk at the suggestion that seemingly senseless matter could possess even the remotest aspect of consciousness. We ask such readers the following: given that you are a material body capable of experience, how can you be so sure that other material things don't share a modicum of your good fortune? It may simply be a matter of perspective. From the vantage point of outer space, we could be viewed as nothing more than an infestation of "planet fungus" entirely void of consciousness. As noted, there is no known marker for determining whether or not something is conscious, therefore any view that asserts realms in which consciousness is necessarily lacking is simply a statement of personal faith and not scientific fact. William James (1902/2002), who arguably said more right things about consciousness than any one else in history, eloquently expressed his view that science could be premature in adopting a strictly impersonal view of physical reality.

> ...The divorce between scientists' facts and religious facts may not necessarily be as eternal as at first sight seems, nor the personalism and romanticism of the world, as they appeared to primitive thinking, be matters so irrevocably outgrown. The final human opinion may, in short, now impossible to foresee, revert to the more personal style, just as any path of progress may follow a spiral rather then a straight line, the impersonal view of science might one day appear as having been a temporarily useful eccentricity rather than a definitively triumphant position which the sectarian scientist at present confidently announces it to be. p 545

Few today would deny either the physical universe or the interior private realm, notwithstanding some prominent philosophers that we have cited above. Once we acknowledge the co-existence of these domains, it seems no more unreasonable to suggest that they exist to varying degrees in all contexts, as dual aspects of a single reality, than to posit (with no hint of an explanation) how either could emerge whole cloth from the other.

Possible Corollaries of Consciousness

An inherent subjectivity to all of matter suggests the possible existence of attributes that intuitively go hand in hand with what we generically refer to as subjectivity, namely: feeling, understanding, intelligence and agency. To evaluate this intuition it may be helpful to consider the lowest level species for which most readers are probably prepared to grant the likelihood of consciousness. For the sake of argument, let us take the standard generally used as the ethical dividing line for research on animal: vertebrates.

If one thinks a simple vertebrate is conscious then one is also likely to think it has (a) feeling – e.g. experiencing pain, (b) at least some rudimentary understanding – recognizing food, danger, etc., (c) appetites/aversions – e.g. desire for food and avoidance of harm, (d) intelligence – capacity to discern a course of action in keeping with its appetites/aversions, and (e) agency – ability to act in keeping with its desires and rudimentary intelligence. Perhaps, some readers are prepared to grant some of these qualities but not others (e.g. feeling but not agency). However, we expect that these are common intuitions of what characterizes beings that are considered conscious. Thus if we are to seriously entertain the possibility that all of matter enjoys some degree of consciousness, then we must also be prepared to acknowledge that this may also entail at least some very primitive element of the likely components of consciousness, including feeling, understanding, intelligence and agency.

The Combination Problem

Once we grant that all of matter may enjoy some primitive level of consciousness, a critical issue, often referred to as the "combination problem" or "boundary problem" Hunt (in press) arises. The combination problem refers to the question: how do individual elements combine to form larger experiential entities? While the constituent elements of a rock are highly unlikely to cohere together into a singular "rock" experience, it seems clear that some of the elements of our brains unite into at least a semi-coherent higher-order experience, which we call our conscious mind. We suggest that the arising of higher-order entities may be a result of nested hierarchies of conscious agents.

It is now widely accepted that evolution entailed a process in which simple organisms combined to form the organelles (e.g. mitochondria) of more complex eukaryote cells, which in turn combined to become multi-cellular organisms (Margulis and Sagan 1995). We suggest that just as life evolved the capacity to integrate independent living creatures into more complex singular life forms, it may have similarly developed the capacity to integrate subjective experiences into nested hierarchies of higher-order experiencers. Indeed, such a hierarchical view of consciousness represents the basis of the neuroscientist Zeki's (Zeki and Bartels 1999; Zeki 2003) theory of how consciousness manifests in the brain.

Drawing on differences in the processing rates of different areas of the visual system, Zeki suggests that the brain engages in a nested hierarchy of distinct conscious experiences leading to a final unified experience. He proposes three hierarchical levels at which consciousness takes place in the brain: micro-consciousness corresponding to the different levels of the visual system that process distinct attributes (e.g. V4 processes color where as V5 processes motion), macro-consciousness that integrates multiple attributes of a system (e.g. binding color to motion), and unified consciousness corresponding to the experience of the perceiving person. Zeki further suggests that each of these nested levels of consciousness occur in a distinct temporal order, with the lower order levels being ahead of and feeding into the higher order levels. Zeki describes his model as follows.

> It thus becomes possible to distinguish three hierarchical levels of consciousness: the levels of micro-consciousness, of macro-consciousness, and of the unified consciousness. Of necessity, one level depends upon the presence of the previous one. Within each level, one can postulate a temporal hierarchy. This has been demonstrated for the level of micro-consciousness, because colour and motion are perceived at different times. It has also been demonstrated for the level of the macro-consciousnesses, because binding between attributes takes longer than binding within attributes…Micro- and macro-consciousnesses, with their individual temporal hierarchies, lead to the final, unified consciousness, that of myself as the perceiving person. (p. 217)

Although Zeki only describes three levels, in principle we could imagine that there could be many additional lower level micro-consciousness corresponding to the individual receptor cells of the retina, and ultimately down to the level of photons. Thus Zeki's model provides a manner of conceptualizing how reality may entail a hierarchical nesting of conscious observers. Accordingly, the non-organic world may involve only the most micro-level conscious observers. In contrast, life may have evolved the capacity to develop hierarchies of conscious observers within observers, with each level subsuming a more macroscopic perspective, leading ultimately to the highest level at which the unified experience of the organism occurs. Additionally, Zeki's approach offers a critical way of distinguishing these levels, namely, by the temporal order in which they occur, with higher order experiences occurring temporally downstream. In other words, Zeki's view suggests that the different conscious observers in the brain may experience the same events at different times, with the final unified consciousness entailing the longest lag.

The perspective on consciousness that emerges from considering Zeki's model in the context of panpsychism is both elegant and daunting. It requires thinking about consciousness in ways that are somewhat alien to our prevailing conceptualizations so it is worth carefully unpacking the elements.

1. *Nervous systems are composed of a hierarchy of distinct conscious observers each enjoying their own unique experience.* The tension between the unity and disunity of consciousness is a long-standing challenge. On the one hand, there is the phenomenal intuition of a unity of experience, a place in the mind where it all comes together, what Dennett (1991) refers to as the "Cartesian theatre". On the other hand there is the empirical fact that the various elements of our

ostensibly unified experience occur in different parts of the brain at different times. The fractionated nature of the processes that contribute to subjective experience is the principle source of Dennett's contention that the Cartesian theatre, i.e., the unity of consciousness, is an illusion. In its stead, Dennett's proposes a multiple drafts theory (1991) in which conscious reports are viewed as the disjointed amalgam of distinct and often competing information processing systems that never come together into a singular unified experience.[1] However, rather than concluding that there is no Cartesian theatre in which these strands of information processing come together, the alternative perspective suggested by Zeki's analysis, and the panpsychist perspective more generally, is that human consciousness entails a hierarchically organized "multiplex" of Cartesian theatres. From this perspective, every nervous system is actually a society of observers or conscious entities (Loftus and Schooler 1985; Minsky 1986), each organizing the information that they are presented with into distinct conscious experiences, and then passing those experiences on to the next level. In this model a final unified experience emerges as the highest order observer that enjoys the collective informational organization afforded by all of the prior separately conscious observers.

Various other lines of research also suggest that our brains may entail multiple distinct conscious experiences. Corpus callosotomy (split-brain) patients, for whom the primary neural bridge between the two brain hemispheres is severed, evidence a variety of behaviors suggesting that their two hemispheres maintain distinct streams of consciousness (Gazzaniga 2005). For example, when a word (e.g. bike) is presented to the left visual (right hemisphere) the patient reports not having seen anything. However, when the left hand is given the opportunity to draw what was presented it nevertheless draws a bicycle. Similarly, in the case of blindsight (Weiskrantz and Weiskrantz 2009), patients report not seeing anything when an object is placed in a particular area of their visual field, yet when given the opportunity they will reach directly for the object they had just claimed not to have seen. These and other dissociations can be readily understood by the notion that one part of the brain is experiencing a particular type of consciousness that is not available to the higher-level (or simply different) consciousness associated with the capacity for verbal reports.

2. *The various observers within a nervous system experience the same event at different times.* In Zeki's model, elements of the nervous system enjoying micro-consciousness and macro-consciousness experience different aspects of the same stimulus at successive moments in time. As Zeki observes:

[1] How phenomenal experience fits into this model has been a source of some consternation. In some places Dennett seems to suggest that phenomenal experience is just an illusion (Dennett 1991) and indeed many have interpreted him in this way (e.g. Searle 1990). In other places, however, he insists that we are conscious (Dennet 1997). Given that Dennett consistently denies that subjective experience offers any privileged information a critical question arises for him, namely, how it is that he knows that he is conscious.

> Because we become conscious of colour before we become conscious of motion, it follows that the micro-consciousnesses generated by activity at two distinct cortical sites are distributed in time as well. From this it follows that micro-consciousnesses are distributed in time and space, and that there is a temporal hierarchy of micro-consciousnesses, that for colour preceding that for motion. (p. 215)

The notion that different observers in the brain experience the same stimuli at different times also helps to account for various empirical conundrums. For example, when individuals view two lights flashing in close proximity and succession to one another, the phenomenal experience is of apparent motion in which an illusory light appears to move between the first flash and the second. If the first light is blue and the second light is yellow, the apparent motion appears to evolve from blue to green and then finally to yellow in the intervening locations between the two lights. The challenge of this illusion is how does the brain know to make the intervening value green before the yellow light has been perceived? In other words, phenomenally the order of events is blue light, green apparent motion, yellow light, whereas objectively the green apparent motion can only have been generated after the yellow light has been observed. Positing the existence of multiple experiencing observers that process their respective aspects of the event at different times helps to reconcile this apparent paradox (see Dennett 1991 for a related albeit staunchly different resolution). Accordingly, the micro-consciousnesses responsible for individual colors perceive the two flashing lights before the macro-consciousness that binds the two events together infers the illusory intervening color in between. The model that emerges from this analysis of consciousness is that the various observers entailed in the nervous system generate a layered construction of reality, with each layer passing an increasingly integrated and informationally rich experience back to the layer behind it.

3. *The grain size of temporal moments must be larger for observers that are higher in the hierarchy.* The notion that various observers in the brain pass organized experiences forward to later observers, which then integrate those experiences into yet more informationally rich experiences, necessarily requires a courser temporal grain size at the higher levels. (By temporal grain size we mean the duration of time constituting an individual psychological moment. As will be discussed further, temporal moments can be considered much like the perception of stills from a movie in which each static frame is integrated into a continuous experience. In this view, temporal grain size corresponds to the duration of time that the information entailed in any single still extends over). Consider again the case of apparent motion, in order for a macro-consciousness to infer the apparent motion and color shift between two individual flashes of light, the macro-consciousness cannot simply be delayed relative to the micro-consciousness, otherwise it would never be able to integrate the first event with the second event. In order to integrate, it must maintain both events in a larger temporal unit that extends between the two individual flashes. In short, in order to integrate distinct experiences in time, higher-order observers must necessarily have a larger temporal grain size, i.e., their psychological moments must span across longer durations than the observers that contributed to them.

4. *The impact of integrative experience is bi-directional.* One aspect of the relationship between micro, macro and unified consciousness that is not explicitly discussed in Zeki's model but which seems highly likely to be the case is that direction of information flow is both bottom up and top down. There is increasing evidence that the information processing states of higher order regions routinely work their way back to earlier levels impacting how those earlier levels process incoming information. Both the situational context (e.g. Kingstone et al. 2004) and attentional allocation (e.g. Kam et al. 2011) assessed at higher levels of processing routinely impact lower levels. It thus seems likely that not only do the low level observers pass their experiential state up to higher levels, but that the interpretations of higher level observers impact on the experience at the lower levels.

The view of consciousness that emerges from the above analysis is one in which all of matter enjoys an iota of conscious experience, but where life forms are able to construct increasingly complex experiences through hierarchies of distinct yet interrelated conscious observers. These hierarchies are akin to a society of minds, in which each observer organizes and passes experiences forward to higher-order observers. At each level of the hierarchy observers interpret the experiences they receive into an integrated and informationally richer experience that occurs later in time and subsumes a larger psychological moment. The interpretations created at these higher levels may then work their way back down to lower levels, in a recursive exchange of experiences.

The realm of information[2] – The notion that all matter (e.g. our brains) must be associated with both an inner experience and an outer condition seems to require the postulation of two sets of coordinates in order to fully describe any matter. That is, external coordinates corresponding to that matters' physical dimensions, and internal coordinates corresponding to the subjective information state that it enjoys. Thus an apparent implication of panpsychism is an additional dimensional space corresponding to the internal experience. We can call this additional dimensional space "information space." (Chalmers 1996). As the opening quote by Watts intimated, the recognition of an inside and an outside to all of reality suggests two complementary realms: the external physical realm and the internal subjective realm, or what may be described as physical space–time and information.

The neuroscientist Giulio Tononi (2008) has recently developed a sophisticated formalization of the nature of the information space that may characterize this inner realm of consciousness. Though he does not explicitly suggest the existence of two complementary realms Tononi does posit that consciousness is a fundamental aspect of the universe associated with "integrated information," which is defined as "the amount of information generated by a complex of elements, above and beyond the information generated by its parts" (p. 216). Tononi's information based theory of consciousness nicely complements many aspects of Zeki's neurocognitive theory, and the more general view presented here. Zeki's and Tononi's models assume that

[2]Hunt parts ways with the Schoolers on the ideas in this section due to its postulation of an ontological distinction between information and physical/reality, see Afterword.

the brain entails nested hierarchies of separate consciousnesses that sometimes exist in coordination and other times in exclusion of one another. Tononi's approach complements this analysis by providing a formal quantitative way of characterizing the information space that consciousness inhabits (see Hunt in press for a related approach). Critically however, neither Zeki nor Tononi fully come to terms with the implications of their complementary approaches, namely that they suggest two distinct realms of reality – the physical realm of space-time and subjective realm of experience and information. As will be seen, when we consider the challenge of conceptualizing movement in time, we again see the possible advantage of positing distinct yet inter-related realms of physical space-time and subjectivity.

The Flow of Time as a Wave of Experience[3]

As noted throughout this chapter, in addition to being unable to adequately account for subjective experience, current scientific notions offer little help in explaining the flow of time or the privileged status of the present moment, despite their phenomenological self-evidence. This challenge arises because time itself and movement in time are understood only with respect to the experience of a conscious observer, and consciousness generally has no explicit status in today's physics. A reasonable starting point for a solution is to introduce consciousness as a fundamental aspect of the universe by which movement in time is defined.

We propose that the present moment is akin to a wave of consciousness moving through physical space-time. As noted earlier, there is nothing in physics that corresponds to the flow of time. Physicist Linde (1990) observes "Thus we see that without introducing an observer, we have a dead universe which does not evolve in time" (p. 25). Yet it is self-evident that we, as conscious observers, experience change in time. If we are to maintain the self-evident fact that experience entails the passage of time, and if we maintain the Einsteinian notion that time has properties akin to a physical dimension, then it follows that the observer "moves" in relationship to time. Characterizing the collective movement of all observers in relationship to time as a wave of consciousness thus constitutes a reasonable, if not logically necessary, characterization of the flow of time.

Importantly, the claim that the ever-changing present entails in some sense the flow of consciousness through physical space-time highlights a central paradox about the experience of time whose resolution may require some further rather dramatic refinements to our current conceptualization of reality. The paradox arises from the following question: On what metric can the flow of time be gauged? The seeming answer would be time itself, but this becomes tautological, as a metric

[3]Hunt parts ways with the Schoolers on this section due to its endorsement of a block universe model of physical reality, and its suggestion of an ontological distinction between subjective and objective time (see Afterword).

cannot be both the measure and the object of measurement. The physicist Davies (2002) puts this paradox succinctly:

> But what meaning can be attached to the movement of time itself? Relative to what does it move? Whereas other types of motion relate one physical process to another, the putative flow of time relates time to itself. Posing the simple question "How fast does time pass?" exposes the absurdity of the very idea. The trivial answer "One second per second" tells us nothing at all. (p. 34)

Davies' solution to this paradox is to conclude that the flow of time is an illusion, but as we have argued such a claim is so counter to experience that it seems untenable, or at least deeply deserving of alternatives. Another option is to postulate that there may be a subjective realm of reality against which movement in objective time can be understood. Put another way, it seems possible (and perhaps even a mathematical necessity) that in order to move through space-time, there needs to be at least one additional dimension to provide the degree of freedom necessary to enable such movement, giving rise to objective time and subjective time.

Although the postulation of additional dimensions of reality should not be taken lightly, it is not without precedent. In physics, string theory has postulated seven additional spatial dimensions beyond the three dimensions of space and one dimension of time that are customarily acknowledged (Greene 2004). If there can be multiple dimensions of space, then might there not also be additional dimensions of time? Indeed, some physicists have argued that an additional dimension of time might be very useful for conceptualizing various issues in physics (Bars et al. 1998). If the postulation of an additional dimension of subjective time could also resolve the paradox of time and provide a realm for subjectivity then surely that would also warrant its consideration as a possibility.

We are not the first to suggest that the failure of objective time as it is currently conceptualized to afford the flow of time or inner experience requires the postulation of an additional subjective dimension (or dimensions) in which the observer moves relative to physical space-time. Noting the inability of current theories of physics to account for the flow of time or the existence of subjective experience, physicist Linde speculates that dimensions of consciousness may be required to provide the necessary degrees of freedom. Linde (2004) observes:

> Is it possible that consciousness, like space-time, has its own intrinsic degrees of freedom, and that neglecting these will lead to a description of the universe that is fundamentally incomplete? What if our perceptions are as real (or maybe, in a certain sense, are even more real) than material objects? What if my red, my blue, my pain, are really existing objects, not merely reflections of the really existing material world? Is it possible to introduce a 'space of elements of consciousness'….? (p. 451)

The neuroscientist/neurophilosopher John Smythies (2003) similarly observes the challenge of current conceptualizations of time for accommodating the flow of time, noting:

> If one wants to account for our psychological impression that there is a 'now' in time and moreover that time in some way flows, we must look elsewhere than contemporary physics, whether Newtonian or Relativity, to find it. (pp. 53)

Like Linde, Smythies (2003) concludes that in order for time to flow, consciousness must move in real time through both physical space/time and additional dimensions of phenomenal space. Concluding:

> So the new formulation of reality might consist of the following ontologically equal partners — (A) physical space-time (10 or more dimensions) containing physical matter (protons, electrons, etc.); (B) phenomenal space (3 more dimensions of a parallel universe) containing mind stuff (sensations and images); and (C) real time (time 2). A and B are in relative motion along the time 1 axis *in* time 2. Their contents are in causal relations via the brain. The psychological 'now' of time marks the point of contact of the two systems. (p. 55).

We remain agnostic regarding precisely how many additional dimensions may be required in order to provide the degrees of freedom necessary for time to flow and matter to have an experiential inside as well as an outside. Indeed we are not even committed to the notion that such a realm must necessarily be thought of as possessing all of the mathematical formalities of spatial dimensions. Our point is simply that the current material reductionist model of reality has left no "room" for time to flow or for matter to have internal experience. It is as if modern physics has built a pendulum clock but left no space for the pendulum to swing. In statistics there always must be one more degree of freedom than the total number of subjects and conditions so as to leave the freedom for variables to vary. We believe that such degrees of freedom are similarly required to enable the flow of time and a realm of experience. This space of subjectivity, which we believe could be formalized in much the same way that physical reality is formalized (e.g. Tononi 2008), needs to be given its due and recognized as a genuine aspect of reality requiring its own degrees of freedom.

Again consider the flow of time from an inside out perspective, i.e. from how each of us experiences it. What is evident is that each of us is moving both through objective time – clocks always tick – and information. At each moment in time we are exist as a slightly different informational state, a changed understanding of the situation. Conceiving of a subjective realm that changes with respect to, yet is distinct from, an objective realm provides a way of thinking about the flow of time that may enable the integration of two longstanding alternative views. One view, labeled by the philosopher McTaggart (1908) the "A theory of time" and stemming back as far as the Greek philosopher Heraclitus, characterizes reality as an ever-changing process of constant flux. In this view, the past is fixed, the present is the point at which reality continuously manifests, and the future is indeterminate. A second view, termed the "B theory of time" and stemming back to the Greek philosopher Parmenides, characterizes reality as fixed and existing in its entirety. According to this view, also entailed by today's standard block universe account, the past, present, and future all exist and are equally real. The view that we are suggesting here is a compromise between these two accounts, whereby the past and the future exist equally in the realm of physical space-time. In contrast, the present is realized exclusively in the subjective realm of experience. The general view of time that we are speculatively proposing has three key elements

1. *Physical space-time exists as continuous extension between past, present, and future.* This claim is largely in keeping with the notion of a block universe, in

which the past present and future are all equally real despite the fact that observers only perceive the present. However, in contrast to the naïve realism often associated with the materialist perspective (in which it is assumed that physical reality is directly experienced), in the present view observers never actually experience physical/space-time directly. Rather they construct representations of it in the subjective realm (discussed below). Importantly, the dimensions of physical space-time are assumed to be continuous, lending it to the mathematics of Newtonian physics and relativity theory..

2. *Subjective reality exists as a process of changing discrete states.* In stark contrast to continuous space-time, the subjective realm manifest as a series of discrete moments. Each observer moves in discrete intervals between successive values in space-time. The values encountered at each sampling are translated into experienced information. Subjectivity entails the sewing together of distinct informational states into the experience of a continuous flow of time. Because it entails discrete changes, the subjective realm is suited to the mathematics of quantum mechanics and information theory.[4]
3. *Given that physical space-time is continuous, every discrete subjective moment entails alternative potential instantiations.* Just as there are numerous ways that a loaf of bread can be sliced, so too the quantization of continuous space-time into corresponding discrete subjective moments affords multiple alternative instantiations of each subjective moment. In this context, the set of possible alternative instantiations of reality correspond to what is referred to in quantum physics as the probability cloud. The collapse of the probability cloud involves the measurement process, whereby specific values of space-time are extracted into a particular subjective moment. The upshot of this construal is that the subjective slicing up of objective space-time affords multiple possible ways in which experienced reality can unfold.

A Rudimentary Framework of Reality[5]

With these conjectures regarding the nature of consciousness and time in hand, we can now state our thesis succinctly – we suggest that *physical space-time exists as a multidimensional continuum, while subjectivity quantizes space-time*

[4]In this regard is notable that information theory specifically requires that observations proceed in discrete steps rather than being continuous. As Turing observes "It is easy to show using standard theory that if a system starts in an eigenstate of some observable, and measurements are made of that observable N times a second, then, even if the state is not a stationary one, the probability that the system will be in the same state after, say, one second, tends to one as N tends to infinity; that is, that continual observations will prevent motion …" Alan Turing quoted in Teuscher 2003 (p. 54)

[5]Hunt parts ways with the Schoolers on the ideas in this section because it builds on the previously noted disputed assumptions (see Afterword).

in the process of moving from one moment to the next.[6] An intuition of our thesis can be gleaned by considering the metaphor of making a movie. Rather than extracting a continuous representation of reality, a movie camera takes a series of stills, thereby converting a (seemingly) continuous stream of events into a discrete set of static images. However, when viewed, the movie is not perceived as a set of stills but rather as a moving picture. In fact, the speed at which each frame is presented relative to consciousness is so fast that were any frame presented alone, and sensory memory prevented, the frame would not be experienced at all. In effect, when we watch a movie, we are not seeing the actual physical frames at all. What we are seeing is the extended unpacking of the frames in our subjective experience. We suggest that the relationship between the subjective realm and physical space-time may be very much like the process of creating and watching a movie. Each observer moves in relationship to physical space-time, sampling moments (taking stills) as it goes. However, these individual moments are not experienced as such but rather inform the generation of a dynamic subjective representation (watching the movie). Much like Plato's cave metaphor, we don't observe objective reality itself, but rather the reflections of reality as they unfold in our subjective experience.

Closely akin to the snapshot/movie metaphor is the old-fashioned flip book, in which the flipping of pages creates a dynamic animation. We suggest that every entity engages in a process that is equivalent to taking two-sided snapshots of physical reality. One side – the out-facing side is then visible to all other observers. The other side – the in-facing side, corresponds to the observers' own personal perspective. Moreover, rather than taking full pictures of reality, like the layering of a cartoon, each observer does not necessarily provide an opaque frame but rather creates the equivalent of semi-transparencies in cartoons, thereby enabling the layered psychological moments of higher order consciousness. In this manner, each entity in the hierarchy is alternately audience and artist, as it perceives the layers ahead of it, and selects its own particular layer to add to the stack.

Now consider this model from the inside out perspective of the observer. When we attend to our visual experience, in effect we are seeing the experience of earlier entities in our sensory systems, the semi-transparent layers of the metaphorical cartoon flipbook that they have laid down. How many frames have been laid down before we add our own subjective viewpoint on reality depends on how far back in the temporal hierarchy one resides. Micro-consciousnesses experience fewer layers than macro-consciousnesses, which in turn receive fewer layers than the unified experience of the human self. Accordingly, the thickness (number of separate

[6] We note that this account shares some at least superficial similarity with the physicist Lynd's (2003) discussions of the relationship between subjective and physical instantiations of time. Lynd observes "there is not a precise static instant in time underlying a dynamical physical process… it is the human observer who subjectively projects, imposes and assigns a precise instant in time upon a physical process, for example, in order to gain a meaningful subjective picture or "mental snapshot" of the relative position of a body in relative motion" (p. 2).

frames an observer experiences) is inversely related to how early in the set of layers the observer exists. The simplest units of matter – strings, subatomic particles, or whatever units physicists decide are ultimately the tiniest units – have the thinnest duration, while representing the least amount of information in each slice. Consistent with our previous discussions, life may have taken advantage of this fundamental characteristic of the universe, and evolved hierarchical systems of distinct conscious observers that organize experienced moments of increasing numbers of layers, with ever longer (thicker psychological moments), including ever greater amounts of information.

This view also suggests that organisms may have developed increasingly layered systems that are able to amass more and more information, but with thicker and thicker experiential moments. For example, insects may have a relatively short hierarchy of layers leading their unified experience to have experiential moments that quickly processes the environment, but represent relatively little information between one moment and the next. The extended hierarchical structure leading to the higher-order thought associated with the human cortex, may enable us to represent vast amounts of information during each experiential moment, but with each moment corresponding to markedly longer durations. This may explain why it's so hard to swat flies: to them it is as if we are moving in slow motion.[7]

A crucial aspect of the present characterization is that it affords a meaningful way of conceptualizing the source of the unity of conscious experience. Accordingly, each observer maintains its own subjective distinctiveness by virtue of its unique sampling rate, the unique coherence that it maintains between its constituents from one snapshot of physical space-time to the next. Potentially, this sampling rate may correspond to the EEG synchronization that is observed between the neurons contributing to a coherent experience of consciousness (Hameroff 2010). An accumulating body of research suggests that the best neural marker of consciousness is the synchronization of the fluctuation of electrical activity associated with neurons.[8] Intriguingly, although EEG synchrony is one of the most promising neural signatures of consciousness (Hameroff 2010), at present there exists no complete explanation either for the source of this synchronization or the instantaneity by which it occurs across distributed brain regions. For example, Freeman and Vitiello (2006) examined the rate of EEG synchronization between the primary sensory and limbic areas in rabbits and cats. They found rates of

---

[7]We thank Rachel Schooler for this observation.

[8]Striking evidence from binocular rivalry studies (in which competing images are presented to each eye) suggest that while the magnitude of such synchronizations is greatest for the reported percept, synchronization is also associated with the image that is not currently being reported (Srinivasan et al. 2009). This suggests that in binocular rivalry experiments, a conscious experience of both images is maintained in the visual system and reflected by distinct coherent patterns of oscillation associated with each. However, at any one time only one of those conscious experiences is integrated into the higher order experience associated with the observer that is capable of verbal report.

resynchronization between these regions that exceeded the rates permitted by neural propagation mechanisms. As they observed:

> The dominant mechanism for neural interactions by axodendritic synaptic transmission should impose distance-dependent delays on the EEG oscillations owing to finite propagation velocities and sequential synaptic delays. It does not. (p. 93)

The difficulty of standard neuronal mechanisms in accounting for the instantaneity of EEG synchrony potentially represents the type of anomaly (Kuhn 1962/1996) that could leverage serious consideration of the kind of accounts being offered here. Indeed, others have similarly suggested that the synchronization of consciousness may be mediated by processes that are either not currently acknowledged (e.g. Libet's 2004 notion of a "conscious mental field") or that link processes that are currently viewed as entirely disparate (e.g. Hameroff and Penrose 2001 suggest that conscious awareness is associated with quantum collapse). Our suggestion is that it is their connectivity in a subjective realm that allow seemingly distinct elements to develop a synchrony of oscillations such that their temporal gap between one objective moment and the next is identical. From this perspective, then, it would be predicted that disparate brain regions corresponding to a singular subjective experience would evidence a speed of synchronization that would continue to defy recognized modes of transduction in the brain. Clearly, future research needs to carefully follow up on claims that EEG synchronization exceeds the maximum speed of electrochemical signaling, as this could prove a promising avenue for exploring the ideas proposed here.

Other Implications of the Present Approach[9]

The suggestion that the flow of time entails the process of translating segments of continuous space-time into a subjective informational realm that represents those segments as a series of experienced moments, affords a possible way of thinking about several other thorny issues, including (1) a substrate in which to construe meaning; (2) the postulation of laws of consciousness that parallel those of matter (3) an opportunity for postulating genuine free will. It is well beyond both the scope of this chapter or the capacity of the authors to fully defend these claims but allow us to make a few comments with respect to each.

The Domain of Meaning

There is a long tradition of thinkers including Plato (transl. 2008), Kant (1781/1896), Whitehead (1929), and more recently Penrose (1989), and Baumeister (2008a, b) who have argued that the domain of meaning is distinct from the material world.

[9] Hunt also does not agree with some of the ideas in this section, again because they build on disputed ontological distinctions (see Afterword).

From lofty mathematical theorems to simple perceptual interpretations, the understanding of information is, it is argued, is aligned with but distinguishable from the contents of the physical world. Baumeister (2008b) puts this idea as follows:

> Thus, what brains and bodies and other physically real things do conforms to the rules of nature. But they evolved (physically) to take input from another kind of reality that is invisible and not itself made of molecules. These realities include abstract concepts such as justice, credit limits, plausible deniability, floating exchange rates, identification with a religious sect and doctrine, and limited money-back guarantees. None of these things are made of molecules, but molecules are moved because of them. The proposition of a subjective realm in which consciousness extends would afford a landscape in which to represent the "kind of reality" that entails the understanding of information. (p. 37)

The existence of a subjective realm provides a domain of reality in which the understanding of such information is realized. Communication, and in particular language may dramatically enhance this realm by enabling the development of an inter-subjective domain of shared information. Although represented in, and an extension of each individuals private experience, this domain of shared understanding may serve as the foundation of culture and the repository of accumulated wisdom.

Laws of Consciousness

The hypothesis that consciousness represents a fundamental aspect of the universe suggests that consciousness may be guided by lawful rules that parallel physical systems (Schooler 2010, 2011) For example, coherent conscious entities may vary in their "size" in a manner akin to physical mass. Similar to the way in which physical solids are distinguished from gases by virtue of the synchronized movement of their atoms through physical time, higher order mental entities may exist by virtue of the synchronized sampling rate of physical time. In this sense, higher order mental entities might be thought of as possessing a greater "mental mass". Likewise the movement of mental entities through time may be likened to inertia, where the greater the mental mass the more force is required to shift the trajectory of the agent. Such "mental inertia" may play a significant role in constraining the opportunity for genuine free will (J.N. Schooler 2010, 2011). However, limited free will might still be possible if we add the further speculation (suggested previously and developed more below) that an intrinsic aspect of consciousness is its capacity for choice. Accordingly, in addition to being impacted by outside forces, mental agents may introduce their own internal source of impetus that is a function of the rate at which they sample physical time.

Free Will

One of the most disheartening implications of the standard block universe model of time is that it seems to rule out any genuine form of free will. If the future is already determined then in it is not possible for one to choose otherwise. And if one can't choose otherwise, then in what sense can one's decisions really be said to be free? Importantly, the notion of alternative possible futures follows naturally from our suggestion that subjective experience samples discrete moments from continuous space-time. Because there are a potentially infinite number of ways to divide a continuous dimension the quantizing of continuous space-time into discrete subjective moments potentially affords multiple alternative instantiations of each subjective moment. Returning to the flipbook metaphor, depending on the timing of the snapshots the nature of the flip book could be quite different. For example, if a flip book of static images was created corresponding to snapshots of a serial alternation of two different patterns, the pattern that actually appeared in the flip book would depend on the timing of the snapshots relative to the timing of each patterns' respective appearance. If space-time entails continuous dimensions that are segmented by subjective moments, then how those moments are segmented may determine how they are perceived to unfold. Thus the present model potentially provides a way in which the future could remain indeterminate.

The possibility of alternative futures is a necessary but not sufficient condition for genuine free will. For free will to be meaningful, consciousness must have some role in how alternative futures manifest. One way that consciousness might exert control in the unfolding of the future is by discerning the precise intervals of physical space-time contributing to each psychological moment. As discussed above, if observers could control the intervals associated with every subjective moment they could then impact on how their flip book was constructed. In effect they could tune in to the frequency of space-time that was most desirable, decipherable or whatever criteria they were concerned with. In this regard, it is notable that a fundamental characteristic of sensory systems is the process of entrainment, whereby they come to oscillate at rhythms mirroring those associated with the stimulus to which they are attending. Such a process is exactly what one would expect if such systems corresponded to conscious observers attempting to select temporal intervals that provided maximum information regarding the variables they are concerned with. In short, conscious control of how reality unfolds could be implemented in the current framework simply by varying the rate at which observers sampled physical space-time.

Ultimately, however, the essence of free will depends on whether top down processing can be genuinely causal, or whether the only causal direction is (as commonly assumed in physics) bottom up. In other words, even if conscious observers could impact on the unfolding of reality by varying the frequency with which they sampled space/time, it would not necessarily follow that they were exerting genuine causal control. Their behavior could simply be the necessary consequence of bottom up prior causal factors. We acknowledge that the present view remains consistent with an ultimately impotent view of consciousness. However, we make a case for genuine downward causation by returning again to the inside out perspective we have advocated throughout this chapter.

Next to subjectivity and change, the third most self-evident aspect of experience is choice. When one chooses to raise a hand we typically do. More generally, the intentions and actions that we bring to consciousness are markedly more likely to come to pass than those that we don't. In a recent extensive review of studies investigating the relationship between consciousness and action (Baumeister et al 2011) conclude:

> The evidence for conscious causation of behavior is profound, extensive, adaptive, multi-faceted, and empirically strong. However, conscious causation is often indirect and delayed, and it depends on interplay with unconscious processes. (p. 1)

Thus, phenomenologically it certainly seems as if consciousness causally contributes to action, and the empirical data can certainly be interpreted in this manner. The only question is whether this intuition must be abandoned based on a commitment to the view that causality can only proceed upward.

One approach for enabling downward causation is to suggest that it is an emergent property that somehow in the chain of being a new direction of causality became possible where it did not exist before (e.g. Campbell 1974). We have already argued strenuously that the claim that a completely unprecedented element emerges where there was no hint of it before is nothing more than a reiteration of "then a miracle occurs". In our view, if one is going to posit entirely inexplicable elements to a system, they should posit them along with all the other inexplicable elements, as the initial axioms. We do not try to explain why gravity or mass exist, we simply take these and a host of other essential elements of the universe as fundamental and assume that they play out at all levels. We have already argued that the same seems appropriate for consciousness. Having acknowledged the potential fundamentality of consciousness it seems appropriate to consider whether along with consciousness may come the property with which it is most often linked namely – agency. Perhaps consciousness is just along for the ride and offers no causal impact of its own. But it certainly does not feel that way, none of us, not even hard nose determinist act this way, and within the broader context of the metaphysical approach articulated here, a coherent scientifically grounded world view does not demand that we accept that consciousness is impotent. While it may be true that there is only upward causation, and the seeming impact of consciousness is just an illusion, we feel it is not unreasonable to speculate that agency is an inherent aspect of consciousness, and that both are inherent aspects of the very fabric of the universe. Accordingly, all levels of the universe may entail a bi-directionality of causality. Upward causality may be a consequence of the inherent structure of physical-space time. Downward causality may be a consequence of the unfolding of subjective time, whereby consciousness exerts its control in the process of segmenting physical time into discrete subjective moments. At a minimum, such an account seems to keep the door open on the possible existence of genuine free will.

Flatland – An Allegory

We recognize that we have introduced a plethora of ideas that are likely to jolt many readers. There is a natural tendency when faced with alien ideas to recoil, to dismiss them out of hand. Indeed, when worldviews are challenged, individuals will often experience dissonant reactions, and respond by more stridently asserting ideas that affirm their core values (Proulx and Heine 2006). Even scientists, who often try to present themselves as more rational than the average person, will vehemently cling to their existing paradigm long after facts suggest a change may be warranted (Kuhn 1962/1996). Undoubtedly, many readers are experiencing just such a response to the ideas presented here. Indeed, such reactions are both natural and appropriate. We believe we have reviewed some potentially serious problems with the current metaphysical assumptions of mainstream science, and some potential alternatives that might begin to redress these issues. However, we fully recognize that the alternative suggestions that we have proposed are far from fully fleshed out. They are merely presented as a springboard for opening up discussion about possible ways of resolving some of the core limitations in the prevailing metaphysics. Readers are unlikely to abandon long-held views, but perhaps we can instill some doubt regarding views that previously were considered unassailable, and introduce the beginning of speculations about alternative promising approaches. Indeed, to be a true skeptic one must question not only those ideas towards which one is disinclined, but also those that are appealing. Toward this goal, we invite readers to reflect on the ideas presented here in light of the wonderful allegorical tale of *Flatland*, written by Edwin Abbott (1885) more than a century ago.

Flatland depicts a two-dimensional world that is visited by a three-dimensional being. The protagonist of the story is a square who has a dream in which he meets the king of a one-dimensional world ("lineland"). From his two-dimensional perspective the square is able to see right into the middle of the king, as well as all of the subjects to his right and left. The king, who can hear the square but not see him, is incapable of conceiving of the square's claim of the existence of another dimension. Later the square encounters an anomaly in his waking life: a circle that appears out of nowhere, grows, shrinks, and finally disappears. Although no longer visible, this anomalous being informs the square that he is a three-dimensional sphere who has just passed through flatland. The square is initially as disbelieving of the sphere's claims of three dimensions as the king of lineland had been of the existence of a second dimension. However, the sphere proceeds to pull the square out of flatland, providing him with a view that he never before could have imagined. Enthralled by his newfound understanding of the existence of a higher dimension, the square asks the sphere about the possible existence of yet a fourth dimension, exclaiming:

> But just as there was the realm of flatland, though that poor puny Lineland Monarch could neither turn to left nor right to discern it, and just as there was close at hand and touching my frame the land of Three Dimensions, though I, blind senseless wretch had no power to touch it, no eye in my interior to discern it, so of a surety there is a Fourth Dimension, which my Lord perceives with the inner eye of thought. … I ask therefore, is it, or is it not the fact that ere now your countrymen also have witnessed the descent of Beings of a higher order than their own, entering closed rooms, even as your Lordship entered mine. (p. 135–136)

Although confirming the square's conjecture that the inhabitants of his three-dimensional world had reported such sightings, the sphere dismisses them, observing:

> It is reported so. But men are divided in opinion as to the facts. And even granting the facts, they explain them in different ways. And in any case, however great may be the number of different explanations, no one has adopted or suggested a theory of the fourth dimension…. Most people say that these visions arose from the thought… from the brain; from the perturbed angularity of the Seer. (p. 137–138)

Following the square's persistent suggestion of the possibility of higher order dimensions, the Sphere becomes infuriated and throws the square back into flatland where he is promptly convicted of blasphemy.

The story of flatland offers a number of useful lessons for the present discussion. First it provides a powerful metaphor for thinking about the existence of additional dimensions of reality. Long preceding relativity theory, which treats time like a fourth dimension, or string theory, which currently posits the existence of up to seven additional spatial dimensions (Greene 2004), Abbot's tale introduces us to the concept of higher order dimensions. Flatland describes how additional dimensions can be both embedded in and yet simultaneously transcend what we know. Facts that seem anomalous in n dimensions may be entirely resolved in n + 1 dimension. The parallels to consciousness are striking. When the square moves to the third dimension he suddenly sees inside the objects of flatland. Like consciousness, movement in an additional dimension enables the perception of an inside where none could otherwise be possible. Like consciousness's relationship to reality, an additional dimension intersects with the lower dimensions and yet is distinct from them. And like the recognition of an additional dimension, positing consciousness as a fundamental feature of physical reality resolves anomalies that otherwise seem to be incapable of naturalization into the broader theoretical framework of modern physics.

In addition to the explicit demonstrations of how beings could be oblivious to the higher dimensions in which they are embedded, the story of flatland also includes more subtle insights about the relationship between dimensions that may be relevant to the present discussion. Although the beings in flatland had no concept of up/down, upon reflection they should have been able to infer it. This is because in order to have any existence at all they would have had to have a modicum of thickness in the up/down dimension. If they had no thickness at all, they wouldn't exist at all. Even if they were unable to move in this dimension, their very existence could have led to the inference that this dimension must exist. In principle, the same argument may hold for our experience of time. In order to experience time at all we must have some extent in it. In much the same way that the residents of flatland required another dimension of space to exist in the two dimensions of flatland, we and everything else may require another dimension of time to exist in time.

A further implication of the story of flatland comes from the similarity between the added perspective that the square provided to the monarch of Lineland, and that the sphere provided to the square from Flatland. The square infers, based on these parallels alone, that there must be a realm with dimensions even greater than that enjoyed by the sphere. In each of these cases, the higher order beings are able to perceive relationships between parts that the lower order beings were not. This depiction is strikingly reminiscent of the levels of consciousness suggested in this paper. The jump from micro-consciousness to macro-consciousness, or from macro-consciousness to unified experience is potentially similar to the realization of higher order dimensions.

The final critical parallel between flatland and discussions of consciousness is how the beings at every dimensional level respond to the suggestion that there could be dimensions higher than their own. The monarch of lineland, the residents of flatland, and even the enlightened sphere of spaceland, all recoil at the suggestion of higher dimensions. These claims are viewed as preposterous, supernatural, and even blasphemous. The parallels to present views regarding consciousness are unmistakable. Just as the authorities of flatland viewed the notion of a third dimension that transcends the known principles of their two dimensional world as undeserving of consideration, so too the scientific establishment of our day often disdains the notion that consciousness might have any properties that transcend those ascribed to the third person perspective of material objects. The hubris that the beings of flatland evidenced in their assurance that nothing beyond the world as they construed it might exist characterizes well the prevailing physics and metaphysics of today.

Ultimately, the most important lesson from the allegory of flatland is the need to maintain humility regarding what is possible and what is impossible. Experiences or claims that may seem ludicrous or supernatural can, from a more informed perspective, turn out to be accurate. This has been demonstrated many times in the history of human thought. The realization of such truths may require a major shift in worldview but certainly not an abandonment of reason (Kuhn 1962/1996). In a very illuminating passage about meteorites – objects falling from space to Earth – we see the hubris that is present in all ages with respect to the prevailing understanding of nature (Krinov, quoted in Koestler 1978):

> During the period of vigorous scientific development which took place during the eighteenth century, scientists came to the conclusion that the falling of meteorites upon the Earth is impossible; all reports of such cases were declared to be absurd fiction…even the well-known chemist Lavoisier signed a memorandum in 1772 with scientists of the Paris Academy of Sciences, which concluded … that 'the falling of stones from the sky is physically impossible.' (p. 322)

Eventually, of course, the prevailing understanding of physics needed to be changed in order to accommodate the genuine existence of meteorites. Such accommodation may have been substantial but it did not require abandoning reason or the scientific method. Similarly, although the prevailing metaphysics is forced to treat subjectivity as inconsequential and the flow of time as illusory in order to reconcile them with scientific facts, this does not mean that all metaphysical frameworks must be so constrained. Ultimately, there may be ways to conceptualize extant and future scientific findings within a metaphysical framework that does not force us to ignore or abandon the subjectively self-evident facts that experience exists and time flows.

Afterword

As noted earlier, the Schoolers and Hunt differ on a number of central speculations in the second half of the paper. The following is a brief summary of a few of the key issues of contention, presented from the perspective of Hunt and the Schoolers in turn.

Hunt – I respectfully part ways with the Schoolers in the above treatment of time and information. I have fleshed out my views on time and information, which combine the ideas of subjective and objective time into a single dimension of time that is quantized at the most fundamental level, in Hunt 2010, in press. I am uncomfortable with the ontological dualism suggested by the Schoolers in their postulation of distinct subjective and objective realms, primarily due to the same problems that are raised by all dualist ontological theories, including Descartes' well-known dualism. That is, how do these distinct realms interact?

The difference in my view is that I do not distinguish between objective time and subjective time in an ontological sense. Rather, there is one ontological (objective) time that may be experienced at different rates by each subject, but this is merely an epistemological, not ontological, difference. The block universe concept, which leads to the problematic objective/subjective time dichotomy, arises primarily from Einsteinian relativistic physics. There is a broad consensus today that relativistic physics requires accepting a block universe because of the "relativity of simultaneity," as described in the body of this paper. However, the relativity of simultaneity is not a logical, philosophical or scientific necessity, as its proponents often suggest. Rather, there is a lengthy history of debate in this area in the philosophy of physics that demonstrates the feasibility – and from my point of view the desirability – of absolute simultaneity.

This alternative view, known as the "conventionality of simultaneity" argument, asserts that Einstein's special theory of relativity assumes a constant speed of light for all observers, as a stated postulate, but that this postulate is a mere "convention," as Einstein himself states. In other words, there are many other possible postulates regarding the speed of light that lead to the same empirical results. This is the case with Lorentz's "ether theory," a competitor to Einstein's theory of special relativity that Hendrik Lorentz, a Nobel Prize winner from Holland, developed in the early part of the twentieth century before Einstein's competing theory. Lorentz's ether theory asserts that the relativistic effects of length contraction and time dilation are caused by interaction with a non-material ether akin to Newton's absolute space. A key feature of Lorentz's view of time dilation, however, is that time dilation refers only to how clocks track time, which is independent of the background "absolute time."

The general view of physicists and philosophers today who track this debate is that the Einsteinian special relativity and Lorentzian relativity theories (Lorentz 1895, 1899) are empirically indistinguishable (they use the same mathematical formulas, known as the "Lorentz transformations") but that Einsteinian relativity is preferred because it is simpler. I disagree with this conclusion for a variety of reasons, not least because of the havoc Einsteinian relativity wreaks on the ontological validity of the passage of time and of free will, which leads to all the problems regarding the nature of consciousness and time that are described in the present paper. Many other philosophers and physicists have supported the "conventionality of simultaneity" arguments over the decades, including Hans Reichenbach, Adolf Grunbaum, Franco Selleri, and many others, as cited in Max Jammer's (2006) *Concepts of Simultaneity*, an excellent primer on these issues.

Moreover, beyond "mere" philosophical arguments regarding alternatives to Einsteinian relativity and the block universe view, we have the very compelling empirical results from various entanglement experiments in quantum physics. These experiments, led by Alain Aspect and his team in France in the 1980s (Aspect et al. 1982) and most recently with Daniel Salart's work in Switzerland in 2008, demonstrate that some kind of causal influence travels between particles far faster than the speed of light – at least 10,000 times the speed of light, according to Salart and his team. This evidence alone demonstrates the invalidity of Einsteinian relativity as a necessary statement about the ontology of the universe. The debate on the ultimate nature of time and free will is, accordingly, still very open.

With respect to information, I also view information as an epistemological concept and not an ontological concept. We can describe reality as particles, energy, etc., or we can describe reality as more fundamentally simply "information." And there are of course many other possibilities with respect to these and other concepts, which mean only what we choose them to mean (as Humpty Dumpty said so aptly). Regardless of our terminology, we are led to the same insights about reality – which is ultimately only the collection of sense-data each of us receives about the world "out there," and nothing more. But we can't use both sets of terminology at the same time. So, from my point of view, if we describe reality as consisting of information, then the same rules and philosophical considerations apply to this information-based reality as we would otherwise ascribe to a matter/energy-based reality. It's just terminology. Thus information has no ontological status above and beyond matter/energy.

A final note is warranted: though I part ways with the Schoolers in using "information" as something ontologically distinct from matter/energy (the physical universe), I do agree that there is more than just the four dimensions of space and time in the totality of reality. There is not space to flesh out this view here, but I have done so in my other works, detailing how the "new ether" or "ground of being" concept is a necessary concept in physics, psychology and spirituality. There is, thus, something beyond, behind, under or above the four dimensions of physical reality, which plays a key role in determining how each chronon and how each actual entity is instantiated. This may be described as ether, ground of being, *Brahman*, *apeiron,* or any of many other terms. As I and the Schoolers continue to flesh out our views on these highly interesting, complex and important ideas I believe we may re-converge in our views – as the creative advance works its magic.

The Schoolers respond: A key source of contention between our perspective and that of Hunt's is the age-old question of how to conceive of the divide between subjective and objective. We believe that both subjective and objective attributes of reality can be conceptualized within a single system, that they nevertheless represent fundamentally distinct aspects of reality. Hunt characterizes our view as essentially dualist, and therefore susceptible to the same criticisms that have haunted dualist perspectives since Descartes, namely, how it is that the two realms interact, if they are distinct. Critically, it is our claim that domains can be distinct and yet nevertheless interact in meaningful ways. As mentioned earlier we believe that the distinction between dimensions may mirror the relationship that consciousness may have with physical space time. Just as objects possess distinct coordinates in space and time, so too may they have distinct coordinates in experienced information space. Similarly, just as matter and space are fundamentally different yet we can understand how matter moves through space, so too we can conceive of consciousness as distinct from physical space/time and yet still capable of moving through it. Like matter and space, consciousness and objective space/time may be qualitatively different in their properties yet nevertheless inseparable.

We concur with the perspective of the opening quote of Alan Watts in which he suggests that although objective and subjective aspects of everything are different that they nevertheless go together. Hunt also agrees with this sentiment, so it seems that the crux of the issue boils down to what "different" means. From our perspective all matter may indeed have both an outside form that it presents to others and an inside state that it experiences, but these two aspects, though necessarily interdependent, have distinct natures.

Consider for example the difference between matter/energy and information states. Admittedly, when one speaks about matter/energy we are necessarily referring to our informational states about matter/energy. Nevertheless we believe there remains a fundamental difference between the presumed physical states of an objective universe whose characteristics can only be inferred and the experienced informational states of conscious beings. To be sure, our knowledge of physical states exclusively entails our inferences about the characteristics of a physical world. Nevertheless, to the degree that we take the existence of a physical world seriously we can make a fundamental distinction between the presumed states of a physical world that we infer from our observations, and the known informational states that we consciously experience. In our view, the fact that we know information but can only infer the physical world illustrates the fundamental nature of their distinction.

A similar argument can be made for the distinction between objective and subjective time. Objective time corresponds to the time of clocks. It has a precise measurable quantity that can be defined in terms of physical events. Subjective time in contrast is defined exclusively with respect to the experience of the conscious observer. Both, however, carry comparably significant yet distinctly different meanings. While objective time can be physically measured, what is of value from the vantage of the experiencer is not how much time has actually passed, but how much time felt like it passed. Neither objective time nor subjective time can be dismissed, nor can either be reduced to the other, leading to the seemingly inevitable conclusion that they are fundamentally distinct constructs.

Hunt also takes issue with the notion of the block universe, and the related premise that consciousness can be thought of as a wave moving through physical time. He rejects the block universe notion in favor of absolute time and an ontologically real present. We think that the idea of absolute time has much to commend it as it offers a more meaningful way of conceptualizing the privileged present. From our view, however such a framework is not inconsistent with the notion of consciousness as wave. Rather, from this perspective the wave of consciousness would correspond to the cusp of the progression of absolute time moving through some type of absolute ether. Of course all of these suggestions are highly speculative. Our point in offering them is not to necessarily be correct, but rather to encourage bolder speculations about alternative ways of conceptualizing the physical and the mental within a single overarching framework.

Acknowledgments We are grateful to many individuals for their input and assistance with this chapter. Ben Baird provided comments on numerous drafts and assisted with references. The ideas presented here were helpfully refined by discussions with many people, none of whom should be assumed to endorse the resulting product. These include: Daniel Povinelli, Merrill McSpadden, Carmi Schooler, Lael Schooler, Rachel Schooler, Nina Schooler, Roy Baumeister, Daniel Gilbert, Daniel Dennett, Mark Laufer, Jonathan Smallwood, Christine Tipper, Dianne Tice, Thomas Nadelhofer, and Harvey Bottelsen. The writing of this chapter was supported by grants from the James Bower Foundation and the Bial Foundation.

References

Abbott, E. A. (1885). *Flatland: A romance of many dimensions*. Boston: Roberts Bothers.

Aspect, A., et al. (1982). Experimental realization of Einstein-Podolsky-Rosen-Bohm Gedanken experiment: A new violation of Bell's inequalities. *Physical Review Letters, 49*, 91.

Bars, I., Deliduman, C., & Andreev, O. (1998). Gauged duality, conformal symmetry, and spacetime with two times. *Physical Review, D58*, 066004.

Baumeister, R. F. (2008a). Free will, consciousness, and cultural animals. In J. Baer, J. C. Kaufman, & R. F. Baumeister (Eds.), *Are we free? Psychology and free will*. Oxford: Oxford University Press.

Baumeister, R. F. (2008b). Social reality and the hole in determinism. *Journal of Consumer Psychology, 18*(1), 34–38.

Baumeister, R. F., Masicampo, E. J., & Vohs, K. D. (2011). Do conscious thoughts cause behavior? *Annual Review of Psychology, 62*, 331–361.

Bergson, H. (1912). *Matter and memory*. New York: McMillan (Original Work Published 1896).

Berkeley, G. (1734/1971). *A treatise concerning the principles of human knowledge*. Menston: Scolar Press.

Bloom, P. (2004). *Descartes' baby: How the science of child development explains what makes us human*. New York: Basic Books.

Campbell, D. T. (1974). "Downward causation" in hierarchically organized biological systems. In F. J. Ayala & T. Dobzhansky (Eds.), *Studies in the philosophy of biology* (pp. 179–186). London: Macmillan.

Chalmers, D. J. (1995a). Facing up to the problem of consciousness. *Journal of Consciousness Studies, 2*(3), 200–219.

Chalmers, D. J. (1995b). The puzzle of conscious experience. *Scientific American, 273*(6), 91–100.

Chalmers, D. J. (1996). *The conscious mind: In search of a fundamental theory*. New York: Oxford University Press.

Crick, F. (1994). *The astonishing hypothesis: The scientific search for the soul*. New York: Scribner.

Davies, P. (2002). That mysterious flow. *Scientific American, 287*(3), 40–43.

Dawkins, R. (2006). *The god delusion*. Boston: Houghton Mifflin.

de Broglie, L. (1959). A general survey of the scientific work of Albert Einstein. In P. A. Shilpp (Ed.), *Albert Einstein: Philosopher-scientist*. New York: Harper.

Dennet, D. C. (1997). *Kinds of minds: Towards an understanding of consciousness*. New York: Basic Books.

Dennett, D. C. (1991). *Consciousness explained*. Boston: Little, Brown and Co.

Dennett, D. C. (2003). *Freedom evolves*. New York: Viking.
Descartes, R. (1956). *Discourse on method*. New York: Liberal Arts Press (Original Work Published 1637).
Dyson, F. J. (1979). *Disturbing the universe*. New York: Harper & Row.
Einstein, A. (2001). *Relativity: The special and the general theory* (Reprint of 1920 Robert W. Lawson Ed. & trans.). London: Routledge.
Feynman, R. (1999). *The pleasure of finding things out: The best short works of Richard P. Feynman*. Cambridge: Perseus Books.
Fitch, W. T. (2008). Nano-intentionality: A defense of intrinsic intentionality. *Biology and Philosophy, 23*(2), 157–177.
Freeman, W. J., & Vitiello, G. (2006). Nonlinear brain dynamics as macroscopic manifestation of underlying many-body field dynamics. *Physics of Life Reviews, 3*(2), 93–118.
Gazzaniga, M. S. (2005). Forty-five years of split-brain research and still going strong. *Nature Reviews Neuroscience, 6*(8), 653–659.
Greene, B. (2004). *The fabric of the cosmos: Space, time, and the texture of reality*. New York: A.A. Knopf.
Griffin, D. R. (2007). *Whitehead's radically different postmodern philosophy: An argument for its contemporary relevance*. Albany: State University of New York Press.
Hameroff, S. (2001). Time, consciousness and quantum events in fundamental space-time geometry. In R. Buccheri, M. Saniga, & W. Stuckey (Eds.), *The nature of time: Geometry, physics and perception*. Amsterdam: Kluwer Academic Publishers.
Hameroff, S. (2010). The "conscious pilot"-dendritic synchrony moves through the brain to mediate consciousness. *Journal of Biological Physics, 36*(1), 71–93.
Hameroff, S. R., & Penrose, R. (1996). *Conscious Events as Orchestrated Space-Time Selections, Journal of Consciousness Studies*, *3*(1), pp. 36–53.
Harris, S. (2006). *Then a miracle occurs* [Cartoon]. New Yorker.
Hiley, B. J., & Peat, F. D. (1987). *Quantum implications: Essays in honour of David Bohm*. Routledge, London.
Hoffmann, B., & Dukas, H. (1972). *Albert Einstein, creator and rebel*. New York: Viking.
Hofstadter, D. R. (2007). *I am a strange loop*. New York: Basic Books.
Hunt, T. (2010). *Mind, world, god*. Unpublished manuscript, UC Santa Barbara, Santa Barbara.
Hunt, T. (in press). Kicking the psychophysical laws into gear: A new approach to the combination problem, part 1. *Journal of Consciousness Studies*.
Husserl, E. (1980). *Phenomenology and the foundations of the sciences*. The Hague: M. Nijhoff.
James, W. (1898). *Human immortality: Two supposed objections to the doctrine*. Boston: Houghton, Mifflin and Co.
James, W. (1909). *A pluralistic universe*. New York: Longmans, Green, and Co.
James, W. (1917). *"Human Immortality" Ingersoll lectures, 1897–1898*. London: J. M.Dent & Sons, Ltd.
James, W. (1918). *The principles of psychology*. New York: Henry Holt and Company. (Original work published 1890)
James, W. (2002). *Varieties of religious experience: A study in human nature*. New York: Random House (Original Work Published 1902).
Jammer, M. (2006). *Concepts of simultaneity: From antiquity to Einstein and beyond*. Baltimore: Johns Hopkins U.P.
Kam, J. W. Y., Dao, E., Farley, J., Fitzpatrick, K., Smallwood, J., Schooler, J. W., & Handy, T. C. (2011). Slow fluctuations in attentional control of sensory cortex. *Journal of Cognitive Neuroscience, 23*(2), 460–470.
Kant, I. (1781/1896). *Critique of pure reason* (F. M. Muller Trans.). London: Macmillan.
Kingstone, A., Tipper, C. M., Ristic, J., & Ngan, E. T. C. (2004). The Eyes Have It!: An fMRI Investigation. *Brain and Cognition, 55*(2), 269–271.
Koch, C. (2004). *The quest for consciousness: A neurobiological approach*. Englewood: Roberts and Company.

Koestler, A. (1978). *Janus: A summing up*. New York: Random House.
Kuhn, T. (1996). *The structure of scientific revolutions*. Chicago: University of Chicago Press (Original Work Published 1962).
Leibniz, G. (1989). Monadology. In R. Ariew & D. Garber (Eds. & Trans.), *G. W. Leibniz: Philosophical essays*. Indianapolis: Hackett Publishing Company. (Original work published 1714)
Libet, B. (2004). *Mind time: The temporal factor in consciousness*, Cambridge, MA: Harvard University Press
Linde, A. D. (1990). *Particle physics and inflationary cosmology*. Chur: Harwood Academic.
Linde, A. D. (2004). Inflation, quantum cosmology, and the anthropic principle. In J. D. Barrow, P. C. W. Davies, & C. L. Harper (Eds.), *Science and ultimate reality: Quantum theory, cosmology, and complexity*. Cambridge: Cambridge University Press.
Loftus, E. F., & Schooler, J. W. (1985). Information processing conceptualizations of human cognition: Past, present and future. In B. D. Ruben (Ed.), *Information & behavior* (pp. 225–251). New Brunswick: Transaction Books.
Lorentz, H. A. (1895). *Versuch einer Theorie der electrischen und optischen Erscheinungen in bewegten Körpern*. Leiden: E.J Brill.
Lorentz, H. A. (1899). Simplified theory of electrical and optical phenomena in moving systems. *Proceedings of the Royal Netherlands Academy of Arts and Sciences, 1*, 427–442.
Lynds, P. (2003). Time and classical and quantum mechanics: Indeterminacy versus discontinuity. *Foundations of Physics Letters, 16*(4), 343–355.
Margulis, L., & Sagan, D. (1995). *What is life?* New York: Simon & Schuster.
McGinn, C. (1991). *The problem of consciousness: Essays towards a resolution*. Oxford: Blackwell.
McGinn, C. (1999). Consciousness and space. In J. Shear (Ed.), *Explaining consciousness: The hard problem*. Cambridge: MIT Press.
McTaggart, J. E. (1908). The unreality of time. *Mind, 17*(64), 457–474.
Minsky, M. L. (1986). *The society of mind*. New York: Simon and Schuster.
Nagel, T. (1974). What is it like to be a bat? *Philosophical Review, 83*(4), 435–450.
Penrose, R. (1989). *The emperor's new mind: Concerning computers, minds, and the laws of physics*. Oxford: Oxford University Press.
Pinker, S. (1997). *How the mind works*. New York: Norton & Company, Inc.
Pinker, S. (2007). *The blank slate*. New York: Viking Adult.
Plato. (trans. 2008). *The republic* (B. Jowett, Trans.). New York: Cosimo Inc.
Proulx, T., & Heine, S. J. (2006). Death and black diamonds: Meaning, mortality, and the meaning maintenance model. *Psychological Inquiry, 17*(4), 309–318.
Rosenberg, G. (2005). *A place for consciousness: Probing the deep structure of the natural world*. Oxford: Oxford University Press.
Salart, D., et al. (2008). Spacelike separation in a Bell test assuming gravitationally induced collapses. *Physical Review Letters, 100*(22), 1–4.
Schooler, J. W. (2002). Re-representing consciousness: Dissociations between experience and meta-consciousness. *Trends in Cognitive Sciences, 6*(8), 339–344.
Schooler, J. N. (2010). *Mental inertia: Limited free will and determinism*. Paper submitted in fulfillment of UCSB Research Mentorship program, University of California, Santa Barbara.
Schooler, J. N. (2011) *Mental laws: The relationship of mentality and physics*. Unpublished manuscript, San Marcos High School, San Marcos
Schopenhauer, A. (1995). *The world as will and idea: Abridged in one volume*. London: J.M. Dent (Original Work Published 1819).
Searle, J. (1990). Is the brain's mind a computer program? *Scientific American, 262*(1), 26–31.
Shariff, A., Schooler, J. W., & Vohs, K. (2009). The hazards of claiming to have solved the hard problem of free will. In J. Baer & R. Baumeister (Eds.), *Psychology and free will*. Oxford: Oxford University Press.
Skrbina, D. (2005). *Panpsychism in the West*. Cambridge: MIT Press.
Smythies, J. (2003). Space, time and consciousness. *Journal of Consciousness Studies, 10*(3), 47–56.

Spinoza, B. (1985). Ethics. In E. Curley (Ed. & Trans.), *The collected works of Spinoza* (Vol. I). Princeton: Princeton University Press. (Original work published 1677)

Srinivasan, R., Russell, D. P., Edelman, G. M., & Tononi, G. (2009). Increased synchronization of neuromagnetic responses during conscious perception. *Journal of Neuroscience, 19*(13), 5435–5448.

Strawson, P. F. (1974). *Freedom and resentment, and other essays*. London: Methuen.

Strawson, G. (2008). *Real materialism and other essays*. Oxford: Oxford University Press.

Talbot, M. (1991). *The holographic universe*. New York: HarperCollins.

Teuscher, C. (2003). *Alan turing: Life and legacy of a great thinker*. Heidelberg: Springer.

Tononi, G. (2008). Consciousness as integrated information: A provisional manifesto. *The Biological Bulletin, 215*(3), 216–242.

Wallace, B. A. (2010, February 25–27) *Toward a three-dimensional science of the mind*. Paper presented at Neuroscience, Consciousness and Spirituality Meeting 2010: Meditation and Prayer: Neuroscientific Approaches and Philosophical Implications, Freiburg. Available at http://www.uniklinik-freiburg.de/medmed/live/events/Expert-Meeting-Neuroscience-Consciousness-and-Spirituality-2010/Wallace_Three-Dimensional.pdf.

Watts, A. (1991). *Man, nature, and the nature of man, collected public lectures*. New York: Macmillan.

Weiskrantz, L., & Weiskrantz, L. (2009). *Blindsight: A case study spanning 35 years and new developments*. Oxford: Oxford University Press.

Whitehead, A. N. (1929). *Process and reality: An essay in cosmology*. Cambridge: University Press.

Zeki, S. (2003). The disunity of consciousness. *Trends in Cognitive Sciences, 7*(5), 214–218.

Zeki, S., & Bartels, A. (1999). Toward a theory of visual consciousness. *Consciousness and Cognition, 8*(2), 225–259.

Mindfulness Meditation: Deconditioning and Changing View

Henk Barendregt

Abstract Mindfulness interventions and meditation form a mental training towards deconditioning. This paper outlines the mental development during long term mindfulness meditation (vipassanā). It is intended to give researchers in the neurosciences and practitioners of mindfulness based interventions an idea of the phenomenological side of this form of meditation. Mindfulness acts as a separator between the perceived actor in us and the things we cognize and act upon. This makes us more flexible. At the same time our view of self will change: no longer is our 'agency' seen as a fixed 'thing' or 'being' that acts in the world, but as a process of sensory input, appraisal, thinking, acting, depending on various mental states.

The Human Condition

We are conditioned: depending on circumstances our behavior is pushed in certain directions. Already in some unicellular organisms this is the case. If there is food nearby, then the flagellum of the organism starts running in order to move towards the source; in case there is a poison, then movement goes in the opposite direction. In multicellular species and finally in mammals, primates, and homo sapiens the conditioning forces are much more sophisticated, making use of tools like neurons, hormones and long-term memory. Usually the 'possessor' has an evolutionary advantage: a better survival mechanism. For example, the feeling of 'self' of a human being, in particular our feeling of 'agency', with all of its planning and possessing up to egocentricity, forms a powerful way to survive. Thus nature has provided organisms, from the unicellular ones to humans, with the mechanism of desire and

H. Barendregt (✉)
Faculty of Science, Radboud University, Nijmegen, The Netherlands
e-mail: henk@cs.ru.nl

H. Walach et al. (eds.), *Neuroscience, Consciousness and Spirituality*, Studies in Neuroscience, Consciousness and Spirituality 1,
DOI 10.1007/978-94-007-2079-4_12, © Springer Science+Business Media B.V. 2011

aversion. Humans suffer, when they cannot get what they want (and also when they cannot avoid what they resent). This means that we have a strong tendency to arrange things in order to fulfill our desires. Driven by desire humans have constructed cities and cultures. The same drive, however, is also the cause of madness and war. Considerable human misery has resulted.

How is it possible that desire, which is there for our increased chance of survival, sometimes is against us? A first reason why this happens is that circumstances may change. Take for example insects that during flight orient themselves towards the moon. They keep a fixed angle to the 'infinitely distant' source of light and then they fly in a straight line. After humans had invented fires and lamps, this form of conditioning caused the insects to spiral towards the light source and then they burn their feet or worse. But there are 'higher' insects that are able to learn from the heat of fires and lamps and avoid them. Another example of conditioning is initially increasing survival chances, but under some circumstances works against the animal. This can be seen in certain monkeys. The story is that they like bananas so much that after grabbing one they cannot let go of it, until they eat the fruit. If the banana is behind a hole just large enough to pass through their hand, they will grab it. But with the banana in the monkey's fist it cannot get out of the hole. As it is impossible for the monkey to let go of the banana it is stuck. If men come with a cage in order to catch it, then the monkey cannot avoid this. Here we see that a seemingly useful conditioning turns into a monkey-trap. We all have friends that do similar things: holding on to things or situations in such a way that it becomes a hindrance. And if we dare to be honest, we may have to admit that sometimes we also do this ourselves. A final reason for conditioning becoming a cause of suffering is that what we want may simply be impossible.

Science

People have developed science: insight in the phenomena around ourselves. This means that we can explain a great many of these from a small subset of them using a predictive model. This insight can lead to technology, by which we are capable to obtain partial control over some of the phenomena. This makes it possible to fulfill needs in ways that are easier than before. Against the cold we construct a heater, against heat an air-conditioner. Science is an impressive human endeavor towards understanding nature. If applied with wisdom, it can diminish human suffering and bring us a higher standard of living. But it is good to realize that not everything can be controlled. What we want may be expensive, illegitimate, immoral, unecological or sheer impossible. Therefore, in spite of science and technology, there will remain suffering, namely in those cases where control is beyond the possible or that control causes conflicts. People may fight for the same scarce natural resources. Human conditioning is even so strong that one may fight about cherished ideas.

Phenomenology

We may cognitively understand that conditioning causes conflicts with ourselves, with others and with our living environment, the earth. Even then this does not set us free from conditioning. There is another type of investigation that may be useful towards this end. It is the research directed inwards, towards consciousness as it is presented to us. This investigation is called 'phenomenology'.

A highly successful example of phenomenology is the following. I heard the story from colleagues, but see Duck (1988), Ribe and Steinle (2002) and Sepper (2007) for the main parts of it. Around 1800 physicists after Newton knew that colored light forms a one-dimensional (1D) phenomenon. Light comes in various wavelengths and a particular one determines the color of the light. In 1810 the poet Goethe, who was also interested in perception, suggested another theory. He claimed that colors form a 3D phenomenon for the following reason. When we have, say, 125 cubes, each one colored evenly in a different color, then it is not possible to put them in a linear row such that the colors change 'smoothly'; also it is not possible to do this in a plane. But in a cube $5 \times 5 \times 5$ of it is possible to arrange the colors such that in all directions they change smoothly. This observation is phenomenological: it is based on direct perception, independent of rational considerations.

At first physicists maintained their position that colors form a 1D phenomenon. But then the physician Young (1773–1829) and later the physicist Helmholtz (1821–1894) tried to reconcile the observations of Goethe with those from physics. They coined the hypothesis that the human eye has three different receptors for color vision. If this is the case, then a single wavelength induces a triple of reactions in the retina and in the rest of the brain. This implies that colors are 1D at their physical formation, but 3D at their perception. The Young-Helmholtz hypothesis was finally proved as late as 1960, that is 150 years after Goethe's observation. In the meantime the multibillion industry of color photography and color TV-monitors had started, all based on the fact that we have three receptors for color perception. Today in 2009 the idea is still going strong in digital cameras, flat-screens, and video-projectors. Politicians should be happy: fundamental research does lead to economic growth. But in this case one had to wait about 150 years for the spin-off. This is longer than the re-election term, so politicians do not notice it.

Religion

What is the goal of religion? My answer is this: to find threefold peace, within ourselves, with others and with the world. Sometimes we may think: "finding peace within ourselves is easy, the others are being nasty; and as to peace with the world, well we are mortal, but I'm not yet concerned with that." But the problem of death is just a particular case of a more general problem. We are not the absolute boss, neither of the circumstances, nor of our body, nor of our minds. This last fact has been expressed well by St. Augustin:

> imperat animus corpori, et paretur statim; imperat animus sibi, et resistitur. imperat animus ut moveatur manus, et tanta est facilitas ut vix a servitio discernatur imperium: et animus animus est, manus autem corpus est. imperat animus ut velit animus, nec alter est nec facit tamen. unde hoc monstrum? et quare istuc, inquam, ut velit qui non imperaret nisi vellet, et non facit quod imperat? Augustine (1955), Book VIII, Ch. 9, 21.
>
> [The mind commands the body, and the body obeys. The mind commands itself and is resisted. The mind commands the hand to be moved and there is such readiness that the command is scarcely distinguished from the obedience in act. Yet the mind is mind, and the hand is body. The mind commands the mind to will, and yet though it be itself it does not obey itself. Whence this strange anomaly and why should it be? I repeat: The will commands itself to will, and could not give the command unless it wills; yet what is commanded is not done. (The Confessions of St Augustine, translated and edited by A.C. Outler, Dover Thrift Editions, 2002).]

Although the body does not always listen so well as in this example, for example we can get ill at moments that are inconvenient, the meaning of the statement of St. Augustine is well-known. We are not always doing what we think we should be doing; conditioning prevents us.

Through a religious view, for example that of the three monotheistic systems (Judaism, Christianity and Islam), one can find a purpose in life, an ethical code and mentioned threefold peace. A religion helps to find faith, hope, and love. If the assumptions of a religion are formulated carefully, then they do not need to be in contradiction with a scientific worldview. They are, however, metaphysical: assumptions about life after death cannot be verified. In some visions on faith (fides quae), this is not necessary. One can accept the religious dogmas as a collection of axioms, in the mathematical sense, that provide a mental hold. Sometimes it is even claimed that this form of faith is more stable than the one in which one interprets the dogmas as being really true (fides qua): indeed, in that case one can be contradicted.

Meditation

In many religious traditions the preference is to base peace not on faith, but on mental development. Insight (or mindfulness) meditation (vipassanā) coming from classical Buddhism is a prime example showing the possibility of transforming the mind. Meditation can be divided into two main classes. In order to describe these we need to introduce some terminology concerning consciousness. Even if consciousness cannot be properly defined, we have an idea of what is meant by it: an awareness of the things around and within us. Of course, the transition ('consciousness' $\rightarrow$ 'awareness') does not help very much. But the following description of aspects of consciousness delineates it, just like an axiom system explains the meaning of the primitive (undefined) mathematical objects occurring in it. Consciousness has two aspects. First there is the object (or content) of consciousness: it is what one sees, hears, smells, tastes, touches or thinks. Next to the object there is the 'coloring' of consciousness, resulting in mental 'states'. One can observe the same object together with e.g. joy or with boredom or with anger. Other colorings are for example sleepiness, restlessness, equanimity, and compassion. Buddhist psychology, the Abhidhamma (Bodhi 2000), distinguishes 51 such colorings. These 51 are called cetasikas, 'mental factors', and form ingredients of the mental states, called cetas, by combining several of them. There are 89 states, obtained as such mixtures. It is mentioned that there are more mental factors and states than these 51 and 89, but the listed ones are particularly important for the mental development towards deconditioning. Some of the mental factors are called unwholesome (like e.g. hatred, restlessness, and jealousy), some are called beautiful (like e.g. mindfulness, flexibility, and compassion), and finally some are called neutral (like e.g. concentration and joy: these states can be combined with both compassion and hatred).

Now we are ready to describe the two classes of meditation. (1) Concentration (or samādhi) meditation is aimed at increasing positive states of consciousness. (2) Insight (also called mindfulness or vipassanā) meditation is aimed at decreasing and eventually eliminating the negative states of consciousness.

The highpoint of concentration meditation consists of various mystical experiences, including rapture and bliss. Mystical states are described in many cultures, but the theoretical explanation may differ in the various traditions (Staal 1975). The Buddhist mystical states are called jhānas (Pali for 'absorptions', states of high concentration). These states are explained in a language that is understandable in cognitive psychology. The word jhāna has come to the English language as follows: jhāna>dhyana (Sanskrit)>chan (Chinese)>sun (Korean)>zen (Japanese and English). The jhānas form an increasing chain of mystical states, ordered linearly. According to the Abhidhamma jhāna 3, 4 and 5 are distinguished as follows: in jhāna 3 one has rapture, bliss or compassion and equanimity; in jhāna 4 one has bliss or compassion and equanimity; and in jhāna 5 one just has equanimity. State 4 is said to be higher than 3, as rapture gives some restlessness; state 5 is higher than 4 as bliss or compassion give some attachment. It is tempting to relate the states 3 and the two variants of 4 to neuromodulators like dopamine, opioids, and oxytocin (said to correspond to libido, postorgastic bliss and inter-human binding). The situation may be more complex, though.

Because these mystical experiences are very pleasant, there is a tendency to get attached to them, for example in Christianity they are described as the "unity with God". From a Buddhist point of view, but also from that of several Christian mystics, this attachment is to be avoided.

The highpoint of insight meditation is purification or enlightenment that eliminates some or all of the negative states. When these do no longer occur, consciousness has a tendency towards neutral and beautiful states. Metaphorically speaking, one can state that concentration meditation aims at nice weather, while insight meditation aims at a nice climate. The last thing is more difficult to reach, notably because one has to proceed along unpleasant states called 'dark night of the soul' by St. John of the Cross (2008). In the tradition of insight meditation they are called the 'three characteristics' of consciousness: anicca, dukkha, anatta, best translated as 'non-permanence, disgust and loss of control'. This passing along negative states is necessary for the domestication of them. As a pretty high degree of concentration meditation is needed for successful insight meditation, 10 of the 22 chapters of Nanamoli (1976) are devoted to concentration, it takes most people more work to reach enlightenment than to reach mystical states. But the effect of purification is lasting and therefore more worthwhile to be pursued.

The Buddhist tradition distinguishes several types of deconditioned mind: liberated by suppression (vikkhambhana-vimutti), a temporary state, and by insight (paññā-vimutti), which is permanent. The jhānas all belong to the first type, whereas the purified state is permanent.

Mindfulness

One of the beautiful mental factors is mindfulness. During mindfulness one observes the input from the senses in a non-reactive way. This sometimes may happen to us while window-shopping: when we look at nice merchandise that is too expensive to be bought, we may look in a manner that is 'observing', but not 'reacting'. Then we may observe 'there is beauty'. Or if desire comes up, then we even may observe 'there is desire' and still not buy the objects. Another way of explaining mindfulness is the classical metaphor of the money-changer. A child sees money bills as colored pieces of paper. An adult can look at the same money with desire, especially to the higher bills. But a money-changer observes the money with attention and, if all is well, doesn't develop desire, as the money is not hers. This is looking with mindfulness. It is something that happens naturally every now and then, but it also can be trained to make its occurrence intentionally more frequent.

Using developed mindfulness one can work towards detachment from the conditioning mechanisms that nature has bestowed upon us. We already have explained why this is beneficial. In some cases the conditioning brings suffering, like to the insects that fly in the candle or to the monkeys that are being caught while grabbing a banana. In cases there is a free banana, there is no problem to pick and eat it. If a human grabs the banana in the 'monkey-trap', then mindfulness would enable him to observe the object and his desire and then to let go of it. Whether monkeys can learn this I do not know, but humans can.

By systematic training mindfulness can become our second nature and this deconditions our negative traits, like entering ruminations of negative thoughts. To decondition the full mechanism of desire, more work is needed: mindfulness should become our first nature.

Vipassanā

During vipassanā meditation one trains to apply mindfulness to all phenomena in consciousness. If one sees something, then one observes 'seeing, seeing'. If there are notable sounds, then one observes 'hearing, hearing'. This constitutes the so called first foundation of mindfulness directed to the input of the senses. Next to this is the foundation of immediate appraisal: something can be pleasant, unpleasant or neutral. The third foundation consists of mindfulness directed to mental phenomena, like thinking and mental factors and states. Finally the fourth foundation consists of the observations of conditioning, like holding onto something without our really wanting to do this. Increased mindfulness can separate the four foundations. In this separated form of consciousness the mechanisms of conditioning are much looser. This brings freedom.

This separation of the composed consciousness is also related to dissociation, as it is known from psychiatry. In vipassanā one prefers to speak about dis-identification. Our phenomenological world consists of fluctuating patterns of consciousness. These patterns are being reified, by imagining them as being a fixed thing. Think of a wave in the sea that we perceive as an object that moves towards the shore. Note that there is no particle of water that moves from the sea towards the land. It is only the relative height of the particles of water that is being transmitted as a wave from the sea towards the coast. Our mind is used to turn moving patterns into things: the act of reifying. During vipassanā training one learns to be aware of this. This frees us from the manipulating and conditioning effects of our perception and cognition.

One remarkable pseudo-paradoxical aspect of the deconditioning is that in order to be mindful, one has to accept what happens without judgment. Initially one is able to do this. But then our mechanism of desire tries to employ 'mindfulness', making it no longer true mindfulness: there is the hidden agenda to become better by using it. Strong suffering results, as one is doing something apparently without success. If one has enough discipline to continue the training, then after a while one gives up any hope of becoming better by the training. At that moment one regains the right mindfulness and then the method works again and one feels fine. Following an intensive vipassanā retreat this mechanism causes several up and down phases.

Domesticating Existential Fear

Some people may observe dissociations as a result of genetic tendencies or traumatic events; this can happen also without training mindfulness. Then anxiety may result, as one seems to loosen the hold of one's self (depersonalization), of the body (desomatization) or of the world (derealization). If this has happened for a longer period, notably during growing up, then one's personality is often unstable and proper therapeutic aid is advisable, in particular before starting insight meditation.

Often the meditator will also develop temporary fear after the experience of dis-identification. If one works in the safe environment of a meditation center and in the proximity of a skillful and compassionate meditation teacher, this fear will at first be diluted and then one can restart the process of reification in a mindful way. This is the core mechanism of deconditioning. Although the mentioned fear of falling apart is temporary, it may be very strong. In fact it is the mother of all fears and the very cause of suffering. During the process of falling apart one clearly sees that consciousness is in a constant fluctuation; moreover it is unbearable as one goes beyond meaning[1] and arrives in a kind of 'emptiness'; and finally it is beyond our control, also called selflessness. These fundamental qualities of phenomenal consciousness – it is fluctuating, unbearable and selfless – are called *the three characteristics*. The original text about uncovering the three characteristic states is the following (Nanamoli (1976), Ch XXI, 3–4), with some parenthetical remarks by me.

> The characteristic of impermanence does not become apparent because, when rise and fall are not given attention, it is concealed by continuity [due to reification]. The characteristic of suffering does not become apparent because, when continuous oppression [to sitting still; i.e. there is desire to move, scratch one's nose, etcetera] is not given attention, it is concealed by the [holding on to the meditation] postures. The characteristic of not-self does not become apparent, because when resolution [dissolution] into the various elements is not given attention, it is concealed by compactness [the feeling of 'agency']. However, when continuity gets disrupted by discerning rise and fall, impermanence becomes apparent in its true nature. When the postures are exposed by attention to continuous oppression, the characteristic of pain becomes apparent in its true nature. When the resolution of the compact is effected by resolution into elements, the characteristic of no-self becomes apparent in its true nature.

By focusing systematically on input of the physical senses as a mental hold, and by constantly observing the negative aspects of the fear and the corresponding effects on behavior, one can learn to diminish the strong phenomena of the three characteristics and eventually to eliminate them. This can happen by applying mindfulness to the four foundations as they come and go. Mindfulness should be developed such that it is fixed for a while on one of the foundations. After that one develops the mental states of equanimity, calmness and joy. At a certain moment the meditation development is mature for 'surrender'. One sees that the phenomena continue, even if one is dis-identified and subject to the three characteristics of consciousness. In case the meditation is not yet matured, then one of the states of mind will 'glue together' the broken consciousness. And because that state hides the three characteristics, the unbearable feeling disappears. But there will be side-effects of this 'emotion' acting as symptomatic medicine. If the hiding factor is, say, desire, then one becomes greedy. If it is fear, then one becomes phobic (Barendregt 1982). It is far better to reconstruct consciousness using mindfulness, than to glue using emotions. Mindfulness domesticates the three characteristics and sets us free. It should be emphasized that this domestication work can be done only if one is willing to invest the right effort under the right conditions. This happens in so called intensive meditation retreats. One needs to meditate for an uninterrupted period, for example during a 10 day intensive vipassanā retreat. It happens at a place in which one is not disturbed by others, where there is an experienced meditator to help and where the living conditions are appropriate. In short: one

[1] Children sometimes play the following game. They repeat an ordinary word, like 'yellow', many times. After a while they loose the meaning of the word, there is just the sound. This phenomenon is called a 'semantic fateague'. From the point of view of insight meditation a word has a sound and a meaning. When a word is repeated often, the sound and meaning are separated. Actually the child still knows very well what 'yellow' means.

goes to a meditation center for an intensive vipassanā retreat under the guidance of a skillful teacher.

Cover-Up and Purification

During the path of insight meditation one sooner or later discovers the three characteristics. It is a powerful dissociation. The feeling of self ('agency', 'ego', or however you want to call it) is lost. One then notices that this phenomenal experience of the three characteristics is always nearby, but usually we hide it by our feeling, thinking and acting. This is called the 'cover-up' of the characteristics. Although they are always there, usually one does not see the three characteristics. After insight has penetrated to the three characteristics, one also sees clearly that one usually covers them up, including the painful selflessness (Barendregt 1996). This is like dismantling the motor of a car piece by piece. After that the pieces may be purified by mindfulness, i.e. be released from the occurring fear. Usually one is not yet ready for this and the three characteristics are covered-up again: a feeling of agency is created in such a way that one takes it for real, having certain side-effects.

At this stage the meditator feels discouraged to continue the training: a problem that he did not know before has resulted, without there being a known solution. A compassionate meditation teacher encourages the meditator to continue to work by mindfulness and not to cover-up the phenomena. If the meditator works hard and summons the discipline to continue the work, eventually an equanimous, calm and joyful mind is developed. At that moment the meditator becomes ready to surrender. Mindfulness that has become a reliable second nature of the meditator now has to become our first nature. This cannot be done at will, but it helps (and it seems to be essential) if the trainee has the patient intention that this will happen. And then it may happen indeed, but unexpectedly: negative mental factors get eliminated. This means that the negative side-effects of the cover-up like fear or depression to give us agency are no longer necessary for being in balance.

According to tradition the negative states of mind will not all be eliminated at the same time. There are four milestones. At first one stops believing in 'self' or 'agency' as a fixed kernel of consciousness. After that greed and hatred are being diluted. Then at the third milestone these two 'poisons' will be eliminated completely. At the fourth and last milestone it is said that one eliminates (i) restlessness, (ii) sleepiness, (iii) pride, (iv) desire for certain forms of existence and (v) ignorance about unconscious processes that nevertheless do influence us in major ways. In order to be able to do the work of deconditioning, one needs to embrace insight meditation as a life-style: knowing that it is useful to meditate often and then actually doing it! The training to domesticate the three characteristics may be compared to a training for a parachutist or astronaut. An inexperienced person is afraid to fall and gets nauseated by it. But one can domesticate the mental fear and physical disgust for weightlessness by a systematic training. This changes one's view on agency and liberates us from a heavy burden: ego does no longer need to be defended.

Meditation and Research

The claims above are being taught in the living oral tradition of insight meditation, backed by ancient texts, and has been partly verified by the author as a trainee (Barendregt 1988, 1996). Buddha mentioned about such theories the following:

> "Do not believe in anything simply because you have heard it. Do not believe in anything simply because it is spoken and rumored by many. Do not believe in anything because it is found written in your religious books. Do not believe in anything merely on the authority of your teachers and elders. Do not believe in traditions because they have been handed down for many generations. But after observation and analysis, when you find anything that agrees with reason and is conducive to the good and benefit of one and all, then accept it and live up to it." Taken from the Kalama Sutta, AN 3.65, see Gethin (2008).

This witnesses a scientific attitude. Nevertheless is seems worthwhile to verify the claims made by the tradition of insight meditation, not only by personal experience of meditators, but also by a scientific evaluation. It may bring the method of deconditioning known to a wider audience, and thus bring more peace to the world. A second reason is that it may help science to progress, as the mentioned mechanisms of conditioning and deconditioning are plausible and highly relevant. In the author's opinion the following questions are important.

1. Is the experience of the three characteristics (chaos, suffering, selflessness) measurable?
2. Is the symptomatic cover-up with side-effects of the three characteristics measurable?
3. Is purification by means of mindfulness measurable?

By 'measurable' I mean physiologically and/or behaviorally observable. During data-collection for this research it will be relevant to receive feedback from the meditators about the mental state in which they reside. To start with question 3, the Abhidhamma claims that mindfulness is correlated with equanimity, flexibility, good memory, and attention. For more common psychological factors there are validated tests. One could investigate these in meditators before and after a retreat and analyze whether they are increased in similar ways. Also one can compare meditators with non-meditators. Research has reported that years of practicing insight meditation correlates with cortical thickness (Hölzel et al. 2008, see also the chapter by Ott et al., this book). Moreover, attention is increased by a different neuronal strategy (Slagter et al. 2007; van den Hurk et al. 2009). One may conjecture that during mindfulness one uses different neural pathways, not in the limbic system but in the cortex. In question 1 it is asked whether the dis-identification is measurable. A possible hypothesis is that the phenomenon of dis-identification has to do with desynchronized firing of distant neurons. The reason for this is that dis-identification is related to unbinding and that there is the hypothesis (Singer 1999), that binding is related with long-distance synchronization. Question 2 may be addressed by researching the influence of neuro-modulators on mind-states. See Veening et al. (2010, 2010) for work in this direction.

The number of scientific papers on mindfulness training and meditation is growing considerably: presently there are several hundred. It is beyond the scope of this paper to review the present literature. But it seems fair to say that typical aspects of insight meditation, like the experience of the three characteristics, have been neglected in past research. This is understandable, as not all meditators reach this state within a limited period of time. The fundamental fear hampers progress in that right direction: people try to avoid the three characteristics. But in vipassan intensive retreats of 10 days and longer, experiencing them does occur relatively often. There should be collaboration between the researchers and the teacher for the selection of the subjects having reached certain stages. The methodology of this proposal is the following: in this early stage of research one does not yet study the effectiveness of insight meditation as a cure, but one investigates the mechanism of a universal human problem consisting of existential suffering as caused by conditioning. If the neurophysiology of the problem is better understood, this will probably contribute to understanding the solution as presented by insight meditation.

References

Augustine of Hippo. (397) (1955). *Confessions* (A. C. Outler, Trans.). , St. Augustine: Christian Classics Ethereal Library.

Augustine of Hippo (2008). *Confessions*, Translation: H. Chadwick Oxford World's Classics, Oxford University Press.

Barendregt, H. P. (1988). Buddhist phenomenology. In M. dalla Chiara (Ed.), *Proceedings of the Conference on Topics and Perspectives of Contemporary Logic and Philosophy of Science* (pp. 37–55). Bologna: Clueb. URL: www.cs.ru.nl/~henk/BP/bp1.html

Barendregt, H. P. (1996). Mysticism and beyond, Buddhist phenomenology, Part II. In *The Eastern Buddhist* (New series, Vol. XXIX, pp. 262–287). URL: www.cs.ru.nl/~henk/BP/bp2.html

Barendregt, J. T. (2008). *Phobias and related fears*. Unpublished. Available at ftp://ftp.cs.kun.nl/pub/CompMath.Found/JTBarendregtFobias.pdf. (Original work published 1982: Fobiëen en verwante angsten. In J. T. Barendregt *De zielenmarkt, Over psychotherapie in alle ernst* (pp. 163–180). Meppel: Boom).

Bodhi, B. (2000). *A comprehensive manual of Abhidhamma: The Abhidhammattha Sangaha of Ācariya Anuruddha*. Kandy: Buddhist Publication Society, Pariyatti Editions.

Gethin, R. (2008). *Sayings of the Buddha: New translations from the Pali Nikaya*, Oxford World's Classics, Oxford University Press.

Hölzel, B. K., Ott, U., Gard, T., Hempel, H., Weygandt, M., Morgen, K., & Vaitl, D. (2008). Investigation of mindfulness meditation practitioners with voxel-based morphometry. *Social Cognitive and Affective Neuroscience, 3*(1), 55–61.

Kabat-Zinn, J. (2003). Mindfulness-based interventions in context: Past, present and future. *Clinical Psychology: Science and Practice, 10*(2), 144–156.

Ñañamoli, B. (1976). *The path of purification* [Visuddhimagga] (B. Buddhagosa, (approximately 430 A. D.), Trans.). Kandy: Buddhist Publication Society.

Ribe, N., & Steinle, F. (2002). Exploratory experimentation: Goethe, land, and colour theory. *Physics Today, 55*(7), 43–49.

Sepper, D. L. (2007). *Goethe contra Newton: Polemics and the project for a new science of color*. Cambridge: Cambridge University Press.

Singer, W. (1999). Neuronal synchrony: A versatile code for the definition of relations? *Neuron, 24*(1), 49–65.

Slagter, H. A., Lutz, A., Greischar, L. L., Francis, A. D., Nieuwenhuis, S., et al. (2007). Training affects distribution of limited brain resources. *PLOS Biology, 5*(6), e138.

St. John of the Cross (2008). *Dark night of the soul, translated by E. Allison Peers*, Wilder, Radford VA.

Staal, F. (1975). *Exploring mysticism*. Berkeley: University of California Press.

van den Hurk, P. A. M., Giommi, F., Gielen, S. C., Speckens, A. E. M., & Barendregt, H. P. (2009). Greater efficiency in attentional processing related to mindfulness meditation. *Quarterly Journal of Experimental Psychology, 63*(6), 1168–1180. URL: www.informaworld.com/smpp/content~db=jour~content=a915419588

Veening, J. G., & Barendregt, H. P. (2010). The regulation of brain states by neuroactive substances distributed via the cerebrospinal fluid; a review. *Cerebrospinal Fluid Research, 7*(1), 1. URL: www.cerebrospinalfluidresearch.com/content/7/1/1

Veening, J. G., de Jong, T., & Barendregt, H. P. (2010). Oxytocin messages via the cerebrospinal fluid: Behavioural effects, a review. *Physiology & Behavior, 101*(2), 193–210.

Endless Consciousness: A Concept Based on Scientific Studies of Near-Death Experiences

Pim van Lommel

> *As for me, I am more than ever convinced that it is the greatest enemy of scientific progress if one rejects or refuses beforehand to study unknown and seemingly strange facts and findings.*
>
> Frederik van Eeden, MD and author (1860–1932)

Abstract In this chapter a concept of non-local consciousness will be described, based on recent studies on near-death experiences (NDE). Recently several theories have been proposed to explain a NDE. The challenge to find a common explanation for the cause and content of a NDE is complicated by the fact that a NDE can be experienced during various circumstances, such as during severe injury of the brain as in cardiac arrest to a continuum when the brain seems to function normally. Since the publication of several prospective studies on NDE in survivors of cardiac arrest, with strikingly similar results and conclusions, the phenomenon of the NDE can no longer be scientifically ignored. It is an authentic experience which cannot be simply reduced to imagination, fear of death, hallucination, psychosis, the use of drugs, or oxygen deficiency. People appear to be permanently changed by a NDE during a cardiac arrest of only some minutes duration. According to these studies, the current materialistic view of the relationship between the brain and consciousness held by most physicians, philosophers and psychologists is too restricted for a proper understanding of this phenomenon. There are good reasons to assume that our consciousness does not always coincide with the functioning of our brain: enhanced consciousness can sometimes be experienced separately from the body.

P. van Lommel (✉)
Department of Cardiology, Rijnstate Hospital, Arnhem, The Netherlands
e-mail: pimvanlommel@gmail.com

H. Walach et al. (eds.), *Neuroscience, Consciousness and Spirituality*,
Studies in Neuroscience, Consciousness and Spirituality 1,
DOI 10.1007/978-94-007-2079-4_13,

Introduction

A near-death experience (NDE) can be defined as the reported memory of a range of impressions during a special state of consciousness, including a number of special elements such as an out-of-body experience, pleasant feelings, seeing a tunnel, a light, deceased relatives, or a life review. Many circumstances are described during which NDEs are reported, such as cardiac arrest (clinical death), shock after loss of blood, traumatic brain injury or intra-cerebral haemorrhage, near-drowning (children) or asphyxia, but also in serious diseases not immediately life-threatening, during isolation, depression or meditation, or without any obvious reason. Similar experiences to near-death ones can occur during the terminal phase of illness, and are called deathbed visions. Furthermore, so-called "fear-death" experiences are mainly reported after situations in which death seemed unavoidable like serious traffic or mountaineering accidents. The NDE is usually transformational, causing enhanced intuitive sensibility, profound changes of life-insight, and the loss of fear of death (van Lommel 2010). The content of an NDE and the effects on patients seem similar worldwide, across all cultures and all times. However, the subjective nature and absence of a frame of reference for this experience lead to individual, cultural, and religious factors determining the vocabulary used to describe and interpret the experience.

Near-death experiences occur with increasing frequency because of improved survival rates resulting from modern techniques of resuscitation. According to recent random polls in the US and in Germany, about 4% of the total population in the western world experienced an NDE (Gallup and Proctor 1982; Schmied et al. 1999). Thus, about nine million people in the US and about two million people in the UK should have had this extraordinary conscious experience. A NDE seems to be a relatively regularly occurring, and to many physicians an inexplicable phenomenon and hence an ignored result of survival in a critical medical situation. Physicians hardly ever hear a patient tell about his or her near-death experience. Patients are reluctant to share their experience with others because of the negative responses they get. As a doctor you have to be open to hear about a NDE. Patients must feel that you trust them and that you can listen without any comment or prejudice.

How It Started

As a cardiologist I had the privilege to meet many patients who were willing to share their NDE with me. The first time this happened was in 1969. In the coronary care unit the alarm suddenly went off. The monitor showed that the electrocardiogram (ECG) of a patient with a myocardial infarction had become flat. The man had a cardiac arrest. After two electric shocks and a spell of unconsciousness lasting some 4 min, the patient regained consciousness, much to the relief of the nursing staff and attendant doctor. That attendant doctor was me. I had started my cardiology training that year. Following the successful resuscitation everyone was pleased, except the patient. To everyone's surprise he was extremely disappointed. He spoke of a tunnel, of colors, of a light, of a beautiful landscape and of music. He was extremely emotional. The term near-death experience did not yet exist, nor had I ever heard of people having any recollection of the period of their cardiac arrest. Whilst studying for my degree, I had learnt that such a thing is in fact impossible: being unconscious means not being aware, and that applies to people suffering a cardiac arrest or patients in a coma. In the event of a cardiac arrest, patients are unconscious; they have stopped breathing and have no palpable pulse or blood pressure. At such a moment, it should be simply impossible to be conscious or to have memories because all brain function has ceased.

Although I had never forgotten the successfully resuscitated patient from 1969 with his memories of the period of his cardiac arrest, I had never done anything with the experience. This changed in 1986 when I read a book by George Ritchie about his near-death experience with the title 'Return from Tomorrow' (Ritchie 1978). When suffering double pneumonia as a medical student in 1943, Ritchie had experienced a period of clinical death. At the time, antibiotics such as penicillin were not yet widely used. Following a period of very high fever and extreme tightness of the chest, he passed away: he stopped breathing and his pulse had gone. He was pronounced dead by a doctor and covered with a sheet. But a male nurse was so upset by the death of this medical student that he managed to persuade the attendant doctor to administer an adrenalin injection in the chest near the heart – a most unusual procedure in those days. Having been 'dead' for about 9 min, George Ritchie regained consciousness to the immense surprise of the doctor and nurse. It emerged that during his spell of unconsciousness, the period in which he had been pronounced dead, he had had an extremely deep experience of which he could recollect a great many details. At first he was quite unable and afraid to talk about it. Later he wrote his book about what happened to him in those 9 min. And after graduation, he shared his experiences with medical students in psychiatry lectures. One of the students attending these lectures was Raymond Moody, who was so intrigued by this story that he started looking into experiences that may occur during critical medical situations. In 1975 he wrote the book Life after Life (Moody 1975). In this book Moody first used the term near-death experience (NDE).

After reading George Ritchie's book I kept asking myself how someone can possibly experience consciousness during cardiac arrest and indeed whether this is a common occurrence. That is why, in 1986, I started systematically asking all the patients at my out-patient clinic who had ever undergone resuscitation whether they had any recollection of the period of their cardiac arrest. I was more than a little surprised to hear, within the space of 2 years, 12 reports of such a near-death experience among just over 50 survivors of cardiac arrest. Since that first time in 1969, I had not heard any other such reports. I had not enquired after these experiences either. But all these reports I was hearing now roused my curiosity. After all, according to current medical knowledge it is impossible to experience consciousness when one's heart has stopped beating.

Questions

For me it all started with curiosity. With asking questions; with seeking to explain certain objective findings and subjective experiences. The phenomenon of near-death experience raised a number of fundamental questions. A NDE is a special state of consciousness that occurs during an imminent or actual period of death, or sometimes without any obvious reason. But how and why does a NDE occur? How does the content of a NDE come about? Why does a person's life change so radically after a NDE? I was unable to accept some of the answers to these questions, because they seemed incomplete, incorrect or unfounded. I grew up in an academic environment in which I had been taught that there is a reductionist and materialist explanation for everything. And up until that point, I had always accepted this as indisputably true.

Some scientists do not believe in questions that cannot be answered, but they do believe in wrongly formulated questions. The year 2005 saw the publication of a special anniversary issue of the journal Science, featuring 125 questions that scientists have so far failed to solve (Kennedy and Norman 2005). The most important unanswered question 'What is the universe made of?' was followed by 'What is the biological basis of consciousness?' I would reformulate this second question as follows: Is there a biological basis of consciousness (at all)? We can also distinguish between both temporary and timeless aspects of our consciousness. This prompts the following question: Is it possible to speak of a beginning of our consciousness and will our consciousness ever end?

In order to answer these questions, we need a better understanding of the relationship between brain function and consciousness. We shall have to start by examining whether there is any indication that consciousness can be experienced during sleep, general anesthesia, coma, brain death, clinical death, the process of dying and, finally, after confirmed death. If the answers to any of these questions are positive, we must look for scientific explanations and scrutinize the relationship between brain function and consciousness in these different situations. By studying everything that has been thought and written about death throughout history, in all times, cultures and religions, we may be able to form a different or better picture of death. But we may achieve the same on the basis of findings from recent scientific research into near-death experiences. It has emerged that most people lose all fear of death after a NDE (van Lommel 2010). Their experience tells them that death is not the end of everything and that life goes on in one way or another. According to most people with a NDE, death is nothing other than a different way of being with an enhanced and broadened consciousness, which is everywhere at once because it is no longer tied to a body. This is what someone wrote to me after his NDE: 'It is outside my domain to discuss something that can only be proven by death. However, for me personally this experience was decisive in convincing me that consciousness endures beyond the grave. Death turned out to be not death, but another form of life.'

The Dutch Prospective Study on NDE in Survivors of Cardiac Arrest

In order to obtain more reliable data to corroborate or refute the existing theories on the cause and content of a NDE, we needed a properly designed scientific study. This was the reason why in 1988 Ruud van Wees and Vincent Meijers, both psychologists who wrote their doctoral theses on NDE, and I, a cardiologist with an interest in the subject, started designing a prospective study in the Netherlands (van Lommel et al. 2001, p. 2041). This study was carried out under the auspices of Merkawah, the Dutch branch of the International Association of Near-death Studies, IANDS the Netherlands. At that point, no large-scale prospective studies into NDEs had been undertaken anywhere in the world. Our study aimed to include all consecutive patients who had survived a cardiac arrest in one of the ten participating Dutch hospitals. In other words, this prospective study would only be carried out among patients with a proven life-threatening crisis. All of these patients would have died of their cardiac arrest had they not been resuscitated within 5–10 min. This kind of design also creates a control group of patients who have survived a cardiac arrest but who have no recollection of the period of unconsciousness. In a prospective study such patients are asked, within a few days of their resuscitation, whether they have any recollection of the period of their cardiac arrest, i.e. of the period of their unconsciousness. All patients' medical and other data are carefully recorded before, during and after their resuscitation. The advantage of this prospective study design was that all procedures were defined in advance and no selection bias could occur.

We had a record of the electrocardiogram, or ECG, for all patients included in our study. An ECG displays the electrical activity of the heart. In cardiac arrest patients this ECG record always displays a normally lethal arrhythmia (ventricular fibrillation) or an asystole (a flat line on the ECG). In the event of resuscitation outside the hospital we were given the ECG done by the ambulance staff. Following successful resuscitation we carefully recorded the demographic data of all patients, including age, sex, education, religion, foreknowledge of NDE and whether or not they had had an earlier NDE. They were also asked whether they had been afraid just before their cardiac arrest. Likewise, we carefully recorded all medical information, like: what was the duration of the actual cardiac arrest? What was the duration of unconsciousness? How often did the patient require resuscitation? What medication, and in what dosage, was administered to the patient before, during and after resuscitation? We also recorded how many days after resuscitation the interview took place, whether the patient was lucid during the interview and whether his or her short-term memory was functioning well. Within 4 years, between 1988 and 1992, 344 successive patients who had undergone a total of 509 successful resuscitations were included in the study. In other words, all the patients in our study had been clinically dead. Clinical death is defined as the period of unconsciousness caused by lack of oxygen in the brain (anoxia) because of the arrest of circulation, breathing or both, as caused by cardiac arrest in patients with an acute myocardial infarction. If, in this situation, no resuscitation is initiated, the brain cells will be irreversibly damaged within 5–10 min and the patient will always die.

A longitudinal study into life changes was based on interviews after 2 and 8 years with all patients who had reported a NDE and who were still alive, as well as with a control group of post-resuscitation patients who were matched for age and sex, but who had not reported a NDE. The question was whether the customary changes in attitude to life after a NDE were the result of surviving a cardiac arrest or whether these changes were caused by the experience of a NDE. This question had never been subject to scientific and systematic research before. The Dutch study was published in The Lancet in December 2001 (van Lommel et al. 2001).

If patients reported memories from the period of unconsciousness, the experiences were scored according to a certain index, the WCEI, or "weighted core experience index" (Ring 1980). The higher the number of elements reported, the higher the score and the deeper the NDE. Our study found that 282 patients (82%) had no recollection of the period of their unconsciousness, whereas 62 patients – 18% of the 344 patients – reported a NDE. Of these 62 patients with memories, 21 patients (6%) had some recollection; having experienced only some elements, they had a superficial NDE with a low score. And 42 patients (12%) reported a core experience: 18 patients had a moderately deep NDE, 17 patients reported a deep NDE and 6 patients a very deep NDE. The following elements were reported: half of the patients with a NDE were aware of being dead and had positive emotions, 30% had a tunnel experience, observed a celestial landscape or met with deceased persons, approximately a quarter had an out-of-body experience, communication with 'the light' or perception of colours, 13% had a life review and 8% experienced the presence of a border. In other words, all the familiar elements of a NDE were reported in our study, with the exception of a frightening or negative NDE.

Are there any reasons why some people do but most people do not recollect the period of their unconsciousness? In order to answer this question we compared the recorded data of the 62 patients with a NDE to the data of the 282 patients without NDE. To our big surprise we did not identify any significant differences in the duration of the cardiac arrest, no differences in the duration of unconsciousness and no differences in whether or not intubation was necessary for artificial respiration in seriously ill patients who remained in a coma for days or weeks after a complicated resuscitation. Nor did we find differences in the 30 patients who had a cardiac arrest during electrophysiological stimulation (EPS) in the catheterization laboratory and whose heart rhythms were always re-established via defibrillation (an electric shock) within 15–30 s. So we failed to identify any differences between the patients with a very long or a very brief cardiac arrest. The degree or gravity of the lack of oxygen in the brain (anoxia) appeared to be irrelevant. Likewise, it was established that medication played no role. Most patients suffering a myocardial infarction receive morphine-type painkillers, while people who are put on a respirator following complicated resuscitation are given extremely high doses of sedatives. A psychological cause such as the infrequently noted fear of death does not affect the occurrence of a NDE either, although it did affect the depth of the experience. Whether or not patients had heard or read anything about NDE in the past made no difference either. Any kind of religious belief, or indeed its absence in non-religious people or atheists, was irrelevant and the same was true for the standard of education reached. Factors that do affect the frequency of a NDE are an age below 60 and if patients required several resuscitations during their stay in hospital, the chances of a NDE report were greater. Remarkably, we found that patients who had had a NDE in the past also reported significantly more frequent NDEs in our study. A complicated resuscitation can result in a long coma and most patients who have been unconscious on a respirator for days or weeks are more likely to suffer short-term memory defects as a result of permanent brain damage. These patients reported significantly fewer NDEs in our study. This suggests that a good memory is essential for remembering a NDE.

We were particularly surprised to find no medical explanation for the occurrence of a NDE. All the patients in our study had been clinically dead and only a small percentage reported an enhanced consciousness with lucid thoughts, emotions, memories, and sometimes perception from a position outside and above their lifeless body while doctors and nursing staff were carrying out resuscitation. If there were a physiological explanation such as a lack of oxygen in the brain (anoxia) for the occurrence of this enhanced consciousness, one might have expected all patients in our study to have reported a NDE. They had all been unconscious as a result of their cardiac arrest, which caused the loss of blood pressure and the cessation of breathing and all physical and brain-stem reflexes. And it is also well established that people without any lack of oxygen in the brain like in depression or meditation can experience a 'NDE'. Likewise the gravity of the medical situation, such as long-term coma after a complicated resuscitation, failed to explain why patients did or did not report a NDE, except in the case of lingering memory defects. The psychological explanation is doubtful because most patients did not experience any fear of death during their cardiac arrest as it occurred so suddenly they failed to notice it. In most cases they were left without any recollection of their resuscitation. This is borne out by Greyson's study (Greyson 2003), which only collected the subjective data of patients after their resuscitation and showed that most patients did not even realise they had had a cardiac arrest. This is similar to fainting. When people regain consciousness they have no clear idea of what happened. A pharmacological explanation could be excluded as well, as the medication had no effect on whether or not patients reported an NDE.

The later interviews in our Dutch longitudinal study were conducted using a standardised inventory featuring 34 life-change questions (Ring 1984). Among the 74 patients who consented to be interviewed after 2 years, 13 of the total of 34 factors listed in the questionnaire turned out to be significantly different for people with or without a NDE. The second interviews showed that in people with NDE fear of death in particular had significantly decreased while belief in an afterlife had significantly increased. We then compared these 13 factors, which had been so significantly different between the two groups with and without NDE after 2 years, in the same patients after 8 years. It struck us that after 8 years the people without NDE were also undergoing unmistakable processes of transformation. Nevertheless, clear differences remained between people with and without NDE, although by now these differences had become a little less marked. We were also surprised to find that the processes of transformation that had begun in people with NDE after 2 years had clearly intensified after 8 years. The same was true for the people without NDE. In summary, we could say that 8 years after their cardiac arrest all patients had changed in many respects, showing more interest in nature, the environment and social justice, displaying more love and emotions and being more supportive and involved in family life. Nevertheless, the people who had experienced a NDE during their cardiac arrest continued to be clearly different. In particular, they were less afraid of death and had a stronger belief in an afterlife. We saw in them a greater interest in spirituality and questions about the purpose of life, as well as a greater acceptance of and love for oneself and others. Likewise, they displayed a greater appreciation of ordinary things, whereas their interest in possessions and power had decreased. The conversations also revealed that people had acquired enhanced intuitive feelings after a NDE, along with a strong sense of connectedness with others and with nature. Or, as many of them put it, they had acquired 'paranormal gifts'. The sudden occurrence of this enhanced intuition can be quite problematic, as people suddenly have a very acute sense of others, which can be extremely intimidating, and also experience clairvoyance (non-local perception), prophetic feelings (precognition) and visions. This intuitive sense can be quite extreme, with people 'sensing' feelings and sadness in others, or having the sense of knowing when someone will die – which usually proved to be accurate (van Lommel 2010). The integration and acceptance of a NDE is a process that may take many years because of its far-reaching impact on people's pre-NDE understanding of life and value system. Finally, it is quite remarkable to see a cardiac arrest lasting just a few minutes give rise to such a lifelong process of transformation.

Only the large-scale Dutch study allowed for statistical analysis of the factors that may determine whether or not a NDE occurs. It thus ruled out the aforementioned possible physiological, psychological and pharmacological explanations for the occurrence of a NDE. Our study was also the first to include a longitudinal component with interviews after 2 and 8 years, which allowed us to compare the processes of transformation between people with and without NDE. We identified a distinct pattern of change in people with a NDE and revealed that integrating these changes into daily life is a long and arduous process. And we reached the inevitable conclusion that patients experienced all the aforementioned NDE elements during the period of their cardiac arrest, during the total cessation of blood supply to the brain. Nevertheless, the question how this could be possible remained unanswered.

Other Prospective Studies on NDE

Bruce Greyson, who published a prospective study in 116 survivors of cardiac arrest in the USA (Greyson 2003), found that 15.5% of the patients reported a NDE: 9.5% reported a core NDE and 6% a superficial NDE. He writes that

> no one physiological or psychological model by itself could explain all the common features of an NDE. The paradoxical occurrence of a heightened, lucid awareness and logical thought processes during a period of impaired cerebral perfusion raises particular

> perplexing questions for our current understanding of consciousness and its relation to brain function. A clear sensorium and complex perceptual processes during a period of apparent clinical death challenge the concept that consciousness is localized exclusively in the brain (Greyson 2003, p. 275).

The British prospective study by Sam Parnia and Peter Fenwick (Parnia et al. 2001) included 63 patients who survived their cardiac arrest. They found in their study that 11% reported a NDE: 6.3% reported a core NDE, and 4.8% a superficial NDE. They write that the NDE-reports suggest that the NDE occurs during the period of unconsciousness. This is a surprising conclusion, in their view, because "when the brain is so dysfunctional that the patient is deeply comatose, those cerebral structures, which underpin subjective experience and memory, must be severely impaired. Complex experiences as reported in the NDE should not arise or be retained in memory. Such patients would be expected to have no subjective experience, as was the case in the vast majority of patients who survive cardiac arrest, since all centres in the brain that are responsible for generating conscious experiences have stopped functioning as a result of the lack of oxygen." (Parnia et al. 2001, p. 151) Another, frequently cited explanation might be that the observed experiences occur during the early phases of the cessation or during the recovery of consciousness. Parnia and Fenwick, however, claim that "the verifiable elements of an out-of-body experience during unconsciousness, such as patients' reports on their resuscitation, render this extremely unlikely" (Parnia et al. 2001, p. 151).

Over a period of 4 years Penny Sartori carried out an even smaller study into NDE in 39 survivors of cardiac arrest in the UK (Sartori 2006). She found that 23% reported a NDE: 18% reported a core NDE, and 5% a superficial NDE. She concludes that "according to mainstream science, it is quite impossible to find a scientific explanation for the NDE as long as we 'believe' that consciousness is only a side effect of a functioning brain." The fact that people report lucid experiences in their consciousness when brain activity has ceased is, in her view, "difficult to reconcile with current medical opinion" (Sartori 2006, p. 25).

Some Typical Elements of an NDE

Before I discuss some of our current medical and neurophysiological ideas about consciousness and the function of our brain in more detail, I would like to reconsider certain elements of the NDE that were experienced during a transient period of a non-functioning brain during cardiac arrest: an out-of body experience, a holographic life-review, a meeting with deceased relatives, and a conscious return into the body.

First the *out-of-body experience* (OBE): in this experience people have veridical perceptions from a position outside and above their lifeless body. This out-of-body experience is scientifically important because doctors, nurses, and relatives can verify the reported perceptions, and they can also corroborate the precise moment the NDE with OBE occurred during the period of cardiopulmonary resuscitation

(CPR). This proves that OBE cannot be a hallucination, i.e. experiencing a perception that has no basis in "reality", like in psychosis, neither can it be a delusion, which is an incorrect assessment of a correct perception, nor an illusion, which means a misapprehension or misleading image. Should an OBE be considered as a kind of non-sensory perception? This is the report of a nurse of a Coronary Care Unit (van Lommel et al. 2001, p. 2041):

> During night shift an ambulance brings in a 44-year old cyanotic, comatose man into the coronary care unit. He was found in coma about 30 minutes before in a meadow. When we go to intubate the patient, he turns out to have dentures in his mouth. I remove these upper dentures and put them onto the 'crash cart.' After about an hour and a half the patient has sufficient heart rhythm and blood pressure, but he is still ventilated and intubated, and he is still comatose. He is transferred to the intensive care unit to continue the necessary artificial respiration. Only after more than a week do I meet again with the patient, who is by now back on the cardiac ward. The moment he sees me he says: 'O, that nurse knows where my dentures are.' I am very, very surprised. Then the patient elucidates: 'You were there when I was brought into hospital and you took my dentures out of my mouth and put them onto that cart, it had all these bottles on it and there was this sliding drawer underneath, and there you put my teeth.' I was especially amazed because I remembered this happening while the man was in deep coma and in the process of CPR. It appeared that the man had seen himself lying in bed, that he had perceived from above how nurses and doctors had been busy with the CPR. He was also able to describe correctly and in detail the small room in which he had been resuscitated as well as the appearance of those present like myself.

During a holographic *life review* the subject feels the presence and renewed experience of not only every act but also every thought from one's life, and one realizes that in some way we are connected to others and to ourselves such that we influence ourselves as well as others. All that has been done and thought seems to be significant and stored. Because one is connected with the memories, emotions and consciousness of another person, you experience the consequences of your own thoughts, words and actions to that other person at the very moment in the past that they occurred (interconnectedness or entanglement). They understand now what in some religions and cultures is known as the cosmic law that everything one does to others will ultimately be returned to oneself. Patients survey their whole life in one glance; time and space do not seem to exist during such an experience (non-locality). Instantaneously they are where they concentrate upon, and they can talk for hours about the content of the life review even though the resuscitation only took minutes. Quotation (van Lommel 2004, p. 121): "All of my life up till the present seemed to be placed before me in a kind of panoramic, three-dimensional review, and each event seemed to be accompanied by a consciousness of good or evil or with an insight into cause or effect". "Not only did I perceive everything from my own viewpoint, but I also knew the thoughts of everyone involved in the event, as if I had their thoughts within me. This meant that I perceived not only what I had done or thought, but even in what way it had influenced others, as if I saw things with all-seeing eyes. And so, even your thoughts are apparently not wiped out. And all the time during the review the importance of love was emphasised. Looking back, I cannot say how long this life review and life insight lasted, it may have been long, for every subject came up, but at the same time it seemed just a fraction of a second, because I perceived it all at the same moment. Time and distance seemed not to exist. I was in all places at the same time, and sometimes my attention was drawn to something, and then I would be present there."

Also a preview (flash-forward) can be experienced, in which both future images from personal life events as well as more general images from the future occur. Again it seems as if time and space do not exist during this preview (non-locality).

If *deceased acquaintances* or relatives are encountered in an otherworldly dimension, they are usually recognized by their appearance, and communication is possible through what is experienced as thought transfer. Thus, it is also possible to come into contact with fields of consciousness of deceased persons (interconnectedness). Quotation (van Lommel 2004): "During my cardiac arrest I had an extensive experience (...) and later I saw, apart from my deceased grandmother, a man who had looked at me lovingly, but whom I did not know. More than 10 years later, at my mother's deathbed, she confessed to me that I had been born out of an extramarital relationship, my father being a Jewish man who had been deported and killed during the second World War, and my mother showed me his picture. The unknown man that I had seen more than 10 years before during my NDE turned out to be my biological father."

Some patients can describe how they *returned into their body*, mostly through the top of the head, after they had come to understand that "it wasn't their time yet" or that "they still had a task to fulfil." This conscious return into the body is experienced as something very oppressive. They regain consciousness in their body and realize that they are "locked up" in their damaged body, meaning again all the pain and restriction of their disease. Quotation (Van Lommel 2004): "And when I regained consciousness in my body, it was so terrible, so terrible...that experience was so beautiful, I never would have liked to come back, I wanted to stay there..... and still I came back. And from that moment on it was a very difficult experience to live my life again in my body, with all the limitations I felt in that period."

Theories About NDE

With our current medical and scientific concepts it seems indeed impossible to explain all aspects of the subjective experiences as reported by patients with an NDE during a transient loss of all functions of the brain. Scientific studies into the phenomenon of NDE highlights the limitations of our current medical and neurophysiological ideas about the various aspects of human consciousness and the relationship between consciousness and memories on the one hand and the brain on the other. The prevailing paradigm holds that memories and consciousness are produced by large groups of neurons or neural networks. For want of evidence for the aforementioned explanations for the cause and content of a NDE the commonly accepted, but never proven concept that consciousness is localised in the brain should be questioned. After all, how can an extremely lucid consciousness be experienced outside the body at a time when the brain has a transient loss of all functions during a period of clinical death, even with a flat EEG? Furthermore, even blind people have described veridical perceptions during out-of-body experiences at the time of their NDE (Ring and Cooper 1999). Another theory about NDE holds that NDE might be a changing state of consciousness (transcendence, or the theory of continuity), in which memories, self-identity, and cognition, with emotion, function independently from the unconscious body, and retain the possibility of non-sensory perception. Obviously, during NDE enhanced consciousness is experienced independently from the normal body-linked waking consciousness, during the period of cardiac arrest, during the period of apparent unconsciousness. But how do we know that the EEG is flat in those patients with cardiac arrest, and how can we study this? Through many studies with induced cardiac arrest in both human and animal models cerebral function has been shown to be severely compromised during cardiac arrest, with complete cessation of cerebral blood flow (Gopalan et al. 1999), causing sudden loss of consciousness and of all body reflexes, but also with the abolition of brain-stem activity with the loss of the gag reflex and of the corneal reflex, and fixed and dilated pupils are clinical findings in those patients. And also the function of the respiratory centre, located close to the brainstem, fails, resulting in apnoea (no breathing). The electrical activity in the cerebral cortex (but also in the deeper structures of the brain in animal studies) has been shown to be absent after 10–20 s (a flat-line EEG) (De Vries et al. 1998; Clute and Levy 1990; Losasso et al. 1992; Parnia and Fenwick 2002). In acute myocardial infarction the duration of cardiac arrest in the Coronary Care Unit is usually 60–120 s, in an out-of-hospital arrest it even takes much longer. Thus, all 562 survivors of cardiac arrest in the four prospective studies on NDE should have had a flat EEG because no patient had been resuscitated within 20 s.

The quite often proposed objection that a flat line EEG does not rule out any brain activity, because it is mainly a registration of electrical activity of the cerebral cortex, misses the mark. The issue is not whether there is any brain activity of any kind whatsoever, but whether there is brain activity of the specific form, and in different neural networks, as regarded by contemporary neuroscience as the necessary condition of conscious experience. And it has been proven that there is no such specific brain activity at all during cardiac arrest. A flat line EEG is also one of the major diagnostic tools for the diagnosis of brain death, and in those cases the objection about not ruling out any brain activity whatsoever is never mentioned. Moreover, although measurable EEG-activity in the brain can be recorded during deep sleep (no-REM phase) or during general anesthesia, no consciousness is experienced because there is no integration of information and no communication between the different neural networks (Massimini et al. 2005; Alkire and Miller 2005; Alkire et al. 2008). So even in circumstances where brain activity can be measured sometimes no consciousness is experienced. A functioning system for communication between neural networks with integration of information is essential for experiencing consciousness, and this does not occur during deep sleep or general anesthesia, let alone during cardiac arrest. So there are good reasons to assume that our consciousness does not always coincide with the functioning of our brain: enhanced consciousness can sometimes be experienced separately from the body.

Consciousness and Brain Function

For decades, extensive research has been done to localize consciousness and memories inside the brain, so far without success. Also we should ask ourselves how a non-material activity such as concentrated attention or thinking can correspond to an observable (material) reaction in the form of measurable electrical, magnetic, and chemical activity at a certain place in the brain by EEG, MEG and PET-scan, and in the form of increased blood flow by fMRI. Neuro-imaging studies have shown these aforesaid activities, with specific areas of the brain becoming metabolically active in response to a thought or feeling. However, although providing evidence for the role of neuronal networks as an intermediary for the manifestation of thoughts (neural correlates), those studies do not necessary imply that those cells also produce the thoughts. A correlation doesn't elucidate anything about cause or result. And how should "unconscious" matter like our brain "produce" consciousness, while the brain only is composed of atoms and molecules in cells with a lot of chemical and electrical processes? Direct evidence of how neurons or neuronal networks could possibly produce the subjective essence of the mind and thoughts is currently lacking. We cannot measure what we think or feel (van Lommel 2010). There are no known examples of neural-perceptual matches, and hence reasons to doubt the truth of the "matching content" doctrine. The assumption in the "matching content" doctrine is that following activation of special neuronal networks you always will have the same content of thoughts or feelings. This seems extremely unlikely, because neural activation is simply neural activation; it only reflects the use of structures. This could be compared with a radio: you can activate the radio by turning it on, and you can activate a certain wavelength by tuning in on a special channel, but you will not have any influence on the content of the programme you are going to hear. Activating the radio does not influence the content of the programme, and neural activation alone does not explain the content of emotions or sensations.

In the last decade many articles and books have been published about consciousness, but up to now there are no uniform scientific views about the relationship between consciousness and the brain (Chalmers 1996). Most of the people who carry out research into consciousness, such as neuroscientists, psychologists, psychiatrists and philosophers, are still of the opinion that there is a materialist and reductionist explanation for consciousness. The well-known philosopher Daniel Dennett believes, and many with him, that consciousness is nothing other than matter (Dennett 1991), and that our subjective experience that our consciousness is something purely personal and differs from someone else's consciousness is merely an illusion. According to these scientists, consciousness originates entirely from the matter that constitutes our brain. If this were true then everything we experience in our consciousness would be nothing but the expression of a machine controlled by classical physics and chemistry, and our behaviour the inexorable outcome of nerve cell activity in our brain. Obviously the notion that all subjective thoughts and feelings are produced by nothing other than the brain's activity also means that it is an illusion to believe in free will. This viewpoint has serious implications for concepts such as moral responsibility and personal freedom. To many others consciousness seems to resist a materialistic explanation, so besides several materialistic concepts there is also an 'interactionistic-dualistic' model, where consciousness and the brain are totally different entities with different fundamental properties but somehow are able to interact with each other. And finally there is a concept of 'phenomalism' or 'immaterial (or neutral) monism', which is also called 'panpsychism' or 'idealism'. According to this model all material, physical systems should have a form of subjectivity at a fundamental level, and intrinsic properties of the physical world should be themselves phenomenal properties. If so, then consciousness and physical reality are deeply intertwined. This view acknowledges a clear causal role for consciousness in the physical world, and so consciousness should be regarded as a fundamental property of the universe (Chalmers 2002).

Summary of Conclusions from Research on NDE, Consciousness and the Brain

- At present more and more experiences are reported by serious and reliable people who to their own surprise and confusion have experienced an enhanced consciousness independently of their physical body. These experiences have been reported in all times, in all cultures and in all religions (van Lommel 2010).
- In several prospective empirical studies it has been proven that an enhanced and clear consciousness can be experiences during the period of cardiac arrest (clinical death) (van Lommel et al. 2001; Greyson 2003; Parnia et al. 2001; Sartori 2006).
- Based on these well documented prospective studies about NDE in survivors of cardiac arrest one has to come to the conclusion that current scientific views fail to explain the cause and content of a NDE (van Lommel et al. 2001; Greyson 2003; Parnia et al. 2001; Sartori 2006).
- It seems scientifically proven that during cardiac arrest no activity of the cortex and the brainstem can be measured, and also the clinical findings point out the transient loss of all functions of the brain (De Vries et al. 1998; Clute and Levy 1990; Losasso et al. 1992; Parnia and Fenwick 2002).
- All scientists who performed prospective studies on NDE came to the same conclusion: lack of oxygen by itself cannot explain the cause and content of NDE (van Lommel et al. 2001; Greyson 2003; Parnia et al. 2001; Sartori 2006). And this view is also supported by the fact that a NDE can be reported by people who did not have life threatening illnesses but were in fear of death or in depression (van Lommel 2010).
- In studying the function of the brain it has been proven that under normal daily circumstances, during deep sleep and during general anesthesia a functioning network and a co-operation between many different centres of the brain is a prerequisite for the experience of our waking consciousness (Massimini et al. 2005; Alkire and Miller 2005; Alkire et al. 2008). This is never the case during a cardiac arrest (De Vries et al. 1998; Clute and Levy 1990; Losasso et al. 1992; Parnia and Fenwick 2002).

- It is scientifically proven that consciousness can change the structure and function of the brain. This is called neuroplasticity (Schwartz and Begley 2002; Beauregard 2007; Davidson et al. 2003; Benedetti et al. 2005).
- In many respects consciousness as well as the function of the brain is still a great mystery.

About Concepts in Science

When empirical scientific studies discover phenomena or facts that are inconsistent with current scientific theories, so-called anomalies, these new facts must not be denied, suppressed or even ridiculed, as is still quite common these days. In the event of new findings the existing theories ought to be developed or adjusted, and if necessary rejected and replaced. We need new ways of thinking and new kinds of science to study consciousness and acquire a better understanding of the effects of consciousness. Some scientists, such as the philosopher Chalmers, are more receptive and take consciousness seriously (Chalmers 1995, p. 200): 'Consciousness poses the most baffling problems in the science of the mind. There is nothing that we know more intimately than conscious experience, but there is nothing that is harder to explain.' Chalmers has specialised in the problem of consciousness and has written a review of the various theories that seek to explain the relationship between consciousness and the brain (Chalmers 2002).

In the past, too, new kinds of science developed when prevailing scientific concepts could no longer explain certain phenomena. At the start of the previous century, for instance, quantum physics emerged because certain findings could no longer be accounted for with classical physics. Quantum physics upset the established view of our material world. The slow acceptance of the new insights provided by quantum physics can be attributed to the materialist worldview we have been raised with. According to some quantum physicists, quantum physics even assigns to our consciousness a decisive role in creating and experiencing the physical world as we perceive it. This not yet commonly accepted interpretation holds that our picture of reality is based on the information received by our consciousness. This transforms modern science into a subjective science with a fundamental role for consciousness. The quantum physicist Werner Heisenberg (1901–1976) formulates it as follows (Heisenberg 1958, p. 21): 'Science no longer is in the position of observer of nature, but rather recognizes itself as part of the interplay between man and nature. The scientific method ... changes and transforms its object: the procedure can no longer keep its distance from the object.'

For me science means asking questions with an open mind. Science should be the search for explaining new mysteries, rather than stick with old concepts. He, who has never changed his mind because he could not accept new concepts, has rarely learned something. We desperately need a real paradigm shift in science, and I sincerely hope that quantum physicist Max Planck was wrong when he said in 1934 (Planck 1948, p. 33–34): "A new scientific truth does not triumph by convincing its opponents and making them see the light, but rather because its opponents eventually die, and a new generation grows up that is familiar with it.". In my opinion, current science must reconsider its hypotheses about the nature of perceptible reality, because these ideas have led to the neglect or denial of significant areas of consciousness. Current science usually starts from a reality based solely on objective, physical phenomena. Yet at the same time one can (intuitively) sense that besides objective, sensory perception there is a role for subjective aspects such as feelings, inspiration and intuition. As stated before, current scientific techniques are incapable of measuring or demonstrating the content of thoughts, feelings and emotions. A purely materialist analysis of a living being cannot reveal the content and nature of our consciousness.

Non-local Consciousness

So it is indeed a scientific challenge to discuss new hypotheses that could explain the reported interconnectedness with the consciousness of other persons and of deceased relatives, to explain the possibility to experience instantaneously and simultaneously (non-locality) a review and a preview of someone's life in a dimension without our conventional body-linked concept of time and space, where all past, present and future events exist, and the possibility to have clear and enhanced consciousness with memories, with self-identity, with cognition, with emotion, with the possibility of perception out and above the lifeless body, and even with the experience of the conscious return into the body, which is experienced as something very oppressive. And it is important to mention that until now it has been impossible to induce a real out-of-body experience with veridical perception from a position out and above the body by any method whatsoever, despite incorrect suggestions about this possibility in the medical literature (van Lommel 2010).

In some articles (van Lommel 2004, 2006) and in my recent book (van Lommel 2010) I describe a concept in which our endless consciousness with declarative memories finds its origin in, and is stored in a non-local space as wave-fields of information, and the brain only serves as a relay station for parts of these wave-fields of consciousness to be received into or as our waking consciousness. The latter belongs to our physical body. These informational fields of our non-local consciousness become available as our waking consciousness only through our functioning brain in the shape of measurable and changing electromagnetic fields. Could our brain be compared to the TV set, which receives electromagnetic waves and transforms them into image and sound? Could it as well be compared to the TV camera, which transforms image and sound into electromagnetic waves? These waves hold the essence of all information, but are only perceivable by our senses through suitable instruments like camera and TV set. The function of the brain should be compared with a transceiver, a transmitter/receiver, or interface. Thus there are two complementary aspects of consciousness, which cannot be reduced one to the other, and the function of neuronal networks should be regarded as receivers and conveyors, not as retainers of consciousness and memories. This view is highly compatible with the concept of 'phenomalism' or 'immaterial (or neutral) monism' (Chalmers 2002). In this concept, consciousness is not rooted in the measurable domain of physics, our manifest world. This also means that the wave aspect of our indestructible consciousness in the non-local space is inherently immeasurable by physical means. However, the physical aspect of consciousness, which originates from the wave aspect of our consciousness through collapse of the wave function, can be measured by means of neuro-imaging techniques like EEG, fMRI, and PET-scan. This non-local aspect of consciousness, which can also be called our higher consciousness, divine consciousness or cosmic consciousness, could be compared to gravitational fields, of which only the physical effects throughout the universe can be measured, but the fields themselves are not directly demonstrable. In trying to understand this concept of interaction between the invisible non-local space and our visible, material body, it seems appropriate to compare it with modern worldwide communication. There is a continuous exchange of objective information by means of electromagnetic fields for radio, TV, mobile telephone, or laptop computer. We are not consciously aware of the vast amounts of electromagnetic fields that constantly, day and night, exist around us and even permeating us, as well as permeating structures like walls and buildings. At each moment we are invaded by hundreds of thousands of telephone calls, and hundreds of radio and TV programmes. We only become aware of these electromagnetic informative fields at the moment we use our mobile telephone or by switching on our radio, TV or laptop. What we receive is neither inside the instrument, nor in the components, but thanks to the receiver, the information from the electromagnetic fields becomes observable to our senses and hence perception occurs in our consciousness. The voice we hear over our telephone is not inside the telephone. The concert we hear over our radio is transmitted to our radio. The images and music we hear and see on TV are transmitted to our TV set. Internet with more than a billion websites can be received at about the same moment in the USA, in Europe and in Australia, and is obviously not located inside our laptop.

One cannot avoid the conclusion that endless consciousness has always existed and will always exist independently from the body, because there is no beginning nor will there ever be an end to our consciousness. There is a kind of biological basis of our waking consciousness, because during life our physical body functions as an interface or place of resonance. But there is no biological basis of our whole, endless, or enhanced consciousness because it is rooted in a non-local space. Our enhanced and non-local consciousness resides not in our brain and is not limited to our brain. So our brain seems to have a facilitating, and not a producing function to experience consciousness. It is quite interesting to mention that this conclusion is in striking concurrence with the view of the philosopher and neuroscientist Alva Noë, who, based on totally different neuroscientific research, writes in his recent book 'Out of our heads' (Noë 2009): 'All scientific theories rest on assumptions. It is important that these assumptions be true. I will try to convince the reader that this startling assumption of consciousness research that consciousness is a neuroscientific phenomenon and that it happens in the brain is badly mistaken. Consciousness does not happen in the brain. What determines and controls the character of conscious experience is not the associated neural activity. That is why we have been unable to come up with a good explanation of its neural basis'. Noë proposes that the brain's job is that of facilitating a dynamic pattern of interaction among brain, body and world.

Conclusion

By making a scientific case for consciousness as a non-local and thus ubiquitous phenomenon, this view can contribute to new ideas about the relationship between consciousness and the brain. I am aware that this concept can be little more than a stimulus for further study and debate, because at present we lack definitive answers to the many important questions about our consciousness and the relationship between consciousness and the brain. I have no doubt that in the future, too, many questions about consciousness and the mystery of life and death will remain unanswered. However, faced with extraordinary or abnormal findings we must question a purely materialist paradigm in science. A near-death experience is one such extraordinary finding. Although consciousness remains a huge mystery, new scientific theories based on research into NDE appear to be helping us answer some of the questions about this mystery.

Scientific studies on NDE challenge our current concepts about consciousness and its relation with brain function, and its conclusions are important for many aspects in healthcare, because this view of consciousness as a non-local phenomenon might well induce a huge change in the scientific paradigm in western medicine. It could have practical implications in actual medical and ethical problems such as the care for comatose or dying patients, euthanasia, abortion, and the removal of organs for transplantation from somebody in the dying process with a beating heart in a warm body but with a diagnosis of brain death. Such understanding also fundamentally changes one's opinion about death, because of the almost unavoidable conclusion that at the time of physical death consciousness will continue to be experienced in another dimension, in which all past, present and future is enclosed. As someone with an NDE wrote to me: "Death is only the end of our physical aspects". But we should acknowledge that research on NDE cannot give us the irrefutable scientific proof of this conclusion, because people with a NDE did not quite die, but they all were very close to death, and without a functioning brain. But, as I have explained, it has been scientifically proven that during NDE enhanced consciousness was experienced independently of brain function. Quoting from a recent death announcement: "All what you have, falls into decay, but what you are, lives on, beyond time and space". So we have a body and we are conscious. Without a body we still can have conscious experiences. Recently someone with a NDE wrote me: 'I can live without my body, but apparently my body can not live without me'. The conclusions seems compelling that endless consciousness has and always will exist independently from the body. There is no beginning nor will there ever be an end to our consciousness. For this reason we indeed should seriously consider the possibility that death, like birth, can only be a transition to another state of consciousness, and that during life the body functions as an interface or place of resonance. This view of a non-local consciousness also allows us to understand a wide variety of special states of consciousness, such as mystical and religious experiences, deathbed visions (end-of-life experiences), peri-mortem and post-mortem experiences (nonlocal interconnectedness), heightened intuitive feelings (nonlocal information exchange), prophetic dreams, remote viewing (non-local perception) and the effect of consciousness on matter (nonlocal perturbation).

A NDE is both an existential crisis and an intense lesson in life. People change after a NDE as it gives them a conscious experience of a dimension in which time and distance play no role, in which past and future can be glimpsed, where they feel complete and healed and where they experience unlimited knowledge and unconditional love. These life changes mainly spring from the insight that love and compassion for oneself, for others and for nature are major prerequisites for life. Following a NDE most people realize that everything and everyone is connected, that every thought has an effect on both oneself and the other, and that our consciousness continues beyond physical death. Regarding what we can learn from people who are willing to share their NDE with others, I would like to quote Dag Hammerskjöld (Hammerskjöld 1964): "Our ideas about death define how we live our life". Because as long as we believe that death is the end of everything we are, we will give our energy towards the temporary and material aspects of our life. We should recognize that our view of the world is wrong, because we do not realize that the world, as we see it, only derives its (subjective) reality from our consciousness. Because it is only our consciousness that is determining how we see this world. If we are in love, the world around us is beautiful, when we are depressed our world is like hell, and when we are frightened (made terrified by politicians and by the press) our world will be full of terror. 'The mind in its own place, and in itself, can make a heaven of hell,' wrote John Milton even in 1667 in his poem 'Paradise Lost'.

People with a near-death experience have been my greatest teachers. My many conversations with them and my in-depth study of the potential significance of an NDE have changed my views on the meaning of life and death. I feel more connected with nature. I realize that everything originates from consciousness. I understand now that people create their own reality based on their consciousness and also create the intention one lives his life. So for me there was much to learn from the insights acquired through an NDE, and according to me it is evident that one doesn't need his own near-death experience to gain new insights into life and death.

References

Alkire, M.T., & Miller, J. (2005). General anesthesia and the neural correlates of consciousness. *Progress in Brain Research, 150*, 229–244.

Alkire, M.T., Hudetz, A.G., & Tononi, G. (2008). Consciousness and anesthesia. *Science, 322*(5903), 876–880.

Beauregard, M. (2007). Mind does really matter: Evidence from neuroimaging studies of emotional self-regulation, psychotherapy, and placebo effect. *Progress in Neurobiology, 81*(4), 218–236.

Benedetti, F., Mayberg, H.S., Wager, T.D., Stohler, C.S., & Zubieta, J.K. (2005). Neurobiological mechanisms of the placebo effect. *The Journal of Neuroscience, 25*(45), 10390–10402.

Chalmers, D.J. (1995). Facing up to the problem of consciousness. *Journal of Consciousness Studies, 3*(1), 200.

Chalmers, D.J. (1996). *The conscious mind: In search of a fundamental theory*. New York/Oxford: Oxford University Press.

Chalmers, D.J. (2002). Consciousness and its place in nature. In D. Chalmers (Ed.), *Philosophy of mind: Classical and contemporary readings.* New York/Oxford: Oxford University Press. Also at: http://consc.net/papers/nature.html

Clute, H., & Levy, W.J. (1990). Electroencephalographic changes during brief cardiac arrest in humans. *Anesthesiology, 73*(5), 821–825.

Davidson, R.J., Kabat-Zinn, J., Schumacher, J., Rosenkrantz, M., Muller, D., Santorelli, S. F., et al. (2003). Alterations in brain and immune function produced by mindfulness meditation. *Psychosomatic Medicine, 65*(4), 564–570.

De Vries, J.W., Bakker, P.F.A., Visser, G.H., Diephuis, J.C., & Van Huffelen, A. C. (1998). Changes **in cerebral oxygen uptake and cerebral electrical activity during defibrillation threshold testing**. *Anesthesia and Analgesia, 87*(1), 16–20.

Dennett, D. (1991). *Consciousness explained.* Boston/London: Little, Brown and Co.

Gallup, G., & Proctor, W. (1982). *Adventures in immortality: A look beyond the threshold of death.* New York: McGraw-Hill.

Gopalan, K.T., Lee, J., Ikeda, S., & Burch, C.M. (1999). Cerebral blood flow velocity during repeatedly induced ventricular fibrillation. *Journal of Clinical Anesthesia, 11*(4), 290–295.

Greyson, B. (2003). Incidence and correlates of near-death experiences in a cardiac care unit. *General Hospital Psychiatry, 25*(4), 269–276.

Hammerskjöld, D. (1964). Markings (L. Sjöberg & W.H. Auden, Trans.). London/New York: Faber and Faber/Knopf. (Originally published in 1963: Vägmärken. Stockholm: Bonniers)

Heisenberg, W. (1958). *Physics and philosophy*. New York: Harper and Row. (Original published in 1955: Das Naturbild der heutigen Physik)

Kennedy, D., & Norman, C. (2005). What we don't know. *Science, 309*(5731), 75.

Losasso, T.J., Muzzi, D.A., Meyer, F.B., & Sharbrough, F.W. (1992). Electroencephalographic monitoring of cerebral function during asystole and successful cardiopulmonary resuscitation. *Anesthesia and Analgesia, 75*(6), 1021–1024.

Massimini, M., Ferrarelli, F., Huber, R., Esser, S. K., Singh, H., & Tononi, G. (2005). Breakdown of cortical effective connectivity during sleep. *Science, 309*(5744), 2228–2232.

Moody, R.A., Jr. (1975). *Life after life*. Covington: Mockingbird Books.

Noë, A. (2009). *Out of our heads. Why you are not your brain, and other lessons from the biology of consciousness*. New York: Hill and Wang, A division of Farrar, Straus and Giroux.

Parnia, S., & Fenwick, P. (2002). Near-death experiences in cardiac arrest: Visions of a dying brain or visions of a new science of consciousness. Review article. *Resuscitation, 52*(1), 5–11.

Parnia, S., Waller, D.G., Yeates, R., & Fenwick, P. (2001). A qualitative and quantitative study of the incidence, features and aetiology of near death experience in cardiac arrest survivors. *Resuscitation, 48*(2), 149–156.

Planck, M. (1948). ***Scientific autobiography and other papers*** (F. Gaynor, Trans., pp. 33–34). New York: Philosophical Library (1949).

Ring, K. (1980). ***Life at death: A scientific investigation of the near-death experience***. New York: Coward, McCann & Geoghegan.

Ring, K. (1984). *Heading toward omega: In search of the meaning of the near-death experience*. New York: Morrow.

Ring, K., & Cooper, S. (1999). *Mindsight. Near-death and out-of-body experiences in the blind*. Palo Alto: William James Center for Consciousness Studies.

Ritchie, G.G. (1978). *Return from tomorrow*. Grand Rapids: Chosen Books of The Zondervan Corp.

Sartori, P. (2006). The incidence and phenomenology of near-death experiences. *Network Review (Scientific and Medical Network), 90*, 23–25.

Schmied, I., Knoblauch, H., & Schnettler, B. (1999). Todesnäheerfahrungen in Ost- und Westdeutschland. Eine empirische Untersuchung. In H. Knoblauch & H.G. Soeffner (Eds.), *Todesnähe: Interdisziplinäre Zugänge zu Einem Außergewöhnlichen Phänomen* (pp. 65–99). Konstanz: Universitätsverlag Konstanz.

Schwartz, J.M., & Begley, S. (2002). *The mind and the brain: Neuroplasticity and the power of mental force*. New York: Regan Books.

van Lommel, P. (2004). About the continuity of our consciousness. In C. Machado & D.A. Shewmon (Eds.), *Brain death and disorders of consciousness. Advances in Experimental Medicine and Biology* (volume 550, pp. 115–132). New York/Springer

van Lommel, P. (2006). Near-death experience, consciousness and the brain: A new concept about the continuity of our consciousness based on recent scientific research on near-death experience in survivors of cardiac arrest. *World Futures, The Journal of General Evolution, 62*(1/2), 134–151.

van Lommel, P. (2010). *Consciousness beyond life: The science of the near-death experience*. New York: Harper Collins. (Original published 2007: Eindeloos Bewustzijn: Een wetenschappelijke visie op de bijna-dood ervaring. Kampen: Ten Have)

van Lommel, P., Van Wees, R., Meyers, V., & Elfferich, I. (2001). Near-death experiences in survivors of cardiac arrest: A prospective study in the Netherlands. *The Lancet, 358*(9298), 2039–2045.

The Hard Problem Revisited: From Cognitive Neuroscience to Kabbalah and Back Again

B. Les Lancaster

Abstract The dialogue between cognitive neuroscience and spirituality/mysticism has largely entailed measuring the neural and cognitive effects of spiritual practices. Such research follows from the spiritual traditions' teachings about the intended psychological effects of practice. The ontologically more challenging postulates of spiritual traditions (e.g., mind beyond brain, 'higher' or 'ultimate' realities) are ignored when focusing in this way on measurable concomitants of practice. In this chapter I argue that the dialogue should be widened to include some of the ontologically more challenging concepts, where these involve references to the brain and psychological states. A specific example is examined in some detail: the kabbalistic worldview posits a *correspondence* between higher and lower levels in the cosmos ('macrocosm' and 'microcosm'), and includes notions of unconscious thought arising in 'brains' in the Godhead. I demonstrate that the macrocosmic principles advanced in kabbalistic literature display a degree of concordance with the results of current research into the neural correlate of consciousness. I explore the implications of this concordance for the light it may cast on the enduring *hard problem* of consciousness.

The Challenge of Explanation

Recent interest in the interrelationships between cognitive neuroscience and spiritual traditions has focused in two areas: the correlates of specific experiential states (Aftanas and Golocheikine 2002; Beauregard and Paquette 2006, 2008; Lutz et al. 2004; Newberg et al. 2001a; for recent reviews, see Cahn and Polich 2006;

B.L. Lancaster (✉)
School of Natural Sciences & Psychology, Liverpool John Moores University, UK
e-mail: b.l.lancaster@ljmu.ac.uk

H. Walach et al. (eds.), *Neuroscience, Consciousness and Spirituality*,
Studies in Neuroscience, Consciousness and Spirituality 1,
DOI 10.1007/978-94-007-2079-4_14,

Lutz et al. 2007, 2008) and the evolutionary psychology of belief (Boyer 2001; Boyer and Bergstrom 2008; Persinger 1987). Thus, for example, studies have examined the spatial and temporal characteristics of brain activity associated with meditative or prayerful states, the therapeutic benefits of mindfulness, and the ways in which religious beliefs fit into the overall cognitive and social organisation of the human mind. The picture of religion that emerges from this recent interest inevitably sees religion as 'naturalistic,' meaning that religious phenomena can be integrated into our current scientific worldview with its materialistic bias. According to this approach, some experiential states may have beneficial effects in terms of physical and mental health, but, from the evolutionary standpoint, religion is generally seen as a mundane consequence of the normative quest for explanation and security that characterises our psychological makeup. Its central postulates are, at best, illusory.

The question I wish to address in this chapter concerns the extent to which such observations do adequate justice to the complexities of spiritual and mystical traditions. My concern is not simply that our view of the traditions may be impoverished if we effectively take from them only those aspects which sit easily with the prevailing scientific worldview. Were that the only issue then one could rightly retort that it is for theologians and devout practitioners to present the noble insights of religion; psychologists and neuroscientists should not muddy the waters of their own disciplines in the process of extolling spiritual teachings. No, my concern lies rather with the knowledge base that is growing where spirituality and psychology come together. The knowledge base will be compromised if we draw on spiritual ideas with, as it were, one hand tied behind our backs.

The point may be expressed another way by emphasising that we should not be approaching spiritual teachings and practices merely in order to fit them into the explanatory categories that are currently *de rigueur* in neuroscience and psychology. Such an approach means that one is inevitably blinded to new insights that may arise when those categories are transcended. As Rothberg argues, "To assume that the categories of current western epistemology are adequate for interpreting spiritual approaches is to prejudge the results of [a profound encounter with practices of spiritual inquiry], which might well lead to significant changes in these categories" (Rothberg 2000, p. 176).

I am concerned here with the unbridled *psychologising* of spiritual concepts. My emphasis should not be taken as implying that I am less concerned over what may be construed as 'smuggling in' religious or spiritual categories to the discourse of psychology. Of course such *fundamentalism* is problematic. However, in introducing my chapter I want to draw attention to the bias that has been present in recent conceptualising of the relationship between cognitive neuroscience and religion. I believe that this bias has ensued from an unquestioning acceptance of the hegemony of psychological science as it has held sway over recent decades, and has compromised the potential outcomes of the relationship.

It seems to me that there is an intellectual path that fills the area of overlap between psychology and religion, and which, crucially, is not reducible to either of those disciplines. The topic of enquiry on that path is the mind as it extends beyond the immediate confines of personal consciousness. From this perspective, the subject matter includes what we perhaps rather simplistically call the *unconscious* as well as those spheres of mind that might be deemed *transpersonal*. Transpersonal realms have been identified using terms such as *self* (when this is distinguished from the personal ego, e.g., Jung's concept of the Self), *transpersonal self*, and *higher unconscious* (Assagioli 1993; Firman and Gila 2002).

It is of course one thing to coin a term; it is another to demonstrate that the construct proposed has any validity. This is by no means a trivial issue; indeed, to my mind it goes to the heart of psychology as an empirical discipline. Psychological constructs are heuristically valuable, but the ontological status of even as ubiquitous a construct as 'I' (or 'ego') may be questioned (Lancaster 1991, 1997; Metzinger 2003). From this perspective there is little difference between the empirical basis of transpersonal psychology and other branches of psychology. If we posit a 'transpersonal self', it is because it is useful in synthesising and organising data drawn from experience; no-one is suggesting that one can objectively observe such an entity.

This is a crucial point when it comes to studying the psychological value of mystical texts. The majority of writings classed as 'mystical' do not focus on experience per se, but rather are directed to understanding the ways in which the "mystery out of which everything arises" (Ferrer and Sherman 2008, p. 18) unfolds into our spatial-temporal realm. The logic of such writings entails acceptance of entities ('celestial intelligences', 'souls') that enable the scholarly community within the given tradition to synthesise and organise data, just as occurs in psychology. The insights from such writings may be valuable irrespective of the ontological status of the entities, just as psychological insights into self-related activity are valid irrespective of the reality of 'I'.

Given that the focus of most mystical writings is not directly on experience, psychological studies focussing on correlates of meditative, or mystical, experience are exploring a selective feature of the spiritual traditions. In itself this selectivity is not problematic. The study of experience is clearly a legitimate approach for psychology, and the fact that the experiences may be extraordinary, *even within the traditions of mysticism*, is no reason to ignore them. Moreover, exploring the neural correlates, and psychological impact, of such experiences raises no major ontological issues. Whilst some may challenge *materialistic* explanations of the data (Beauregard and O'Leary 2007), the data themselves are neutral with regard to the prevailing worldview of psychological science.

I want to suggest, however, that this narrow focus on the concomitants of spiritual experiences generates an inadequate foundation for our interest in the connections between mysticism and cognitive neuroscience. And the inadequacy comes about for two reasons. Firstly, the seeming distinction between mystical writings that are reflective of direct experience and writings which attempt to explain spiritual realms from a more hermeneutic approach (e.g., exegesis and commentary on revealed scripture) is by no means clear-cut. I shall substantiate this point shortly, but just to note here the consequence that asking participants in a study to engage in a specific practice, or to re-live moments of mystical union, is not doing justice to the range of experience relevant to our interest. The second reason why I believe that we must engage fully with the seemingly more 'theoretical' material in the mystical traditions concerns the interpretations we place on experience. Experience is undoubtedly structured by the categories established through enculturation, and to de-code experiential referents in mystical material requires a rich understanding of the categories involved. The problem here is that those of us coming from a psychological background tend to operate with a highly simplified dichotomous approach when it comes to interpreting mystical experience: either the experiences are fantasies[1] or they represent contact with an Ultimate of which nothing further can be said. The phrase "devil and the deep blue sea"[2] comes to mind! It is the "of which nothing further can be said" that stymies us. And the fact is that plenty has been said, especially of those intermediary realms which, I purport, are crucial for the debate between psychology and mysticism.

Let me elaborate on the points I have raised in this last paragraph, with specific reference to the tradition of Kabbalah, which, as I shall explain later, seems to me peculiarly important for our present purposes on account of its psychological orientation and elaboration of the intermediary realms between the Ultimate and the human mind.

Kabbalistic texts in which an individual mystic describes a more-or-less ineffable experience are the exception. Most writings are exegetical and explore the *concealed teachings* which give shape to the intermediary realms between human and God. However, in line with my first point above, it is evident to all scholars of Kabbalah, that exegetical material is both framed through the lens of experience and acts as a spur to mystical experience. Wolfson (1994) uses the term "inspired exegesis" to capture the point that kabbalistic insights invariably arise between the two poles of the revealed scriptural text and the mystic's own experience. As he writes in relation to the central text of Kabbalah, the *Zohar*, "visionary experience is a vehicle for hermeneutics as hermeneutics is a vehicle for visionary experience" (Wolfson 2004, p. 113).[3]

[1] Again, the reductive approach can cover a number of viewpoints, not all of which are necessarily dismissive of mystical practice. *Mindful* experiences, for example, may be viewed as constructive simply in terms of health benefits. However, my term 'fantasies' is meant to convey the fact that in the vast majority of the world's spiritual traditions, such experiences include transcendent categories ('soul', 'godhead', 'world soul', etc.) which are dismissed as irrelevant in psychological discourse.

[2] The 'deep blue sea' may be a sacrilegious term to apply to the Ultimate, but I claim dispensation from Rabbi Meir's aphorism that, 'blue resembles the colour of the sea, and the sea resembles the colour of the sky, and the sky resembles the colour of the *Throne of Glory*' (Talmud, *Menakhot* 43b)!

[3] Wolfson is writing here about the mystical experience of *light*, and therefore his interest lies with *visionary* experience. The point stands that more generally for the Jewish mystic, hermeneutics is intertwined with all forms of mystical experience.

The *Zohar* has assumed canonical status within Judaism as being the teaching *par excellence* of the 'secrets of the Torah.' Kabbalah holds that the Torah comprises revealed and concealed teachings, the latter pertaining to the nature of God and the ways to align oneself with God in order to promote divine beneficence to flow to the world. Thus, the mysticism that pervades the *Zohar* is one that promotes 'ascent for the sake of descent,' as Hellner-Eshed (2009, p. 317) characterises it. The *Zohar* comprises a corpus of writings, many of which are presented as esoteric commentary on the biblical text, with others cryptically elaborating the structure of the Godhead. Its authorship has been the subject of extensive scholarly debate. The major part of the corpus is seen by most scholars as having been penned in thirteenth-century Spain by Moses de León. This view is opposed by religious authorities who regard the *Zohar* as having been revealed through miraculous means to the second-century Shimon bar Yohai in the Land of Israel.

One example will suffice to make this point. The twelfth-century *Sefer haBahir* explores the meaning of the prayer of the biblical Habakkuk in which the prophet writes, "Lord, I have heard thy hearing [otherwise translated, "teaching", but literally "hearing"] and feared" (*Habakkuk* 3:1):

> What did he understand that he should fear? He understood God's thought. Just as [human] thought has no end, for even a mere mortal can think and descend to the end of the world, so too the ear also has no end and is not satiated. It is thus written [Ecclesiastes 1:8] "The ear is not satiated from hearing" (Abrams 1994, p. 149 [*Bahir* para 53]).

The writer is evidently asserting that real fear brings about a dissolution of the normal boundaries of self, such that thought itself is no longer dominated by 'I'-associations. The idea of "hearing God's *hearing*" is reminiscent of the Islamic scripture, "When I [Allah] love my servant … I become the hearing with which he hears, the seeing with which he sees, the hand with which he grasps …" (*Hadith, Bukhari* 81:38*)*. The insight that there is ultimately no subject other than the divine is a fundamental of all theistic traditions. Whilst the *Bahir's* exegesis is supported by the scriptural text, it is undoubtedly the writer's experience of thought in mystical states that underpins the meaning. In the *Zohar*, which circulated in the century after the *Bahir*, the subtleties of such transpersonal thought are explored in relation to the *sefirotic* pleroma[4] that mediates between the transcendent mystery and the divine presence in our world. The "mystery of the concealment of thought" (*Zohar* 1:21a) is traced from its origin in the infinite divine essence until it manifests as outward speech. Although the codified format by which the various stages in this unfolding of thought are presented makes it difficult to capture succinctly, it is clear that the pattern broadly entails the progression from unconscious roots to outer expression (Lancaster 2005). Moreover, there is a reciprocity between the 'downward' movement as seen in the process of Creation (from 'above' to 'below') and the 'upward' movement in mystical states of consciousness (from 'below' to 'above'). The ascent, then, entails moving towards the unconscious roots of thought: "Human thought has the ability to strip itself and to ascend to and arrive at the place of its source. Then it unites with the supernal entity whence it comes, and it [the thought] and it [its source] become one entity" (anonymous thirteenth-century Kabbalist, cited in Idel 2005a, p. 219).

To capture my argument in this chapter succinctly, I think our interest in bringing neuroscience, consciousness, and spirituality together will be served by paying more attention to these discussions of 'higher realms'. At present, the project seems to be focused on the concomitants of experiential states, and, to elaborate my earlier point, we are presented with two alternatives. Either the spiritual experience under examination is basically *nothing but* the distinctive brain state associated with it, or the brain state is mediating contact with something beyond itself, designated by terms such as "Absolute Unitary Being" (Newberg et al. 2001b, p. 171) or the "divine Ground of Being" (Beauregard and O'Leary 2007, p. 293). The first

[4] This term refers to the intermediary realm between the unknowable transcendent God and the natural world. The *sefirot* are the emanations of God.

alternative essentially leads up a blind alley as far as the goals of mysticism are concerned. At best, it may be claimed that the spiritual state has some value in promoting mental efficiency and increased well-being. But study of the brain states is tangential to those benefits. The second alternative also seems to me to be somewhat limited. There may be some 'Absolute' beyond the brain, but without further detail of its 'hither' dimension (as William James might have put it) this is simply a matter of belief and assertion.

Earlier I referred to the kinds of constructs ('higher self' etc.) psychologists with a transpersonal inclination have employed to explain not only experiences that seem to go beyond the normal realms but also the notion that the individual may grow into higher levels or stages of development (Wilber 2006). In order to link this aspect of contemporary psychology with the classical worldview of the Kabbalah, let me propose that these constructs are heirs to the Aristotelian term, *Active Intellect*, which was used during late antiquity and the middle ages by scholars and mystical writers of the Abrahamic traditions to convey the notion of a realm of mind bridging human and divine (see Lancaster 2004, for more detail). According to Aristotle, the active intellect comes into the soul from outside, and thus brings into human psychology an element of the divine and the potential to reach the divine. Whilst some psychologists would resist aligning this term and one such as 'higher self' for reasons of ontology, the challenge of accounting for transcendent experiences and attainment of higher levels of being ("synthesising and organising data", as I put it earlier) unites the kabbalist and the transpersonal psychologist.

All explanation requires use of appropriate structures and processes that can be seen as the underpinning and cause of the matter to be explained. On account of its presumed extra-bodily, divine source, the active intellect served to explain mystics' experiences of intellectually participating in a divine realm. In our day, when the connotations of a term such as 'active intellect' are no longer seen by many as ontologically acceptable, the challenge for those – like myself – who wish to hold onto the sense of a "mystery out of which everything arises" is to embrace the data relating spirituality to cognitive neuroscience in non-reductive ways.[5] But it is not simply a matter of generating appropriate terminology; we need to conceptualise mystical states in *operational* terms.

This challenge is *one and the same* as the challenge to explain consciousness in non-reductive ways. The cognitive neuroscience of consciousness is a burgeoning field, yet its rich database does nothing to explain the core mystery of consciousness, its *phenomenality*. I use this term to convey the subjective element, the fact of

[5] I make no defence for my "wish to hold onto" this sense of mystery. The key point is that my sense of the mystery is independent of data in cognitive neuroscience. The *data themselves do not justify any overarching belief for explanation, be it materialistic or transcendental.* I hold onto the sense of mystery for a host of reasons, mainly relating to what I consider to be meaningful values and goals in life. In this chapter I am not attempting to 'prove' the truth of kabbalistic, or any other, insights. My point is simply that those insights are worthy of exploration for the relationship they have with observations in the cognitive neuroscience of consciousness.

qualia, or what it is to have experiences. Brain science can yield copious data about neural processes involved in signalling the presence of certain sensory details, for example, but it cannot explain this inner quality whereby we experience the *blueness* of light of a given wavelength or the scent of lavender when certain chemicals interact with the olfactory system. We need to be clear about "what brain science can, and what it cannot, tell us about the mind," as I put it many years ago (Lancaster 1991, p. 5). The *explanatory gap* (Levine 1983) between our understanding of neural dynamics and the nature of experience remains as much of a gap as it ever was, and our approach to bridging it depends more on our beliefs than it does on the objective data available (Barušs 2001; Lancaster 2004).

Given the parallel between the two challenges – to understand consciousness and to explain spiritual and mystical states – it follows that the division that has been made between the "hard problem" and the "easy problems" (Chalmers 1995), which applies to one, will in turn pertain to the other. Such a division has indeed informed my research into *Kabbalistic Psychology* (Lancaster 2005), which forms the substance of this Chapter. In brief, the "easy" aspects include those kabbalistic teachings which can be related to processes known to psychological science (e.g., the effects of concentrative practices in bringing about altered states of consciousness; the role of language in relation to self; associative processes in accessing unconscious material). Whilst the original teachings are inevitably embedded in the language of religion, with its ontological assumptions of a divine realm etc., the teachings themselves are meaningful when extracted from the religious belief system. The "hard" aspects, by contrast, are those which depend upon the ontological assumptions and seem to have no applicability when viewed outside the religious framework (teachings of ontologically separate worlds; microcosm-macrocosm correspondence; the goal of *cleaving* to God). I shall return shortly to elaborate on these "easy" and "hard" aspects of Kabbalah.

The Principles of Kabbalistic Psychology

My specific interest in Kabbalah is not only a result of my own Jewish roots but also stems from the distinctive psychological emphasis found within this mystical path. It is beyond the scope of this chapter to explore fully the reasons for this psychological emphasis, but it is worth noting in passing that three factors are primarily involved.

First is the emphasis in Judaism on *embodied spirituality*. The separation at the inception of the Common Era between nascent Christianity and what became Rabbinic Judaism was fundamentally about *worldview*. The Rabbis taught that the spiritual path is very much *this*-worldly, and that ordained embodied action brings holiness to the world. In perpetuating this emphasis, Kabbalah enters into elaborate discussion of what might be best described as *spiritual anatomy*. As will become clear below, the Godhead is understood in terms of its relation to highly physical features, such as brains and facial structures. Embodied structure becomes the key to grasp 'higher structure'.

This first factor is itself entwined with the second, namely the core teaching that the human is in the image of God (*Genesis* 1:27; 5:1). This reflective principle means that understanding human nature, and especially the human mind, is the key to understanding God. 'Psychology' is therefore essentially spiritual. There can be no such thing as a 'secular' psychology, for any insight into the human mind necessarily pertains to the divine mind. "From the 'I' of flesh and blood you may learn about the 'I' of the Holy One, blessed be He," runs a Jewish *Midrashic* text.[6] The insight that the human mind recapitulates the mind of God is implicit in much kabbalistic literature. Insights into the divine mind – God's thought, His will to create – illumine deep aspects of the human mind (the unconscious determinants of thought, creativity). Indeed, this notion of correspondence becomes an explicit *isomorphism* between God and man (Shokek 2001). As Wolfson puts it, "In seeing God, one sees oneself, for in seeing oneself, one sees God" (Wolfson 2005, p. 39).

A relationship with human psychology is evident wherever kabbalistic texts discuss thought processes in the Godhead. In the *Zohar*, the initial stirrings of creation are repeatedly described in these psychological terms. To take just one example of the *Zohar*'s teaching about the primordial moments of creation: "From the midst of thought a desire arose to expand, and it spread from the place where thought is concealed and unknown until it expanded to settle in the larynx" (*Zohar* 1:74a). The larynx is the source of spoken language and in this passage symbolises the creative voice of God preparing for the great "Let there be…" of *Genesis*. For our purposes, the key notion in this extract is that God's initial thought, which precedes vocalisation, arises in the "place where thought is concealed," and that its development is grasped through the embodied experience of using the human larynx. This "place where thought is concealed" parallels the human unconscious, in which the thinking that occurs is indeed unknown. The earliest explicit reference to a human unconscious in kabbalistic literature is found in the work of Dov Baer, the Maggid of Mezeritch (1710–1772), although the idea is implicit in much earlier work. In the Maggid's thought, the unconscious is both a dimension of the human mind in which thoughts arise prior to their entering the realm of consciousness, and also an aspect of the 'mind' of God Himself. The unconscious is, accordingly, a region of confluence between the mind of man and the mind of God.

The third factor that underpins the psychological emphasis found in Kabbalah again interrelates with the first two, since it relates to the concealment of thought and the logic of the unconscious. This principle concerns the multiplicity of meaning inherent in the divine text, the Torah. The Talmud conveys the point as follows: "In Rabbi Ishmael's School it was taught: '…And like a hammer that shatters the rock' (*Jeremiah* 23:29). Just as this hammer produces many sparks, so also may one Biblical

[6] *Genesis Rabbah* 90:1; *Leviticus Rabbah* 24:9. The quote is given in the name of Rabbi Levi. *Midrash* refers to a corpus of Jewish literature, dating from the second to the twelfth centuries C.E., and still of the utmost importance to the practice of Judaism today. The style of Midrash is largely homiletical, and frequently draws on word play to derive a teaching from a scriptural passage. Jewish mysticism largely draws its intellectual roots from the Midrashic imagination (Idel 1993).

verse convey many meanings" (Talmud, *Sanhedrin* 34a). The quintessence of Jewish spirituality lies in the unpacking of the meaning of sacred scripture, and a whole armoury of techniques for decoding those meanings is central to all branches of Judaism, with the most elaborate being found in the Kabbalah. As has been pointed out by many authors, these techniques parallel those used by Freud and others in their attempts to discern unconscious meanings in outward thought, speech and action (see Lancaster 2004 for elaboration of these parallels and for bibliography). Again, the isomorphism between man and God is crucial here, for it implies that the exegetic path to uncovering deeper meaning in the mind of God – the Torah – must apply in parallel fashion to the human mind. In other words, the implication is that concealed meanings in our own thoughts and actions may be made conscious by applying rules of hermeneutic analysis that parallel those used in scriptural exegesis.

These three factors – embodied spirituality, isomorphism between human and divine, and the logic of discerning 'unconscious' meaning – ensure that my term, 'Kabbalistic Psychology' penetrates to the core of the mystical enterprise within Judaism. For Kabbalah, the unconscious is very much the domain for the most profound human-divine encounter, and many kabbalistic practices are directed towards engaging with the *unknown* roots of thought.[7] No wonder, then, that Jung recognised the Maggid of Mezeritch as an illustrious predecessor: He is reported to have said in a 1955 interview, "But do you know who anticipated my entire psychology in the eighteenth century? The Hassidic Rabbi Baer from Meseritz, whom they called the Great Maggid. He was a most impressive man" (Jung 1977, pp. 271–2).

The Hard Problem

> There is no light except that which emerges from darkness.[8]

Earlier I suggested that kabbalistic teachings may be divided into "easy" and "hard" as far as their implications for the links between cognitive neuroscience and mysticism are concerned. The easy teachings can be incorporated into psychological discourse in ways that evade major challenges to our prevailing scientific worldview. That is to say that they may be explored psychologically whilst ignoring any *unmeasurable* factors (e.g., those pertaining to the divine) and any ontological challenges to physicalism. Thus, for example, many kabbalistic practices may be

---

[7] Terminology is always difficult here. Obviously, as soon as one engages with the *unknown* it is no longer 'unknown!' Indeed, the prefix 'un-' in 'unconscious' I find problematic, and for this reason I prefer to use the term 'preconscious,' although this does not fully resolve the problem (for further discussion, see Lancaster 2004). Introspectively, my sense is of a region of mind detached from the everyday imposition of I-centeredness, this detached region being *always already engaged with the 'Mystery'* (to use Ferrer and Sherman's term mentioned earlier). Mystical practice builds bridges between this non-I-centred region of mind and the everyday realm of consciousness (see below).

[8] *Zohar* 2:184a.

understood as employing distinctive linguistic techniques to *deconstruct* normal perceptual categories pertaining to the outer world and the self (Lancaster 2000). Such linguistic deconstruction is a specific case of the *deautomatization* that Deikman (1966) sees as a hallmark of the mystic experience. For the purposes of psychological analysis, the traditional context for such kabbalistic language practices (which holds that the kabbalist is ultimately enjoining with the divine mind in its arcane linguistic activity) may be bracketed off. Similarly, practices that entail breath control, repetitive body movements, chanting, fasting etc. (Idel 1988) may be understood in normative psychophysiological terms as promoting a state of trance (Lancaster 2004). An example would be the intense concentrative practices taught by the thirteenth-century Abraham Abulafia, which have been viewed psychologically in terms of *absorption* and neurologically in terms of activity in the brain's temporo-parietal junction (Arzy et al. 2005).

But what of the *hard problem*? Kabbalistic practices can be viewed alongside those of other traditions as instilling mental discipline, and training skills of attention and associative imagination. For all the "easy" psychology involved here, the central consideration in the texts in which these practices are described is that of *formatting the mind in correspondence with a higher realm*. This touchstone of kabbalistic practice is captured by the thirteenth-century Azriel of Gerona when he writes that all practice comes down to one great principle: "to attune one's thought in faith as if it were attached to the world above" (cited in Tishby 1949/1989, p. 9). It is only by such formatting of one's mind that the mystic becomes a suitable vessel to receive the divine influx from above. In the kabbalistic worldview, when like is attuned to like, then the inner essence flows.

Azriel goes on to specify that such attunement comes about through use of the letters of the divine name, YHVH, for they embody the higher realm. Here we find the core of all kabbalistic practice: "The goal for the kabbalist – indeed what justifies him being called a kabbalist – is to receive the secret of the name, that is, to cleave to YHVH" (Wolfson 2005, p. 2009). The hard problem – transposed from the study of consciousness to the cognitive neuroscience of mysticism – is to account for these ideas of lower and higher realms, and for the principle of correspondence which links them. Indeed, the three factors I enunciated above as underpinning kabbalistic psychology centre on this microcosm-macrocosm relationship. In Wolfson's exquisite language:

> There is a single essence shared by all three [God, human and world], whence follow the corollaries that knowledge of God is equivalent to self-knowledge, and self-knowledge to knowledge of the cosmos. The consubstantiality of self and God in kabbalistic literature … is coupled with the correspondence of the macrocosm and microcosm, the depiction of the world as a "large human" (*adam gadol*) and the human as a "small world" (*olam qaṭan*). God, world, and human are intertwined in a reciprocal mirroring (Wolfson 2005, p. 32).

It bears repeating that these ideas form the very heart of Kabbalah (and of esoteric Christianity, alchemy, and Sufism; indeed they are hardly foreign to many Vedic teachings). Are they simply the dogmatic dross that can be jettisoned in arriving at a cognitive neuroscience of spirituality?

There are two reasons for answering *no* to this question. The first I have already mentioned: the spiritual experiences that we may wish to view positively in psychological terms are typically arrived at in connection with the interpretive framework provided by the ideas of correspondence, microcosm etc. Attempting to explain the experience without reference to the interpretive framework would be like trying to explain the experience of love without addressing the relationship within which it arises. Kabbalistic mystics elevating their core state of mind are the selfsame mystics who are penetrating into a scriptural text understood as explicating some secret of the higher realm. The "enchanted chain" (Idel 2005b) that interconnects lower and higher, is inseparably both the ladder of ascent and the lens through which the mystic's perceptions are focussed. "God, Torah and man share the same structure, and this is the reason why the scholar is able to ascend on high" (Idel 2005b, p. 141).

The second reason for my hesitance in rejecting the microcosm-macrocosm framework central to Kabbalah is that I believe it holds up in the face of recent developments in our understanding of the cognitive neuroscience of consciousness (Lancaster in press). As I have argued elsewhere (Lancaster 2004), there are different levels of explanation through which the mystical material may be examined. At the level of depth psychology, these notions of correspondence might be explored for their *archetypal* significance: a 'correspondence' may underpin the way that images forming in the 'lower' mind recapitulate structures transmitted through a 'higher', collective mind. But such a level of explanation lacks foundation in the substantive research project that addresses the brain's involvement in the mind. I believe that two features of recent developments in the cognitive neuroscience of consciousness provide tentative evidence that the kabbalistic worldview is uncannily grounded in neuroscience. These features are *reflexivity* and *resonance*.

Reflexivity

There is a growing body of evidence to suggest that the neural correlate of consciousness involves *recurrent processing* through which neural activity at 'higher' levels impinges on that at 'lower' levels (Boehler et al. 2008; Dehaene et al. 2006; Edelman and Tononi 2000; Lamme 2003, 2004, 2006). The evidence largely comes from studies of the timing of events in the brain's perceptual systems, using, for example, *trans-cranial magnetic stimulation* to interfere with neural activity in discrete brain areas at specific times (Pascual-Leone and Walsh 2001; see Ro 2010, for review), and from studying paradigms such as *backward masking* (e.g., Supèr et al. 2001; Fahrenfort et al. 2007) and the *attentional blink* (e.g., Sergent et al. 2005). I have argued that the importance of such recurrent processing for consciousness may lie in its role in integrating object-related and self-related processing (Lancaster 1991, 1997, 2004, in press). Such a proposal is supported by recent research demonstrating the role of *reafference* (cf. *recurrent processing*) in bringing about a self-specific perspective in perception (Legrand and Ruby 2009).

Lamme succinctly captures the essence of this principle of reflexivity in his assertion that, "RP [recurrent processing] is the key neural ingredient of consciousness. We could even define consciousness as recurrent processing" (Lamme 2006, p. 499). Whilst agreeing that recurrent processing is the key neural ingredient that correlates with the immediate sense of *access* consciousness, I would refrain from defining consciousness in this way. As argued more fully elsewhere (Lancaster 2004), a definition of consciousness requires that we recognize different *dimensions* of consciousness. Recurrent processing appears to be the key ingredient in the brain mechanisms involved with the dimensions of *intentionality* and *accessibility*. This form of processing does not, however, account for the fundamental dimension of *phenomenality*; it does not answer the hard problem of consciousness.

Resonance

Many have stressed the importance for understanding the neural correlate of consciousness of the *binding problem* (Crick and Koch 1990; Treisman and Schmidt 1982; Treisman 1996). The problem concerns how the brain registers that certain neural responses should be linked with others in order to establish the presence of whole objects in the world. If I am looking at a pen lying on my desk, the *feedforward* visual sweep (i.e., neural activity preceding recurrent activation) will detect a host of features in the sensory array. The question is, on what basis can the brain determine that a subset of these features (e.g., those relating to the pen) belong together?

The emphasis on recurrent processing leads to a straightforward answer, namely that the feedforward sweep does not itself establish the presence of objects. Recognition of objects requires the contribution of recurrent processing. It is likely that the feed-forward system simply detects the presence of basic features in the visual input. On the basis of these features, higher cortical regions connected with the memory store become activated, with those structures (memory traces, or *schemata*) sharing the greatest number of features with the sensory analysis becoming the most highly activated. The *re-entrant* system (neural pathways from higher to lower areas, giving rise to recurrent processing) then modulates the responses of the feedforward system in an attempt to establish whether or not the most activated schemata can match the current input. Again, considerable research underpins the summary view of Enns and di Lollo that the perceptual system, "actively searches for a match between a descending code, representing a perceptual hypothesis, and an ongoing pattern of low-level activity. When such a match occurs, the neural ensemble is 'locked' onto the stimulus" (Enns and di Lollo 2000, p. 348). The stages hypothesized as being involved in the perceptual process are represented in Fig. 1.

Models of this kind have replaced those favoured some 30 years ago that stressed only the *spatial* aspects of brain organization. Previously, the search was for increasing evidence of *localization of function*, with consciousness being seen as merely the most complex in a hierarchy of functions. Over recent years, however, there has been a major shift towards greater emphasis on the temporal dimension of cerebral processing.

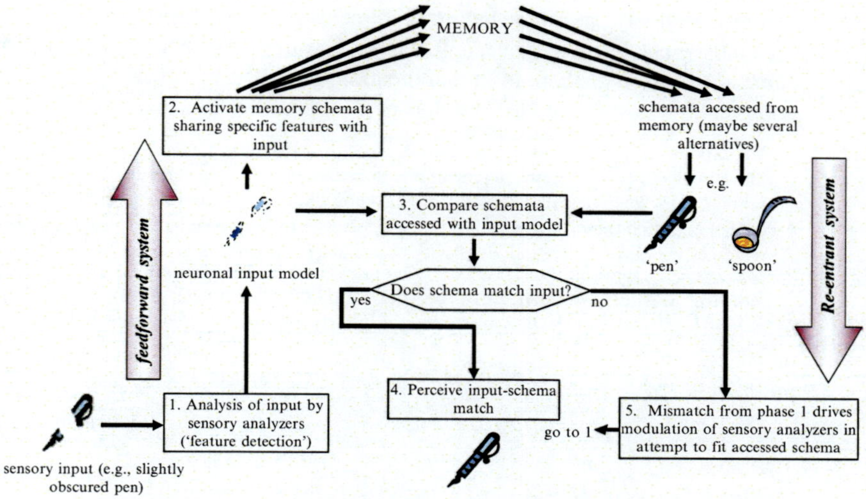

Fig. 1 A psycho-physiological model of stages in perception (based on Lancaster, 2004)

As first proposed by von der Malsburg (1981), it is the temporal dimension of neural signals that underpins the binding of neurons into functional groups. There is thus a principle of *resonance* that seems to be the key for assemblages of neurons to integrate data processing. Much research has demonstrated that phase synchrony, or resonance, in the gamma band (40 Hz approx.) is established between individual neurons, and amongst groups of neurons that are functioning together at a given time (for reviews, see Engel and Singer 2001; Revonsuo 1999; Singer 1999, 2000). Such neural phase synchrony, or *coherence*, is sometimes viewed as a necessary condition for consciousness.

The relation between neural coherence and consciousness is unlikely to be monolithic, however. As Luo et al. (2009) conclude from a recent study, neural coherence "is related to, but not sufficient for, consciousness" (p. 1896). We need to distinguish binding in the feedforward pathway alone, which is unlikely to be the correlate of consciousness, from binding which unifies feedforward and recurrent processing. It is this latter which appears to underlie the brain's relation to consciousness. It is this form of binding which, for example, would be involved in the unification across different orders of cognitive representation that has been proposed as the basis of consciousness (see, for example, Kriegel 2007).[9]

[9] "Conscious states arise from the integration, or unification, of what are initially two distinct representations, a first-order representation of an external stimulus and a higher-order representation of that first-order representation; once the two representations are unified, they form a single representational state with two parts, one directed at the other and the other directed at the stimulus" (Kriegel 2007, p. 899). I would accord the *neuronal input model* in Figure 1 the status of first-order representation, and the schemata accessed from memory, the status of second-order representation.

Introspectively, one of the hallmarks of consciousness seems to be its unity; there is a *oneness* across all the diverse features of awareness at any given time. It should come as no surprise, then, to find unification, signalled by neural coherence, as the brain feature most related to consciousness. As von der Malsburg put it,

> We experience mind states of different degrees of consciousness, and … the difference is made by the difference in the degree of coherence, or order … between different parts of the brain. Let us, then, describe a state of highest consciousness as one characterized by global order among all the different active processes…. A globally coupled state could be one in which all the different [parts] are phase-locked to each other. (1997, pp. 196–197)

Bearing in mind our earlier discussion of the key role of neural reflexivity in relation to consciousness, we must qualify von der Malsburg's assertion with the proviso that the global coupling entails interactions between feedforward and recurrent neural processing.

Reflexivity and Resonance in Mystical Language

> Come and see. Through the impulse from below is awakened an impulse above, and through the impulse from above there is awakened a yet higher impulse, until the impulse reaches the place where the lamp is to be lit and it is lit … and all the worlds receive blessing from it. (*Zohar* 1: 244a)

In this section I shall argue that the above two core principles of brain function related to consciousness accord with kabbalistic teachings. It is for the reader to consider the persuasiveness of the arguments. But, should the parallels between core teachings of Kabbalah and these features of neuroscience be convincing, then we must address their significance. Putting it simply, there would seem to be three alternative ways to explain the parallels: They may be attributable (1) to chance (which I doubt); (2) to the mystics having uncannily accurate insight into brain function (which I also doubt); or (3) to the mystics' grasp of principles finding expression at different levels in the 'created' hierarchy – due to the isomorphism I discussed earlier. It is this latter alternative which will be further explored in the final section of the chapter.

As Scholem noted, the essential idea of the impulse from below stimulating that from above – as portrayed in the above extract – is central to the *Zohar's* narrative (Scholem 1941/1961, p. 233). Activity at the lower ontic level is "magically reflected" at the higher level. If the lower impulse is acceptable, then the response is "blessing," i.e., "light from the supernal lamp," or "oil emerging from *The Holy Ancient One*." There are many metaphors to describe this influx from the higher level. All of them may be best understood in modern terms as concerned with the arising and flow of consciousness.[10]

[10] A number of modern scholars use the term *consciousness* or *awareness* in attempting to render into contemporary language the *Zohar's* symbolic language. See, for example, Giller (2001), Magid (2002), Matt (1995).

The picture emerging from cognitive neuroscience parallels this more cosmic picture. As discussed above, intentionality and access consciousness seem to be dependent on recurrent processing. The impulse from below (detection of elemental features in sensory activity and their integration in a *neuronal input model*) brings about higher activity (memory readout and self-related processing), which acts back on the lower activity, bringing consciousness of the perceptual object. Stated in this crude form, however, the analogy may be less than convincing. A deeper analysis is required to support the substance of the analogy.

We may start by noting that the parallel extends into the *means* whereby recurrent processing is effected. As discussed above, this seems to depend on the binding mechanisms achieving phase synchrony among neurons. Analogously, at the cosmic level, awakening of the higher influx depends upon unification of the lower "limbs": "Whenever all the parts of the body are brought together in a single bond enjoying pleasure and delight from the head, above and below..., then he becomes a flowing river going out of the real Eden" (*Zohar* 1:247b). The "flowing river" from "Eden" is another symbol of the influx from the higher level.[11] Indeed, the theme of sexual intercourse, whereby the disparate parts are bound together *par excellence*, is a favoured image in the *Zohar*'s discourse. Unification below brings about a celestial uniting that eventuates in an outpouring from above that nourishes those below.

Indeed, practices directed at unification are central to the goals of Kabbalah. A description of one of these practices in the *Zohar* strikes a resonant chord with von der Malsburg's reference to "globally coupled" states quoted above:

> "One" - to unify everything from there upwards as one; to raise the will to bind everything in a single bond; to raise the will in fear and love higher and higher as far as *En-Sof* [the limitless essence of God]. And not to let the will stray from all the levels and limbs but let it ascend with them all to make them adhere to each other, so that all shall be one bond with *En-Sof* (*Zohar* 2: 216b)

Further examination of the concepts of *lower* and *higher* in the respective contexts will inform the comparison. Superficially, of course, the neuroscientific and the kabbalistic versions appear highly disparate. What can brain regions and functions possibly have in common with supposed mystical planes of reality? However, a deeper grasp of the kabbalistic symbolism indicates that the *earthly* (lower) and *heavenly* (higher) spheres include features that do bring them into alignment with their proposed neurocognitive counterparts.

The lower level is in touch with the earth – it is the level of human worldly activity. This parallels the lower brain regions whose activity is driven by the "impulse from below," i.e., from the energy of the physical world impinging on sensory receptors. The higher regions, in neurocognitive terms, are those areas responsible for the memory readout which guides the perceptual process. The parallel here is to the psychological role assigned in kabbalistic thought to the sphere of *Ḥokhmah*

---

[11] It is difficult in a short treatment of kabbalistic imagery to substantiate fully my claims about the intended meanings in passages such as this. Indeed, concealment of meaning is one of the hallmarks of the medieval Kabbalah. I have explored this issue at greater length in Lancaster (2005).

("Wisdom"), the highest emanation in the kabbalistic hierarchy attainable by the human mind. This is the sphere to which the Maggid of Mezeritch applied the term *kadmut ha-sekhel*, "pre-" or "un-conscious," (see Hurwitz 1968; Matt 1995; Scholem 1975). The Maggid is alluding to the higher level of (preconscious) thought that functions to format (conscious) thoughts in the human mind:

> **In the vessel which a craftsman makes, you find that wisdom exists in a concealed fashion.** So too with thought, for thought requires a preconscious [*kadmut ha-sekhel*] which is above **the thought that thinks.... Hence we find it written** (*Job* 28:12) that *"Wisdom comes from nothing"*[12] (Maggid of Mezeritch, *Or ha-Emet* 15b. Cited in Scholem 1975, p. 355).

The higher level in both the neurocognitive model and in kabbalistic thought depicts the preconscious as the "concealed" inner process of thought.[13] In Zoharic symbolism, this level is that of the thought that precedes expansion into articulation. Similarly, in the neuro-cognitive model, the preconscious activation of matching memory schemata is clearly a kind of thinking that occurs prior to the meaning which accompanies a match between sensory input and memory readout.

In summary, it seems that the distinctive pattern of operation which Kabbalah discerns in the cosmic scheme is found, as it were in microcosmic form, in the neural systems that relate to consciousness. In this sense, then, it could be argued that recent neurocognitive research lends credence to the core mystical assertion that the microcosm recapitulates the macrocosm.

Which comes first ... above or below...?

Tikkun: Repairing the Split

> The highest wisdom that is concealed in the head of the Holy Ancient One is called the supernal brain, the hidden brain, the brain that is tranquil and silent; and no-one knows it other than He Himself.... When the white brilliance is formed in the light, it distils its **essence into this brain, which is illumined; and there emanates from this precious influence** another brain which expands and shines into 32 paths.

[12] This verse from the book of Job is often translated as "From where may wisdom be found?" The Hebrew translated as "from where," if taken more literally, means "from nothingness." The mystics emphasised this latter meaning since it accords with their understanding that the sphere of Wisdom may be accessed only through annulment of the everyday sense of 'I'; "Transformation comes about only by passing through nothingness," writes The Maggid (cited in Matt 1995, p. 87).

[13] The point may be misunderstood on account of confusion over the appropriate direction in the spatial metaphor applied to notions of consciousness. Freud famously viewed the unconscious as *lower* – the portion of an iceberg under water, the basement of a house, etc. However, as Whyte (1962) pointed out, the unconscious might be thought of as *higher* than the conscious sphere on account of its importance for higher creative and spiritual abilities. It is unfortunate that we are compelled to understand these psychic ideas through spatial metaphor, since we confuse the metaphor with the meaning. There is no spatiality in the psyche. Kabbalistically, higher means *closer to the divine*. But, the parallel with neuro-cognitive terminology arises by virtue of the critical idea that the terms *higher* and *closer to the divine* mean that the process comes *earlier* in the generation of mental content. This is essentially the meaning of Dov Baer's term *kadmut ha-sekhel*, which is why it should be translated as *pre*conscious rather than *un*conscious.

> The light of wisdom expands in its 32 directions emanating from the light that is in the concealed brain. There is hewn from wisdom a light that streams forth and proceeds to water the garden. It enters into the head of the "Small Face," and forms a certain other brain. And from there it is extended and flows into the whole body, and waters all those plants, as it is written: "And a river went out of Eden to water the garden...." (*Genesis* 2:10) (*Zohar* 3:288a, 289b [extracts from *Idra Zuta Kadisha*]) .

The "enchanted chain" of kabbalistic teaching may be thought of as holographic inasmuch as all parts are reflective of the whole. The overarching principles of the whole chain are those we have examined: ***reflexivity***, whereby the lower impulse triggers the higher, which in turn infuses the lower; and *resonance*, which binds the parts into a harmonious whole. That we find these principles at work in the human brain in relation to consciousness provides, I suggest, circumstantial evidence for the potential validity of the kabbalistic scheme. What does it mean if we accept this circumstantial evidence? What may be the implications for a cognitive neuroscience of spirituality? In relation to the three alternatives enunciated earlier, I shall explore the third, namely that the mystics grasped an underlying ground plan which finds expression in all working systems, notably in our case, the system of the human brain.

The above extract from the *Zohar* lays out the ground plan in terms of a hierarchy of "brains" in the godhead, of which our human brain is a lower recapitulation. In terms of the "spiritual anatomy" I referred to earlier, the spinal cord plus 31 pairs of spinal nerves (total = 32) into which the human brain "expands" recapitulates the 32 paths into which the "higher" brain shines. Recurrent processing and phase synchrony amongst neurones recapitulate the relationship between the brains of the godhead (those of the "small face" and the "large face", the latter term being used interchangeably with "Holy Ancient One") in terms of reflexivity and unification. Whilst the physiognomic imagery of the *Zohar* is clearly not to be taken literally, kabbalistic symbolism generally sees physical entities as embodying the higher principles operating in the ***sefirotic*** pleroma. Thus it is implicit in the *Zohar*'s scheme that the human brain functions by the same principles as those operating with the 'higher brains'.

The "white brilliance" of the "light" enters the hierarchy from its highest level, from where it streams down to infuse the whole system with spiritual vitality. "The worlds ... exist in gradations, one atop the other ... until they all ascend to the secret of the awesome faces whence the light emerges," writes Moses de León, reputed to be the author/editor of the *Zohar* (cited in Wolfson 2005, p. 34). I suggest that the "light" refers to what we would describe as the essence of consciousness, namely *phenomenality*. Whilst the mundane contents of consciousness are formed in terms of the interactions between the brain, its memory systems and the physical world, this elusive quality – phenomenality that makes experience what it is – derives from a higher level.

Time to take stock: The evidence accrued in this chapter may be summarised in three statements.

- The hard problem for a psychological approach to Kabbalah concerns the latter's teaching of isomorphism across ontologically distinct levels in the created hierarchy;
- The hard problem for the psychology of consciousness is *phenomenality*;

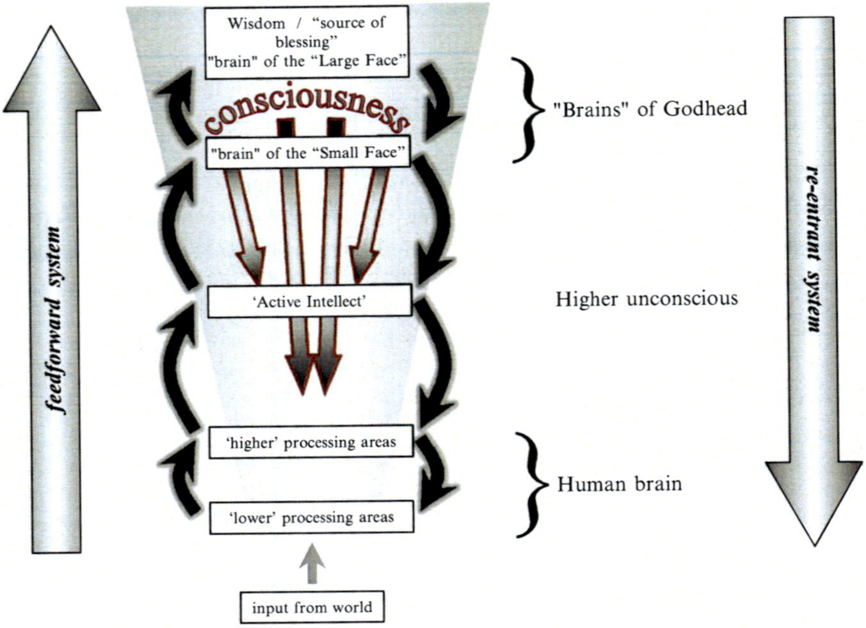

Fig. 2 **An extended view of reflexivity and consciousness**

- The cognitive neuroscience of consciousness lends some credibility to the major principles which are said to operate in the kabbalistic scheme.

These lead me to pose the following question, which I regard as central to the nascent cognitive neuroscience of spirituality:

- Might cognitive neuroscience draw some inspiration from the kabbalistic system in attempting to address its hard problem?

In Fig. 2, I present a stylised depiction of the kabbalistic scheme in which I include the material deriving from the cognitive neuroscience of consciousness. What we know of the workings of the *physical* brain regarding consciousness occupies the two lower levels of the figure; the kabbalistic teachings regarding the "brains" in the Godhead, the upper two levels. Between these I have included the *Active Intellect*. The figure intentionally depicts the isomorphism between brain mechanisms related to consciousness and the kabbalistic scheme whereby the highest levels in the created hierarchy are activated. As stated in the *Zohar* (1:70b) in the name of Rabbi Ḥizkiyah: "The upper world depends upon the lower, and the lower upon the upper."

In the figure I have designated the *Active Intellect* as the 'higher unconscious', the term Assagioli used to depict that sphere from whence our highest intuitions and inspirations derive (Firman and Gila 2002). It is portrayed in the figure as interacting with the human brain in the same reflexive way as operates within the brain's perceptual systems. In the previous section I emphasized the preconsciousness of higher activity, in both cognitive, and kabbalistic, schemes. Just as upward activation in the brain preconsciously activates memories and associations pertaining to the *individual's* prior experience relating to the current sensory array, so the feedforward system may be viewed as activating *collective* resonances in the higher unconscious /Active Intellect. Still higher feedforward resonances open the portal of phenomenality. The final effect is that phenomenality as well as personal and collective projections enter the narrative mind of mundane consciousness via the downward re-entrant system.

A non-reductive operational framework such as this will always beg questions about ontology. Is the hard problem answered by suggesting that phenomenality derives from a higher dimension of reality? Surely the problem is just pushed into the area of faith. My answer to this is that it is already an act of faith to assume that physicalism can answer the problem. The evidence presented here for interrelating cognitive neuroscience and kabbalistic mysticism gives a basis for the extended framework I have presented. Moreover, what exactly the 'godhead' is may be open to discussion. The notion of 'higher brains' is but imagery locked into a particular cultural formulation; today we may be more comfortable with a formulation that places such consciousness beyond the human brain within a naturalistic framework. The microcosmic-macrocosmic correspondences I have discussed may be a further example of the "parsimony of nature" that Hunt sees at work in the way that consciousness recapitulates principles found at the quantum level: "Consciousness appears in the universe as the most hierarchically complex system we know, or possibly could know, and which re-creates, on its new emergent level, principles first manifested on quantum and nonlinear systems levels" (Hunt 2001, p. 36).

Many questions remain. My argument is simply that the explanatory gap seems unbridgeable when we see only the two extremes of matter and phenomenality (or brain and spirit). The gap is made no smaller when, in the name of spirituality, we appeal to an *Absolute Other*. The 'otherness' simply widens the gap! The central point I wish to make is that we make progress towards bridging the gap when we explore the intermediary realms between the extremes, between the unknowable essence of God and the scientifically knowable world of nature. This has been the major interest of most strands of mysticism throughout the world, and, as I have argued here, represents a fruitful area of dialogue with cognitive neuroscience (and other areas of science such as quantum mechanics and astrophysics).

In closing let me emphasise that the dialogue between cognitive neuroscience and spirituality unfolds within a larger context. When we ask questions about the neural and cognitive effects of spiritual practice, it is not simply because we hope to fill a gap in our understanding. It is generally because we intuit that such practice is essentially good for us; we seek *confirmation* through science. In Kabbalah, that larger context is given the name, *Tikkun*, meaning the restoration of a divinely-intended level of harmony to all things. In an outward sense, this restoration involves concern for our world and for the well-being of our fellow humans. More inwardly, *tikkun* is promoted when we embrace the path of unification, recognising the spiritual essence that penetrates the physical realm, and the mirroring relationship that pertains between the physical and the spiritual realms of being. "[F]or kabbalists," quips Wolfson, "the mirror is a medium that renders appearances real and reality apparent" (Wolfson 2005, p. 33).[14] Discerning the spiritual dimensions of brain function inevitably challenges our sense of what is 'real.' In responding to that challenge, and exploring the relationship between the microcosm of the human brain and the macrocosm, we begin to heal the split which Christian thinkers imposed between the physical and spiritual realms. We can contribute to re-enchanting the world, and in so doing perpetuate the work of *tikkun*.

References

Abrams, D. (1994). *The book Bahir: An edition based on the earliest manuscripts*. Los Angeles: Cherub Press.

Aftanas, L. I., & Golocheikine, S. A. (2002). Non-linear dynamic complexity of the human EEG during meditation. *Neuroscience Letters, 330*(2), 143–146.

Arzy, S., Idel, M., Landis, T., & Blanke, O. (2005). Speaking with one's self: Autoscopic phenomena in writings from the ecstatic Kabbalah. *Journal of Consciousness Studies, 12*(11), 4–29.

Assagioli, R. (1993). *Transpersonal development: The dimension beyond psychosynthesis*. London: Thorsons.

Barušs, I. (2001). The art of science: Science of the future in light of alterations of consciousness. *Journal of Scientific Exploration, 15*(1), 57–68.

Beauregard, M., & O'Leary, D. (2007). *The spiritual brain: A neuroscientist's case for the existence of the soul*. New York: HarperCollins.

Beauregard, M., & Paquette, V. (2006). Neural correlates of a mystical experience in Carmelite nuns. *Neuroscience Letters, 405*(3), 186–190.

Beauregard, M., & Paquette, V. (2008). EEG activity in Carmelite nuns during a mystical experience. *Neuroscience Letters, 444*(1), 1–4.

Boehler, C. N., Schoenfeld, M. A., Heinze, H.-J., & Hopf, J.-M. (2008). Rapid recurrent processing gates awareness in primary visual cortex. *Proceedings of the National Academy of Sciences of the United States of America, 105*, 8742–8747.

Boyer, P. (2001). *Religion explained: The evolutionary origins of religious thought*. New York: Basic Books.

Boyer, P., & Bergstrom, B. (2008). Evolutionary perspectives on religion. *Annual Review of Anthropology, 37*, 111–130.

Cahn, B. R., & Polich, J. (2006). Meditation states and traits: EEG, ERP, and neuroimaging studies. *Psychological Bulletin, 132*, 180–211.

Chalmers, D. (1995). Facing up to the problem of consciousness. *Journal of Consciousness Studies, 2*, 200–219.

[14] It is worth noting in passing that isomorphism as presented in Kabbalah is conceptually distinct from cognitivism's *representationalism*, inasmuch as the latter entails an arbitrary relation between the representation and that represented. Kabbalah asserts that the "mirror" that relates two entities (such as God and human) entails an identity of substance. Indeed, it is axiomatic that such identity is critical for any knowledge; man can know God and God can know man only because they share an essential nature. As Wolfson (2005) remarks, this axiom implies ultimately that there is no non-divine reality. And, we might add, if there is no non-divine reality there is no explanatory gap!

Crick, F. H. C., & Koch, C. (1990). Towards a neurobiological theory of consciousness. *Seminars in the Neurosciences, 2*, 263–275.

Dehaene, S., Changeaux, J.-P., Naccache, L., Sackur, J., & Sergent, C. (2006). Conscious, preconscious, and subliminal processing: A testable taxonomy. *Trends in Cognitive Sciences, 10*(5), 204–211.

Deikman, A. J. (1966). Deautomatization and the mystic experience. *Psychiatry, 29*, 324–338.

Edelman, G. M., & Tononi, G. (2000). Reentry and the dynamic core: Neural correlates of conscious experience. In T. Metzinger (Ed.), *Neural correlates of consciousness: Empirical and conceptual questions* (pp. 138–151). Cambridge: The MIT Press.

Engel, A. K., & Singer, W. (2001). Temporal binding and the neural correlates of sensory awareness. *Trends in Cognitive Sciences, 5*(1), 16–25.

Enns, J. T., & di Lollo, V. (2000). What's new in visual masking? *Trends in Cognitive Sciences, 4*(9), 345–352.

Fahrenfort, J. J., Scholte, H. S., & Lamme, V. A. (2007). Masking disrupts reentrant processing in human visual cortex. *Journal of Cognitive Neuroscience, 19*(9), 1488–1497.

Ferrer, J. N., & Sherman, J. H. (2008). Introduction: The participatory turn in spirituality, mysticism, and religious studies. In J. N. Ferrer & J. H. Sherman (Eds.), *The participatory turn: Spirituality, mysticism, religious studies* (pp. 1–78). Albany: State University of New York Press.

Firman, J., & Gila, A. (2002). *Psychosynthesis: A psychology of the spirit*. Albany: State University of New York Press.

Giller, P. (2001). *Reading the Zohar: The sacred text of the Kabbalah*. Oxford: Oxford University Press.

Hellner-Eshed, M. (2009). *A river flows from Eden: The language of mystical experience in the Zohar* (N. Wolski, Trans.). Stanford, CA: Stanford University Press. (Originally published 2005)

Hunt, H. (2001). Some perils of quantum consciousness. *Journal of Consciousness Studies, 8*(9–10), 35–45.

Hurwitz, S. (1968). Psychological aspects in early Hasidic literature (H. Nagel, Trans.). In J. Hillman (Ed.), *Timeless documents of the soul* (pp. 149–239). Evanston: Northwestern University Press.

Idel, M. (1988). *The mystical experience in Abraham Abulafia* (J. Chipman, Trans.). Albany: State University of New York Press.

Idel, M. (1993). Midrashic versus other forms of Jewish hermeneutics: Some comparative reflections. In M. Fishbane (Ed.), *The Midrashic imagination: Jewish exegesis, thought, and history* (pp. 45–58). Albany: State University of New York Press.

Idel, M. (2005a). *Kabbalah and Eros*. New Haven: Yale University Press.

Idel, M. (2005b). *Enchanted chains: Techniques and rituals in Jewish mysticism*. Los Angeles: Cherub Press.

Jung, C. G. (1977). In W. McGuire & R. F. C. Hull (Eds.), *C. G. Jung speaking: Interviews and encounters*. Princeton: Princeton University Press.

Kriegel, U. (2007). A cross-order integration hypothesis for the neural correlate of consciousness. *Consciousness and Cognition, 16*(4), 897–912.

Lamme, V. A. F. (2003). Why visual attention and awareness are different. *Trends in Cognitive Sciences, 7*(1), 12–18.

Lamme, V. A. F. (2004). Separate neural definitions of visual consciousness and visual attention: A case for phenomenal awareness. *Neural Networks, 17*(5–6), 861–872.

Lamme, V. A. F. (2006). Towards a true neural stance on consciousness. *Trends in Cognitive Sciences, 10*(11), 494–501.

Lancaster, B. L. (1991). *Mind, brain and human potential: The quest for an understanding of self*. Shaftesbury: Element Books.

Lancaster, B. L. (1997). On the stages of perception: Towards a synthesis of cognitive neuroscience and the Buddhist *Abhidhamma* tradition. *Journal of Consciousness Studies, 4*(2), 122–142.

Lancaster, B. L. (2000). On the relationship between cognitive models and spiritual maps: Evidence from Hebrew language mysticism. *Journal of Consciousness Studies, 7*(11–12), 231–250.

Lancaster, B. L. (2004). *Approaches to consciousness: The marriage of science and mysticism*. Basingstoke: Palgrave Macmillan.

Lancaster, B. L. (2005). *The essence of Kabbalah*. London: Arcturus.

Lancaster, B. L. (in press). The cognitive neuroscience of consciousness, mysticism and psi. *International Journal of Transpersonal Studies*.

Legrand, D., & Ruby, P. (2009). What is self-specific? Theoretical investigation and critical review of neuroimaging results. *Psychological Review, 116*(1), 252–282.

Levine, J. (1983). Materialism and qualia: The explanatory gap. *Pacific Philosophical Quarterly, 64*, 354–361.

Luo, Q., Mitchell, D., Cheng, X., Mondillo, K., Mccaffrey, D., Holroyd, T., Carver, F., Coppola, R., & Blair, J. (2009). Visual awareness, emotion, and gamma band synchronization. *Cerebral Cortex, 19*, 1896–1904.

Lutz, A., Greischar, L. L., Rawlings, N. B., Ricard, M., & Davidson, R. J. (2004). Long-term meditators self-induce high-amplitude gamma synchrony during mental practice. *Proceedings of the National Academy of Sciences of the United States of America, 101*(46), 16369–16373.

Lutz, A., Dunne, J. D., & Davidson, R. J. (2007). Meditation and the neuroscience of consciousness: An introduction. In P. D. Zelado, M. Moscovitch, & E. Thompson (Eds.), *The Cambridge handbook of consciousness* (pp. 499–551). Cambridge: Cambridge University Press.

Lutz, A., Slagter, H. A., Dunne, J. D., & Davidson, R. J. (2008). Attention regulation and monitoring in meditation. *Trends in Cognitive Sciences, 12*(4), 163–169.

Magid, S. (2002). Origin and overcoming the beginning: *Zimzum* as a trope of reading in post-Lurianic Kabbala. In A. Cohen & S. Magid (Eds.), *Beginning again: Toward a hermeneutic of Jewish texts* (pp. 163–214). New York: Seven Bridges Press.

Matt, D. (1995). Ayin: The concept of nothingness in Jewish mysticism. In L. Fine (Ed.), *Essential papers on Kabbalah* (pp. 67–108). New York: New York University Press.

Metzinger, T. (2003). *Being no one: The self-model theory of subjectivity*. Cambridge: Bradford.

Newberg, A., Alavi, A., Baime, M., Pourdehnad, M., Santanna, J., & D'Aquili, E. (2001a). The measurement of regional cerebral blood flow during the complex cognitive task of meditation: A preliminary SPECT study. *Psychiatry Research, 106*(2), 113–122.

Newberg, A., D'Aquili, E., & Rause, V. (2001b). *Why God won't go away: Brain science and the biology of belief*. New York: Ballantine Books.

Pascual-Leone, A., & Walsh, V. (2001). Fast backprojections from the motion to the primary visual area necessary for visual awareness. *Science, 292*(5516), 510–512.

Persinger, M. A. (1987). *Neuropsychological bases of God beliefs*. New York: Praeger.

Revonsuo, A. (1999). Binding and the phenomenal unity of consciousness. *Consciousness and Cognition, 8*(2), 173–185.

Ro, T. (2010). What can TMS tell us about visual awareness? *Cortex, 46*(1), 110–113. doi:10.1016/j.cortex.2009.03.005.

Rothberg, D. (2000). Spiritual inquiry. In T. Hart, P. Nelson, & K. Puhakka (Eds.), *Transpersonal knowing: Exploring the horizon of consciousness* (pp. 161–184). Albany: State University of New York Press.

Scholem, G. G. (1941/1961). *Major trends in Jewish mysticism*. New York: Schocken Books.

Scholem, G. (1975). *Devarim be-Go*. Tel Aviv: Am Oved. (Hebrew).

Sergent, C., Baillet, S., & Dehaene, S. (2005). Timing of the brain events underlying access to consciousness during the attentional blink. *Nature Neuroscience, 8*(10), 1391–1400.

Shokek, S. (2001). *Kabbalah and the art of being*. London: Routledge.

Singer, W. (1999). Neuronal synchrony: A versatile code for the definition of relations? *Neuron, 24*(1), 49–65.

Singer, W. (2000). Phenomenal awareness and consciousness from a neurobiological perspective. In T. Metzinger (Ed.), *Neural correlates of consciousness: Empirical and conceptual questions* (pp. 121–137). Cambridge: The MIT Press.

Supèr, H., Spekreijse, H., & Lamme, V. A. F. (2001). Two distinct modes of sensory processing observed in monkey primary visual cortex (V1). *Nature Neuroscience, 4*, 304–310.
Tishby, I. (1949/1989). *The wisdom of the Zohar: An anthology of texts.* (D. Goldstein, Trans., 3 Vols.). Oxford: Oxford University Press.
Treisman, A. M. (1996). The binding problem. *Current Opinions in Neurobiology, 6*, 171–178.
Treisman, A. M., & Schmidt, H. (1982). Illusory conjunctions in the perception of objects. *Cognitive Psychology, 14*(1), 107–141.
von der Malsburg, C. (1981). The correlation theory of brain function. Internal report 81–2, MPI biophysical chemistry. Reprinted in E. Domany, J. L. van Hemmen, & K. Schulten (Eds.), *Models of neural networks* II (pp. 95–119). Berlin: Springer (1994).
von der Malsburg, C. (1997). The coherence definition of consciousness. In M. Ito, Y. Miyashita, & E. T. Rolls (Eds.), *Cognition, computation, and consciousness* (pp. 193–204). Oxford: Oxford University Press.
Whyte, L. L. (1962). *The unconscious before Freud.* London: Tavistock.
Wilber, K. (2006). *Integral spirituality: A startling new role for religion in the modern and post-modern world.* Boston/London: Integral Books.
Wolfson, E. R. (1994). *Through a speculum that shines: Vision and imagination in medieval Jewish mysticism.* Princeton: Princeton University Press.
Wolfson, E. R. (2004). Hermeneutics of light in medieval Kabbalah. In M. T. Kapstein (Ed.), *The presence of light: Divine radiance and religious experience* (pp. 105–118). Chicago: University of Chicago Press.
Wolfson, E. (2005). *Language, Eros, being: Kabbalistic hermeneutics and poetic imagination.* New York: Fordham University Press.
Zohar. (1978). Margoliot, R. (Ed.), (6th edn., 3 Vols.). Jerusalem: Mosad ha-Rav Kook.

Towards a Neuroscience of Spirituality

Wayne B. Jonas

Abstract Neuroimaging technology has revitalized the study of the brain and with it the opportunity to examine spiritual phenomena has also emerged. However, what is the difference between neurospirituality and neuropsychology? Distinguishing between brain measurements of spiritual phenomena from the psychological aspects of religion is key to defining the field of neurospirituality. In this article, two paths are suggested to distinguish these fields. One path is to explore the neurological correlates of characteristics that spiritual traditions claim for the Divine such as omnipresence, omnipotence, eternality, omniscience, a causality, and beneficence. A second path is examining neurological changes during spiritual healing practices when the practice involves emersion into Divine love. In both cases the interaction with and alteration of the arrow of time is a key distinguishing factor.

Introduction

In the last 20 years, and especially in the last decade, tremendous advances in technologies for imaging the brain have occurred. These include functional magnetic resonance imaging (fMRI), positron emission topography (PET), superconducting quantum interference device (SQUID), magnetic electroencephalogram (MEG), fast computed topography (FCT), and others. These technologies are advancing rapidly, allowing for real time assessment of various brain structures and functions. They allow for visually imaging of electro-magnetic phenomena, and the chemistry, anatomy and dynamics of different brain structures. The advances are

W.B. Jonas (✉)
Samueli Institute, King Street 1737, Alexandia, VA 22314, USA
e-mail: wjonas@siib.org

H. Walach et al. (eds.), *Neuroscience, Consciousness and Spirituality*,
Studies in Neuroscience, Consciousness and Spirituality 1,
DOI 10.1007/978-94-007-2079-4_15,

such that we can now begin to correlate brain images with what heretofore have been psychological phenomena such as expectation, emotion, perception, and meaning. These technologies open up new opportunities for the examination of spirituality and spiritual experiences.

Spiritual or Psychological Neuroscience?

The overlap between psycho-religious constructs such as altruism, forgiveness, belief, and love are so close that what is often called the *neuroscience of spirituality* is indistinguishable from the neuroscience of emotional and psychological experiences. Emotional and psychological experiences are generated from spiritual behaviors, but they are also generated from social and cognitive behaviors that are devoid of spiritual content. Yet many people and religious traditions make a distinction between spiritual and psychological phenomenon. Thus, specific experiences that might only occur only during spiritual behavior may involve different brain functions. Understanding those differences could also help us develop a better understanding of spirituality itself.

What, however, would distinguish between a neuroscience of spirituality and a neuroscience of psychology? How can the two be distinguished? In this paper I propose two approaches. One approach is to define those phenomena that appear to be characteristics of the "transcendent" and are not associated with social interaction alone. By defining these characteristics of the "Divine" as they are experienced by humans, one could examine the associated neurological phenomena. The second approach would be to explore the neuroscience of spiritual healing; that is, healing phenomena that occur in relationship to the Divine rather than in relationship to a particular device or instrumental placebo treatment. While the neurological phenomena may look similar between these and psychological processes, there also may be differences especially if we begin to look at whether such neurological phenomena deviate from the normal processes involved in psychological space and time or from the processes accompanied by the placebo effect. I describe these two approaches to the development of a neuroscience of spirituality and attempt to distinguish it from neuropsychology, and suggest that the key distinguishing feature is related to time perception.

Neuroscience of Divine Characteristics

Most major religions describe the characteristics of Divine phenomena. These include: omnipresence, omnipotence, eternality, non-material nature, omniscience, acausality, and beneficence. While various religious and spiritual traditions describe these characteristics in different ways, they share some core, and cross-cutting aspects. These include an experience of unity and transcendence not due to social bonding and a lack of sense of a separate identity and ego. One of the distinguishing features of spirituality, as Walach has described elsewhere in this book, involves the experience of transcending one's ego or individual identity (see the chapter by Walach et al. on page 1–21 in this book). Most spiritual traditions do not limit this description to simply a perceptual transcendence, that is, a sense of being one with the universe. Rather, they describe this transcendence as real in the sense that both, metaphorical concepts and physical phenomena arise from the unitary spiritual nature of the human being. Thus an examination of the neurological correlates when these Divine characteristics occur would be one of the first approaches to defining a neuroscience of spirituality.

Defining Spiritual Characteristics

For the purposes of illustration, I define the following as Divine characteristics. *Omnipresence* is the quality of being everywhere at the same time. *Omniscience* is infinite knowledge as evidenced by the ability to know something beyond the normal spacial and temporal boundaries. *Eternal* is existing without or having no beginning or end; forever; in the present moment. Fundamentally this means non-locality in time. *Acausal* is an agent or force producing an effect in neutral or reverse time. Finally, *beneficence* or *love* is a deep, unconditional, affection or appreciation for another, including the love of God towards mankind.

Spiritual literature describes these characteristics and the essential unity of the Divine as core spiritual characteristics. For example, in the spiritual writings of George Fox, founder of the Quaker movement, he describes his experiences in encountering the Divine as an omniscient experience in this way:

> One day, when I had been walking solitarily abroad and was come home, I was taken up in the love of God so that I could not but admire the greatness of his love and while I was in that condition it was opened onto me by the eternal light and the power and I therein saw clearly that all was done and was to be done in and by Christ. (Steere 1984)

Omnipotence

Various religious and spiritual traditions take different paths to accessing this unity and dealing with spiritual experiences. When these experiences occur, the neuroscience of those phenomena then becomes a pathway to defining the neuroscience of spirituality.

Let's take an example of these characteristics and explore how they might be evaluated with new neuroscience technology. First, omnipotence; A large body of research has examined the ability of consciousness to affect events at a distance and at different times. These mind-matter interactions are associated with a body of research in parapsychology developed, over several decades, into a rigorous methodology for defining when such interactions at a distance occur (Samueli Institute 2007). This interaction at a distance is known to occur in the microcosm of quantum physics, but is largely unexplored on the macrocosm of biology and neuroscience. Research has now demonstrated that mind-matter interaction can occur at a distance. This data was reviewed rigorously several years ago (Jonas and Crawford 2003). The largest body of studies examines the effect of the mind on random number generators. These studies show that a small, but highly statistically significant effect occurs when an individual intends to influence chance events of a variety of types. These effects, especially when studying automated and computer driven "random event generators" or REGs, have been replicated in over 516 experiments with over 90 different first authors (Radin and Nelson 2003). While some reviews have questioned the statistical stability of these phenomena, it is still possible to measure when they have occurred in any specific data set (Bösch et al. 2006). It would appear that some aspect of our consciousness is able, at times to influence phenomena on a non-local level, demonstrating the characteristic of omnipotence. Since it is now possible to use neuroimaging technologies to examine the brain changes that occur with intention, the same technologies when applied to the impact of intention on physical phenomena non-locally (such as REGs) could be part of the neuroscience of spirituality.

Eternality

What about the characteristic of eternality or timelessness? A similar approach could be taken to examine brain changes during that experience. Dean Radin and others have looked at using Markoff chain models to examine the direction of influence during intention. They found that under certain conditions, intention follows pathways more compatible with non-temporal paths, illustrating that some aspect of our consciousness can travel independent of time (Radin and Borges 2009). For example, we might image the dynamics of the brain that change when the Markoff chain demonstrates reverse causation from the neuropsychological component of intention when forward causation or no intentionality influence occurs.

Omniscience

Another Divine characteristic, omniscience, could be examined by looking at the experiments done with remote viewers. Remote viewing is a phenomenon that was extensively examined during the Soviet-American Cold War to see if individuals could access information through non-sensory means and independent of time and space, that is, in an omniscient manner. Again, through a series of rigorous and replicated experiments, it was shown that certain individuals were especially skilled at accessing remote information in this way and that this skill was possibly trainable (Utts and May 2003). Thus, one could explore the neuroscience of omniscience when it is demonstrated through remote viewing as another example of the neuroscience of spirituality and distinguish it from studies of local meditative or imaginary behaviors that inhabit the realm of neuropsychology.

Neuroimaging Measurement of Divine Phenomena

These "Divine characteristics" correspond with the reported effects from a number of spiritual practices such as Shamanism; Jewish and Greek Orthodox mystical traditions; descriptions of Hindi Yogi Masters, and non-spiritual practices such as parent-child or friend-friend communication; pets who appear to access non-local information from their masters; and other claims such as those made by fortune tellers, psychics, and prophets. While many of these are not quantifiable, the remote viewing experiments use careful control of the receiver, the target, sensory parameters and blind evaluation scoring. How could these be measured neurologically? A number of attempts to assess non-local phenomena through neuroscience have been made. These attempts include distant EEG changes correlated between two connected individuals (Wackermann et al. 2003) and functional magnetic resonance imaging (fMRI) changes during spiritual healing (Achterberg et al. 2005; Standish et al. 2003). These studies point toward assessment of the brain during spiritual healing. During spiritual healing, Divine characteristics are often invoked and this could provide another fruitful approach to developing a neuroscience of spirituality, as described below.

Spiritual traditions from all cultures and all times describe various types of healing in which Divine presence or spiritual energy is transmitted through an individual, resulting in changes in the physical or psychological health of another without the use of an agent such as a needle, pill, or knife. Often, spiritual experiences accompany these healings which are evoked through a ritual or religious process (i.e. prayer). These rituals are sometimes said to evoke a special "presence" that can exude from the practitioners. Could there be neuro-correlates of this spiritual healing presence and if so could they be objectively measured?

Tapping into Spiritual Healing Models

To examine the first part of this question we brought together 12 healers from different spiritual healing traditions from around the world. Table 1 lists those traditions. Each of these practitioners had their own theories, framework, and methodology for delivering spiritual healing. None of them used an agent such as a pill, knife, or needle. We asked these practitioners to describe the background,

Table 1 Healing traditions sampled (Jonas and Crawford 2004)

Ayurvedic	Christian evangelical
TCM (Traditional Chinese Medicine)	African Shamanism
Greek orthodox	Psychotherapy
Kabalistic	Christian science
Sufi	Spiritual healing
Native American medicine	Brazilian psychic surgery

Table 2 Common components of a healing presence (number of traditions identifying the component) (Jonas and Crawford 2004)

Love (11)	Good intention (11)
Spiritual Grace (6)	Belief (5)
Focused Awareness (5)	Direction of Energy (4)
Openness to Healing (4)	Listening (3)
Creativity (3)	Reconciliation (3)
Imagination (2)	Connectedness (3)
	Relationship (3)

the assumptions and approaches used in their practice. We then used a Delphi process to determine which components this group felt were the essential components of the spiritual healing experience and process.

A list of over 25 components emerged. These were gradually condensed through iterative feedback to the group, using a modified Delphi process producing ranked ordering of those components. What emerged was the identification of two components that were almost universally required across all the spiritual healing traditions. These components were said to be necessary, though not always sufficient, to produce healing. Practitioners were careful to distinguish between these and expectancy or placebo effects. Table 2 lists these common components of spiritual healing identified by the traditions included.

Note that the two characteristics that emerged across all traditions were "love" and "good intentions" (W. B. Jonas and Crawford 2004). Love was generally described as a deep, unconditional emotional connection with the spiritual or the Divine aspect of the other person (similar to Agape) rather than a personal love (Eros). It was rooted deeply within the now and to the eternal aspects of existence, transcending time and space. It was a state that connected the personal and profane to the eternal or "sacred" dimensions of the two persons involved as well as to the Divine. The second equally important component was a positive intention. The general concept was that love was the "engine" or the "force" for healing and that good intention provided the "direction" to achieve healing outcomes. All the traditions except for Chinese QiGong distinguished between the spiritual and mental dimensions of the phenomena. In QiGong, however, the practitioners described Qi as being distinct from the psychological and it had non-local characteristics similar to what other traditions labelled as spiritual or Divine.

A Model of the Whole Person

Figure 1 illustrates a graphic that helps conceptualize what most of these traditions envisioned as "the whole person." A whole individual includes the body, mind, and spirit manifested through matter, energy, and information, respectively. On the body level, individuals might be separated by long distances such as are person 1 and person 3 in Fig. 1. Imaginary and mental characteristics such as memory and imagery could bring individuals closer together through imagination, making them less distant.

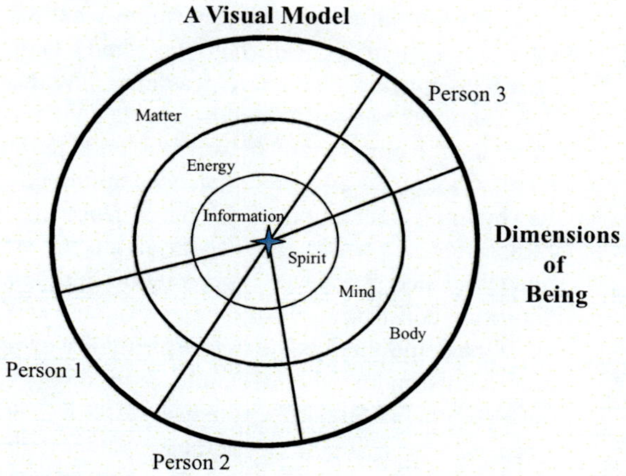

Fig. 1 The whole person (W. B. Jonas and Crawford 2004)

These are cognitive or psychological characteristics. However, these spiritual traditions all identified some point in which a dimension of the person touches all others on a non-local level. These are described as the spiritual aspects of an individual. At that point persons are indistinguishable from other persons although they retain an element of their unique identity on the mental and physical levels. Thus, in Fig. 1, person 1 and person 3 are no further apart in their spiritual aspect than person 1 and person 2. In the same domain as mind/body/spirit I have placed what I consider possible parallel scientific domains of energy/matter/information. While the body clearly consists of matter we should explore if, in a similar fashion whether mind consists of energy and spirit consists of information, the latter having timeless characteristics as well as being also without matter or energy.

From this model, the process of spiritual healing involves acknowledging the reality of the whole person and then becoming more fully aware of their spiritual aspects. Meditative techniques allow the individual to dip down into their spiritual nature, touching all other beings at the non-local point. When this occurs, healing occurs as a return to the full wholeness of the person. One could theoretically distinguish between the mental phenomena of neuropsychology, occurring on the mental level of the individual and the phenomena of neurospirituality occurring on the spiritual or non-local level in this model. In all cases, we would be looking at correlates of brain activity with psychological or spiritual characteristics in neuropsychology and neurospirituality, respectively.

Can We Distinguish Social from Spiritual Neuroscience?

Distinguishing between a neuroscience of love on a Divine level versus the social level might be problematic without being able to measure characteristics of individuals engaged in non-social spiritual activity. A possible pathway to examine this might be in an attempt to determine what aspects or dynamics of the brain and nervous system occur during love when accompanied by non-local and non-social events – spiritual characteristics defined by these spiritual healing traditions. Work on heart rate variability and heart rate coherence indicates that when affection and love are felt emotionally between individuals, enhanced heart rate coherence occurs. This is manifest by a more balanced autonomic and parasympathetic discharge and a rhythmic fluctuation of discharge from the vagus nerve (McCraty et al. 2004). Research on whether this occurs within the brain would allow one to examine the neuroscience of love and appreciation in the brain as both psychosocial and socio-spiritual constructs.

One could then define neurospirituality using two possible approaches. The first would distinguish the love and affection between two people as social, therefore a social psychological phenomenon. If, however, the love and affection were being expressed as a Divine characteristic rather than to another individual, this, could be defined as neurospirituality. A complementary approach would be to explore whether non-local phenomena were occurring during healing when another person is not the target of the affection. Neural patterns that occurred during non social love might be distinguishable from those occurring during social affection. These two complementary approaches then might help to distinguish a neuroscience of spirituality from neuropsychology.

An Alternative (Non-dichotomous) Hypothesis

I have argued that a way to distinguish neurospirituality and neuropsychology may be in a dichotomous way the brain processes time – paralleling what appears to be the dichotomous nature of local and non-local time. However, our brain may have adapted a non-dichotomous approach to processing time instead. For example, recent work on the "default mode network" (DMN) indicates that there are brain regions that are active during basic, non-task performing awareness (Raichle et al. 2001). These same processes may be the structures that modulate our management of time perception (Pagnoni et al. 2008). Meditators, for example, seem to use the brain DMN when changing (expanding and contracting) time perception and reaction rate. Thus, we may find identical or at least overlapping functional areas of the brain used when time shifting is involved in the mental activity. Perhaps it is at the extremes of such DMN variances that local and non-local time phenomenon are detected, like a trombone slide, continuously manipulating sound. One fruitful experiment might be to test whether time shifts during cognitive task processing can be manipulated through changes in consciousness (such as through hypnosis, meditation, suggestion, or biofeedback). If in those experiments we pushed the boundaries of such processing beyond the speed of neural transmission, we could explore how the brain handles both the local and the non-local. When combined with neuro-imaging of the DMN this approach might help us determine how such neural processes are involved in the creation of both cognitive and spiritual experiences.

Conclusions

In this paper I have tried to theoretically circumscribe the differences between a neuroscience of psychology and a neuroscience of spirituality, proposing that a neuroscience of spirituality be defined when phenomena are related to Divine characteristics such as omnipotence, omnipresence, and eternality are manifest and/or when individuals are undergoing spiritual healing and the effects are occurring independent of social expectancy and the placebo effect. A true neuroscience of spirituality will need empirical validation where observed neurological patterns and characteristics have occurred under these differing conditions. If neurological patterns are the same for psychological and spiritual characteristics, this would be evidence that both neuropsychology and neurospirituality are indistinguishable and possibly redundant. If, however, distinct neurological processes occur under these differing characteristics, development of a separate field of neurospirituality would be scientifically justified.

References

Achterberg, J., Cooke, K., Richards, T., Standish, L.J., Kozak, L., & Lake, J. (2005). Evidence for correlations between distant intentionality and brain function in recipients: A functional magnetic resonance imaging analysis. *Journal of Alternative and Complementary Medicine, 11*(6), 965–971.

Bösch, H., Steinkamp, F., & Boller, E. (2006). Examining psychokinesis: The interaction of human intention with random number generators–A meta-analysis. *Psychological Bulletin, 132*(4), 497–523.

Jonas, W.B., & Crawford, C.C. (Eds.). (2003). *Healing, intention and energy medicine: Science, research methods and clinical implications*. London: Churchill Livingstone.

Jonas, W.B., & Crawford, C.C. (2004). The healing presence: Can it be reliably measured? *Journal of Alternative and Complementary Medicine, 10*(5), 751–756.

McCraty, R., Atkinson, M., & Bradley, R.T. (2004). Electrophysiological evidence of intuition: Part 2. A system-wide process? *Journal of Alternative and Complementary Medicine, 10*(2), 325–336.

Pagnoni, G., Cekic, M., & Guo, Y. (2008). Thinking about non-thinking: Neural correlates of conceptual processing during Zen meditation. *PloS One, 3*(9), e3083.

Radin, D.I., & Borges, A. (2009). Intuition through time: What does the seer see? *Explore: The Journal of Science and Healing, 5*(4), 200–211.

Radin, D., & Nelson, R. (2003). Research on mind-matter interactions (MMI): Individual intention. In W. Jonas & C. Crawford (Eds.), *Healing, intention and energy medicine: Science, research methods and clinical implications* (pp. 39–48). London: Churchill Livingstone.

Raichle, M., MacLeod, A., Snyder, A., Powers, W., Gusnard, D., & Shulman, G. (2001). A default mode of brain function. *Proceedings of the National Academy of Sciences, 98*(2), 676–682.

Samueli Institute. (2007). Special issue: The pertinence of the Princeton Engineering Anomalies Research (PEAR) laboratory to the pursuit of global health. *Explore: The Journal of Science and Healing, 3*(3), 191–245.

Standish, L.J., Johnson, L.C., Kozak, L., & Richards, T. (2003). Evidence of correlated functional magnetic resonance imaging signals between distant human brains. *Alternative Therapies in Health and Medicine, 9*(1), 128. 122–125.

Steere, D. (1984). *Quaker spirituality: Selected writings* (p. 66). Mahwah: Paulist Press.

Utts, J., & May, E. (2003). Non-sensory access to information: Remote viewing. In W.B. Jonas & C.C. Crawford (Eds.), *Healing, intention and energy medicine: Science, research methods and clinical implications* (pp. 59–74). London: Churchill Livingstone.

Wackermann, J., Seiter, C., Keibel, H., & Walach, H. (2003). Correlations between brain electrical activities of two spatially separated human subjects. *Neuroscience Letters, 336*(1), 60–64.

Sufism and Healing

Howard Hall

Abstract This chapter examines a neglected phenomenon: rapid wound healing after deliberate bodily damage. Some spiritual traditions, such as a particular school of Sufism, maintain that through immediate contact with the Divine bodily damage may not do harm and any wounds can heal rapidly. During such rituals the body is pierced and wounded by various objects, such as daggers, skewers, glass or razor blades. To investigate this claim extensive field documentations were conducted and a series of well controlled demonstrations under laboratory conditions. These findings are described. The observations and measurements seem to suggest the following: the rapid wound healing phenomenon after deliberate bodily damage is repeatable, even under laboratory conditions and outside the context of the religious ritual; there is clear evidence of penetration of tissue, but no evidence of pain, infection or other lasting damage; concomitant EEG data and the phenomenology do not seem to suggest that this phenomenon is a consequence of hypnosis. Spiritual explanations are discussed.

Introduction

Is there empirical evidence of healing anomalies that challenge our traditional paradigms of normal science that may be initiating a paradigm change or a scientific revolution (Kuhn 1970)? This chapter will review my systematic observations of

H. Hall (✉)
WO Walker Center, Case Western Reserve University, Rainbow Babies and Children's Hospital, University Hospitals of Cleveland, 10524 Euclid Avenue, Suite 3150, Cleveland, OH 44106, USA
e-mail: Howard.hall@uhhospitals.org

H. Walach et al. (eds.), *Neuroscience, Consciousness and Spirituality*, Studies in Neuroscience, Consciousness and Spirituality 1,
DOI 10.1007/978-94-007-2079-4_16,

unusual rapid wound healing practices from the Middle East within a mystical tradition called Sufism, which I have witnessed over the years in field work. These observations may challenge conventional mechanistic explanations employed within our current scientific models. In order to understand this wound healing paradigm better, it is important to know about its role within the Sufi tradition and the wider context of this tradition within Islam. Hence I will introduce this tradition briefly, as it understands itself, and specifically the spiritual tradition called Sufism within it. I will then describe rapid wound healing following deliberate bodily damage and its spiritual role and discuss potential explanations.

Historical Background

The history of Sufism has roots going back about 15 centuries or around 600 CE (Common Era) to the time of the Prophet Muhammad and the birth of Islam. The Prophet Muhammad lived from 570 to 632 CE and was born in the city of Mecca, which is today in Saudi Arabia (Fatoohi 2009). Muhammad would frequently retire for meditation to a cave later known as "Hira" on top of a mountain north of Mecca called "Nur" or light. At the age of 40 in the year 610 CE while meditating in that cave, the Prophet Muhammad received the first of a series of revelations commonly believed to be from God though the angel Gabriel. These revelations continued over a period of around 23 years through this "unlettered" prophet and were memorized and written down by his followers and became the Quran or the holy text of Islam. Tradition holds that the Quran is the last of divinely inspired books from God through chosen Prophets following Moses' book, the Torah, and the Gospels of Jesus. Also, the Quran is held to be divinely protected from being corrupted (15:9) and is also remarkable for its internal consistency, its external agreement with historical and archaeological evidence, as well as providing new information and scientific findings that did not come to light until the nineteenth and twentieth century. Quranic scholars also note that all prophets bring the same divine message calling people to God and also confirming the veracity of the prior messengers (Fatoohi 2010) . For example, human fetal development was vividly described well before the dawn of scientific knowledge of embryology (Quranic verses: 22:5; 23: 12–14; and 40:67). Furthermore, unlike other religious traditions, Muhammad is not worshipped or seen as divine, but is credited with the revelation of a literary masterpiece, along with the founding of a major path to God (Allah). The Sufi musician Hazrat Inayat Khan (1988) provides a description of spiritual religion and a new world power (Armstrong 1992).

Sufism is the mystical spiritual tradition within Islam, which directs humans toward a "nearness" to the direct path development: "The word 'spiritual' does not apply to goodness or to wonder-working, the power of producing miracles, or to great intellectual power. The whole of life in all its aspects is one single music; and the real spiritual attainment is to tune oneself to the harmony of this perfect music."(Khan 1988)

The foundation of Sufism is based on the mystical aspects of the spirituality of the Prophet and the belief in its transforming power (Hussein 1997). During his lifetime, pious individuals from different nations learned under his guidance the spiritual/mystical laws, knowledge, and practices of Islam, that led toward direct experience of the Divine (Angha 1994).

The spiritual leader of a Sufi school is known as a *Shaikh*. The spiritual knowledge of the Shaikh can be traced back to the Prophet Muhammad who later converted his cousin and son-in-law ‘Ali bin abi Talib. ‘Ali is considered a spiritual heir to the Prophet and the one who inherited his spiritual knowledge and power. Thus all Sufi Masters are his students, directly or indirectly, and this is the origin of the title “Shaikh of the Shaikhs.” Through a line of succession, each Shaikh would initiate a successor based upon revelations from Allah. Thus this maintains a direct spiritual link or attachment with the Prophet Muhammad to the present spiritual leader (Chishti 1991). This chain from the Prophet Muhammad down to the present Master of a Sufi school is known as a *silsila* (Hussein 1997). Today there are more than 150 orders or schools of Sufism.

A Shaikh is a mediator or guide to Allah in Sufism in order to help the student draw near to Allah, battle their lower self. This battling of the lower self is, by the way, the true spiritual meaning of jihad and not how the term has been distorted by some Muslims and the media (Fatoohi 2009). The Shaikh also helps channel the spiritual power from Allah to the student in order to perform paranormal events including unusual types of rapid wound healing such as deliberately caused bodily damage. These feats are done to demonstrate to the student and to others this spiritual connection.

Potential Paradigm Shifts in Medicine

The 1990’s were the era of a “silent revolution” in health care for some of the wealthiest, most highly educated mainstream citizens of major industrialized countries of the world, from the United States to Europe. Surveys observed a major increase in the utilization of alternative medicine practitioners at a higher frequency than traditional primary care physician visits for chronic non-life threatening health conditions (Eisenberg et al. 1993, 1998). There is potentially a paradigm shift occurring within mechanistic medicine where Larry Dossey (1993) described medicine moving through three distinct eras. Era I: Mechanical, Material, or Physical Medicine. This was the Newtonian view of the world where the human body was seen, as operating like a machine. Today this is our high tech medicine. Era II: Mind-Body Medicine movement. In the United States, one can place relaxation, meditation, and some Sufi approaches within the alternative medicine movement. Finally, Era III: “spiritual healing”/ “energy healing.” Dossey also called this non-local or transpersonal medicine. The metaphysical healing aspects of Sufism can be viewed within this context.

ERA III Energy Medicine/Spiritual Healing (Non-local, Transpersonal Medicine)

The Silent Revolution or paradigm shift of the 1990's incorporated global healing traditions from other cultures that were thousands of years old. These non-western practices involved whole person approaches to healing. Examples of such global healing traditions included: Ayurveda medicine of India with its various yoga techniques and energy healing practices, Chinese Medicine with Acupuncture, and Qigong (Micozzi 2006). Within Western religious traditions there was intercessory prayer and distant healing intention research (Dossey 1993).

Contemporary western scientific research has documented some of these effects. Daniel Benor addressed the question for the evidence-basis of spiritual healing in "Spiritual Healing: Scientific Validation of A Healing Revolution" (Benor 2001):

> Does spiritual healing work? Does research confirm that healing is an effective therapy?" An impressive number of studies with excellent design and execution answer this question with a "Yes. If we take a broad view, out of 191 controlled experiments of healing, close to two thirds (64.9 percent) of all the experiments demonstrate significant effects. (Benor 2001, p. 371).

In a series of rigorous laboratory energy healing experiments Gary Schwartz (Schwartz and Simon 2007) observed that healers trained in a Japanese energy technique called Reiki, could provide statistically significant protection of growth from heat damage for E. coli cells in a test tube compared to untreated control samples. This effect however was associated with the healer's emotional state. If the healer was in a non stressed self-reported state of mind the cells demonstrated enhanced growth and protection from heat damage in the Reiki treated group. The assessment of this mood variable was done by having all practitioners completed a standardized well-being form when coming into the laboratory. Interestingly those practitioners who came to the study in a stressed mood had laboratory results that went in a negative direction compared to controls. Unlike the increased growth of E. coli in the positive well-being Reiki group, the stressed practitioners had decreased cell growth compared to controls.

In another study in this series, Schwartz (Schwartz and Simon 2007) observed that Reiki energy healing had a protective effect on noise stress-induced microvascular leakage in the gastrointestinal system of rats compared to untreated stressed rats or sham Reiki treated groups. There was also a no noise control condition.

It should be noted that a lot of the research is puzzling, not all results are consistent, and that, in fact, some of the most rigorous experiments have not provided any substantial evidence for replicable effects (Krucoff et al. 2001, 2005; Radin and Yount 2004; Yount et al. 2004; Taft et al. 2005; Zachariae et al. 2005). This raises the question, whether such effects of intention can be classed as causal effects or might not rather be a-causal, non-local effects that defy experimentation.

Rustum Roy (2004) has worked to integrate mind-body-medical-spiritual healing traditions and other non traditional therapeutic approaches and has characterized them as a Whole Person Healing movement. He has worked on an interdisciplinary and international level to bring together and study non traditional healing approaches. Roy has structured these meetings with the same rigor as he brings from the hard sciences with as he noted an: "utter dedication to facts, to data" (Roy 2004).

One global healing tradition from the Middle East that has received very little attention from the West (probably for geographical and political reasons) is the spiritual practice of rapid wound healing from "deliberately caused bodily damage phenomena" (DCBD) from a major Sufi school in Iraq. This has been the focus of my research for over a decade.

The Metapsychology of Sufism and Rapid Wound Healing

The extraordinary instantaneous wound healing from "deliberately caused bodily damage phenomena" (DCBD), has been reported by the Tariqa, Casnazaniyyah School of Sufism in Baghdad, one of the largest Sufi schools in the Middle East (Hussein et al. 1994a, b, c, 1997). Followers (dervishes) of this Sufi school have been observed to demonstrate instantaneous healing of deliberately caused bodily damage (DCBD). For example, dervishes drove a variety of sharp instruments such as spikes and skewers into their body; hammered daggers into the skull bone and clavicle; and chewed and swallowed glass and sharp razor blades without harm to the body and with complete control over pain, bleeding, infection, as well as rapid wound healing within 4–10 s (Hussein et al. 1994a, b, c). The name of this Sufi school is *Tariqa Casnazaniyyah,* an Arabic-Kurdish word, meaning "the way of the secret that is known to no one" (Hussein et al. 1997). Researchers report that such extraordinary abilities are accessible to anyone and not restricted to only a few talented individuals who have developed strong faith and dedication from spending years in special training. In contrast, a unique aspect of this Sufi tradition is how a dervish, once initiated, can instantly demonstrate such rapid healing abilities. These unusual healing phenomena have also been reproduced under controlled laboratory conditions in the Middle East where 28 dervishes inserted sharp objects into their body, chewed and swallowed glass along with other feats with recordings of EEG, skin resistance, skin temperature, and blood pressure (Hussein et al. 1997). My research group has also demonstrated DCBD effects within a laboratory in the United States and found that, like Hussein, this phenomenon is in no way similar to hypnosis (Hall et al. 2001)

Similar isolated observations of DCBD phenomena have been described in various parts of the world in a variety of religious and non-religious contexts (Hussein et al. 1997) from Native American Sun Dance involving inserting skewers into the skin in the jaw, to Indian Mediums beating their bodies with swords. Around the time after the Second World War, Mirin Dajo, the Dutchman provided demonstrations to the public and physicians of him putting swords through his chest (Coppens 2010). Also, trance surgeons in Brazil have employed sharp instruments to cut, pierce, or inject substances into a patient's body for therapeutic purposes (Don and Moura 2000). The Brazilian healing work of Arigo – Surgeon Of the Rusty Knife

(Fuller 1974) in the 60's and the current practice of John of God (Cumming and Leffler 2007) provide dramatic illustrations of anomalous healing observations. Scientific laboratory EEG investigation of trance surgeons have observed that this "state of spirit possession" for the healers was associated with a hyper-aroused brain state (waves in the 30–50 Hz band) (Don and Moura 2000).

Elmer and Alyce Green (Green and Green 1977) have scientifically documented in their psychophysiology and brain laboratory, the ability of exceptional practitioners of "unconventional healing" to control biologic and wound repair processes in their bodies. These have included Swami Rama, Rolling Thunder, and Jack Schwartz who could push a 6 in. steel sailmaker's needle through his biceps with control over bleeding. Given his unusual ability of self-regulation the Greens described Schwartz as being like a Sufi.

Peper et al. (2006) have also conducted neuroscientific research on individuals with the exceptional ability to control pain and bleeding. One such subject included a 63 year old Japanese Yogi with 37 years of experience in his discipline, who could insert a non-sterilized skewer through his tongue. After the skewer was removed there was no bleeding or physical injury to the tongue. Quantitative EEG data suggests that the Yogi during the tongue piercing was able to inhibit areas in his brain that responded to external stimuli.

Despite this scientific research being conducted on anomalous rapid wound healing in the United States and the Middle East, consistent with Kuhn's (1970) analysis these observations have been met with scorn and have even been challenged by so-called skeptic groups (e.g. Committee for the Scientific Investigation of Claims of the Paranormal (CSCIOP)) who say they offer monetary incentives to discredit such claims in unscientific and dangerous settings (Mulacz 1998; Posner 1998). Such claims may threaten the current world view of a certain reading of the scientific knowledge. As such phenomena cannot be explained within the world view, they are being excluded and actively banned. (See Dossey 1999; Fatoohi 1999 for a response).

From a spiritual perspective, this type of healing is described in terms of "healing energies"[1] (Hussein et al. 1994a, b, c). This "higher energy" is alleged to be instantly transferable mediated through a spiritual link from the current Shaikh of the Tariqa Casnazaniyyah Sufi School and through the chain of masters to Muhammad and ultimately from Allah (Hussein et al. 1997). As noted in the Quran: 33:6 "The Prophet is closer to the Believers than their own selves." Followers of this Sufi school describe the ability to accomplish DCBD as an "others-healing phenomena" which goes beyond traditional mechanistic and psychological factors that influence healing.

[1] It is important to note here that the term "energy" is employed metaphorically and not in a strictly physical or material sense, as no known energy would be able to provide an explanation for the phenomena observed.

Empirical Research in Sufi Rapid Wound Healing

How might Western scientists empirically investigate such claims of usual wound healing from the Middle East? To scientifically research these apparently anomalous phenomena I began a serious of systematic observations. This research started with a trip to Baghdad to directly observe and document such claims via video recording. Second, I decided to experience such rapid wound healing in that setting. Third, arrangements were made to have this Middle Eastern demonstration transported to the United States within a traditional medical setting with a dervish from the Middle East as well as studies on myself as the subject. These systematic observations began in 1998 with an invitation from the Shaikh of the Tariqa Casnazaniyyah School of Sufism in Baghdad and support from the Kairos Foundation in Illinois, to travel to Baghdad to meet with the spiritual leader of this group, Shaikh Muhammad al-Casnazani, and witness a group demonstration of DCBD at their major school (Hall 2000). At this meeting, which was professionally video taped, I had the opportunity to examine first hand, the objects that were employed during the DCBD demonstrations, such as the knives, razor blades, and glass, and observe them being inserted into various parts of the body. What I witnessed and recorded was consistent with the extraordinary claims made by this group of rapid wound healing with no apparent pain or infection.

Although I saw no evidence of a ruse, I imagined, however, that some skeptics might question if I had somehow been deluded, even with video footage. Thus, while at this demonstration, I had requested permission to experience DCBD by having my cheek pierced (Hall 2000). After witnessing several demonstrations of DCBD an assistant came up to me and asked if I was ready? I said, "Yes," and he asked me to face the Shaikh to ask permission to allow the spiritual energy for rapid wound healing. The Shaikh nodded indicating that I had his permission. What was most striking was that I did not feel any different, nor was I in an altered state, and my cheek was not numb. The assistant then inserted a metal ice pick from the inside of my left cheek through the outside. It felt like a slight poke, but no pain. I walked around the group circle with the ice pick in my cheek introspecting on how it was not hurting, bleeding, nor numb. I could feel the weight of the object and notice the metal taste in my mouth, but no discomfort. Again, consistent with their reports, my cheek healed rapidly in minutes with only a couple of drops of blood. This personal experience was very compelling to me despite having had much doubt, as well as a strong aversion to pain. Nonetheless, I still imagined that skeptics would question if such practices could be exported and demonstrated outside of this religious context, to the West. Clearly what was needed in this area was a demonstration of such rapid wound healing within a Western medical setting given the scientific implications for such healing. If such spiritually based healing approaches are genuine, they hold much promise for addressing some of today's most serious medical issues.

The investigations of such unusual healing phenomena in the West raise many questions such as what should be measured within a scientific context. Would standard measures of brain activity such as EEG as well as immune responses be associated with rapid wound healing? Would high frequency EEG activity suggesting a hyper-aroused brain state be correlated with healing? Would new technological approaches be needed to detect "fields of consciousness," such as the changes in the output from a random event generator (REG) or measurements of "energy fields" with recordings of the human aura employing the Gas Discharge Visualization measures (GDV) (Korotkov 2002, 2004; Korotkov et al. 2005)? We began our systematic scientific observations by inviting a dervish from the Middle East to a medical facility in Ohio.

Case Report

With the support of the Kairos Foundation of Wilmette, Illinois, a Sufi practitioner (J.H.) was invited from the Middle East, to a local radiology facility in Cleveland, Ohio July 1, 1999, with permission from the Shaikh of the Casnazaniyyah Sufi school to perform a demonstration of rapid wound healing following the insertion of an unsterilized metal skewer, 0.38 cm thick and approximately 13 cm long, while being videotaped by a film crew, in the presence of a number of scientists and healthcare professionals (Hall et al. 2001). This was apparently the first demonstration from this Sufi School in the United States. The practitioner consented to sign a release of liability for the medical facility and personnel against claims from possible injuries that might occur. Emergency medical technicians were present. The major goal of this demonstration was to observe the authenticity of rapid wound healing following a deliberately caused injury within a medical setting. The demonstration was also conducted with radiological, immunological, EEG evaluations, and a zener noise diode random event generator, similar to the one employed at Princeton University by Jahn and colleagues (1987). Based on previous studies in Brazil with healer-mediums engaged in quasi-surgical practices, it was hypothesized that DCBD would be accompanied by alterations in brain waves and effects on random event generators. The alterations in brain waves found with the Brazilian healer-mediums showed statistically significant enhancement of 40 Hz band in the EEG (Don and Moura 2000). A statistically significant deviation from random behavior in random event generators was found, run covertly while the Brazilian healer-mediums were in trance. This methodology was developed by Robert Nelson and Associates at the Princeton Engineering Anomalies Research Laboratory (PEAR) (Nelson et al. 1996, 1998). Such energy fields have been considered as theoretically associated with rapid wound healing (Don and Moura 2000).

Nineteen-channel EEGs were recorded during baseline resting conditions, while the dervish inserted the skewer through his cheek, and immediately after removing the instrument.

A random event generator (REG) was plugged into the serial port and run in the background without informing the dervish. The distribution of binary digits was tested for possible significant deviations from random behavior. Data were acquired before and after the self-insertion, as well as during the skewer insertion condition.

Prior to the insertion of the skewer, and about 1 h after the piercing, blood was collected from the practitioner and three volunteers for an immunological analysis of the percent change in CD4, CD8 and total T cell counts.

Results

Radiological images were done while the skewer was inserted. Axial CT images through the lower mandibular region showed artifact from dental metal. In addition, there was a horizontally oriented metallic bar which elevated the left lateral soft tissues just anterior to the muscles of mastication. There was no associated underlying mass. A single frontal fluoroscopic image showed a presence of EEG leads over the maxilla and mandibular regions. There was a transverse metal superimposed extending from the soft tissues on the right through to the left without interval break.

Because of movement and scalp muscle artifacts throughout the experimental self-insertion condition, it was impossible to assess the EEG for the hypothesized 40 Hz brain rhythms. The frequency spectrum of scalp muscle discharge overlaps the 40 Hz EEG frequency band of interest. The presence of increased theta rhythms after the insertion condition (and a slight decrease in average alpha power) suggests a mild hypo-aroused altered state of consciousness.

The random event generator during baseline periods did not differ significantly from random behavior. However, during the self-insertion condition, there was a trend toward significant non-randomness. The chi-square equalled 3.052, ($df=1$) with $p=.07$.

The immunology testing did not reveal any major difference between the Sufi practitioner and the controls.

Discussion

The behavior of the random event generator was in the predicted direction of non-randomness. This has been interpreted by our and the PEAR laboratories as being associated with states of heightened attention and emotion. Further, PEAR has proposed that a "field of consciousness" is associated with such non-randomness. Unfortunately, the 40 Hz brain wave hypothesis was not testable because of the excessive amount of scalp artifact and so awaits further exploration. The Sufi performing this feat was doing so for the first time. It would seem possible that with further practice or by testing more experienced subjects, it may be feasible to obtain EEG data without large amounts of scalp artifact. Because the subject reported no perceived pain during the self-insertion, preliminary relaxation exercises might eliminate all or most of the artifact. This would enable us to test the 40 Hz hypothesis definitively. Clearly, further work is indicated.

The immunology data suggest that the variation found in the practitioner was not different from normal controls.

The radiological film documented that the skewer had actually penetrated both cheeks, thus addressing skeptic groups that such practices are the result of fakery. Following the removal of the skewer there was a slight trickle of blood, which stopped, with compression of clean gauze to the cheek. The physicians and scientists present documented that the wound healed rapidly within a few moments. The practitioner also reported that there was no pain associated with the insertion or removal of the metal skewer. This demonstration was conducted outside of the traditional religious context, where chanting, drumming, and head movements are generally part of the ceremony when done in the Middle East. Thus, our case study argues against the necessity of a religious context with its accompanying state of consciousness being important for the successful outcome of such a demonstration. This case study also demonstrated that DCBD could be done at such a large distance separating the dervish from the Master (Baghdad and Cleveland). It should be noted that the dervish had obtained prior permission from the Shaikh before doing this demonstration. This would suggest that this is a very robust phenomenon independent of the distance separating its source and the scene where the DCBD phenomena occur.

It should also be pointed out that the skewer stayed in the dervish's cheeks for more than 35 min. This was a longer time frame than I had observed during my field observations at the Major School of Tariqa, which were only a few minutes. Thus, this case study argues against the necessity of a very brief piercing period for a successful outcome of DCBD. Furthermore, the dervish of this demonstration reported that there was no pain associated with this piercing; there was minimal bleeding and no post procedure infection. It should not be overlooked that about a half hour after the completion of the demonstration, the dervish, along with seven other people who witnessed the DCBD event, had dinner together.

Personal Rapid Wound Healing

After witnessing rapid wound healing in the Middle East and experiencing it myself there, I was initiated into the Sufi Order with the ritualistic hand shake taking about 2–3 min. After a subsequent visit with the Shaikh in the United Kingdom in June of 2000, I was given a license to perform rapid DCBD. I first requested permission from the Shaikh in Baghdad to perform a cheek piercing on myself in May at the 2001 World Congress on Complementary Therapies in Medicine in Washington, DC. After lecturing on DCBD, I informed the audience that I needed to take an earlier flight home because of a family medical emergency in Cleveland. I then went right into the cheek piercing for the first time on my own after the lecture. My state of mind was on the family medical crisis back home, but I was instructed to focus on connecting with the Shaikh asking mentally for spiritual energy for rapid wound healing before the piercing. This took about a minute. There was also one physician in the audience who was particularly skeptical, so I invited him to stand right next to me when I did the piercing. I used a skewer from my kitchen drawer for the demonstration which later proved to be not very sharp.

After the 1 min mental connection with the energy from the Shaikh and much nervousness, I pushed a very dull skewer through my left cheek. Yes, I was quite worried about the medical situation at home. The most difficult aspect of this experience was getting this dull object through my cheek. Eventually it went through with no pain. My skeptical medical colleague was very quiet after that. I pulled it out and there were a couple of drops of blood which I blotted with a tissue until the bleeding stopped. From there, I had a friend take me directly to the airport.

The second time I demonstrated DCBD on myself (making this my third experience with DCBD including Baghdad) was at the Fifth World Congress on Qigong in November, 2002 (Hall and Schwartz 2004). Since this was after 9-11-2001 I had to go shopping for hopefully a better piercing instrument. The video camera was then set on a stand on the side of my left cheek. I again focused on connecting with the energy of the Shaikh for rapid wound healing. I did not feel different, but had faith that the connection was there, despite the distance in space. Again, I found that pushing the metal pick through my cheek was very difficult to puncture the cheek. After some effort both physical and mental, it went through. I also spoke on camera about how I was feeling with the object through my cheek. Following the interview, I pulled the pick out and padded a tissue against my cheek with a few drops of blood. The wound closing was also documented on film for the first time. It is interesting to note that I had cut myself shaving early that morning flying out to California and the piercing was the next day. The shaving cut was more noticeable than the piercing after the demonstration. I went out for a late dinner after this demonstration.

The next day after the demonstration I had the opportunity to meet and get evaluated by Dr. Konstantin Korotkov, professor of physics at St. Petersburg State Technical University in Russia, and inventor of the Gas Discharge Visualization technique (GDV) which purportedly measures "human energy fields" like the earlier Kirlian photography (Korotkov 2002, 2004; Korotkov et al. 2005). Dr. Korotkov first took a baseline measure of my "energy field" from my fingers and displayed the results on a screen for the audience. He then asked me to invoke the Sufi energy. I again took about a minute and mentally requested energy from the Shaikh for this demonstration. It should be noted that this demonstration was not planned by me nor had I obtained prior permission from the Shaikh for this energy. After about 1 min I said I was ready for the second (after energy) measure. Dr. Korotkov outwardly expressed surprise at how quickly I had invoked energy. This time when he took energy reading from my hand the computer malfunctioned and another one had to be brought in. After the new computer was in place, the GDV revealed a major increase in my energy field after the quick 1 min invocation.

My fourth experience with DCBD was requested by National Geographic Television (2005). This demonstration was done in collaboration with Gary Schwartz at his Human Energy Systems Laboratory, Center for Frontier Medicine in Biofield Science, University of Arizona, Tucson, AZ (Hall 2004). Again I obtained permission from the Shaikh to conduct another DCBD demonstration and we also explored if there were any changes in brain activity associated with this process or any changes in my "energy field" or "aura" as indicated by the GDV measures. A 19 channel EEG along with GDV recordings were taken before and after piercing my left cheek with a 5 in. ice pick and while being filmed by National Geographic. This meditation took about 90 min to complete connecting with the current Shaikh and other masters who are part of this Sufi school's chain of Shaikhs (i.e. silsila).

During the mediation my brain did reach a more relaxed state as measured by EEG recordings. There was, however, no anomalous neurological activity, such as seizures, sleep, or hyper-aroused brain states. Pre versus post changes on the GDV did reveal, for the first time ever, a selective decrease in the "energy field" where the cheek was pierced revealing a gapping hole in my "aura" in that area. As with my prior experience with DCBD there were a few drops of blood after the ice pick was removed, but the wound healed very quickly and was not noticeable after a few minutes. Again, even with these additional scientific measures, traditional paradigms offered little insight to account for the rapid wound healing phenomenon. Such an observation was consistent with the above case report with the Sufi practitioner from the Middle East where we again failed to find any correlations with DCBD with any of the blood tests or imaging studies. The only hints of associations were the GDV energy measures and the trend of the Random event generator also suggesting some change in the energy field.

Is Rapid Wound Healing Hypnosis?

My review of this rapid wound healing phenomena argues against hypnosis or altered states of consciousness being an explanation for DCBD because of a lack of logical, theoretical, or empirical support (Hall 2000). Neither the EEG research on the 28 dervishes in the Middle East nor my brain wave activity during my piercing showed any evidence of any altered states of consciousness. Dossey (1998) has also argued that the positive clinical outcomes observed in hypnosis do not compare to the dramatic effects observed in Sufi rapid wound healing. The rapidness of healing and the lack of infections in DCBD are quite different from outcomes of hypnosis research. For example, accelerated post-surgical wound healing in hypnosis involved a 7 week postoperative time course (Ginandes et al. 2003). Along similar lines, research of the slowing of wound healing in a group of stressed relatives of Alzheimer's patients had a time courses of 7 weeks for the stressed group versus around 6 weeks for age matched controls (Kiecolt-Glaser et al. 1995). Such a time course in weeks can not be compared to the minutes for the Sufi healing phenomenon.

How Does Sufism Explain How This Can Occur?

Sufism can form a unified theory for mechanistic, mind/body, and spiritual healing. Sufi philosophy suggests that the mechanistic operation of the world can be suspended or superseded by divine intervention.

As explained by Sufi Shaikh Gaylani (Abd al-Qadir et al. 2008, p. 39–40).

> The belief of the followers of the Book and the Sunna of the Messenger of Allah (Salla Allah ta'ala 'alayhi wa sallam) is that the sword does not cut because of its nature, but it is rather Allah ('Azza wa Jall) who cuts with it, that the fire does not burn because of its nature, but it is rather Allah ('Azza wa Jall) who burns with it, that food does not satisfy hunger because of its nature, but it is rather Allah ('Azza wa Jall) who satisfies hunger with it and that water does not quench thirst because of its nature, but it is rather Allah ('Azza wa Jall) who quenches thirst with it. The same applies to things of all kinds; it is Allah ('Azza wa Jall) who uses them to produce their effects and they are only instruments in His hand with which He does whatever He wills.

Thus, according to a Sufi world-view, most of the time the world operates by mechanical laws allowed by the "ultimate reality", called "God" in Islam and theistic religions, but mediation by a Sufi Shaikh based upon the Shaikh's nearness to "the ultimate reality" and through "the ultimate reality" would allow for fire not to burn, or a knife not to cut, etc, thus suspending mechanistic laws. The Quran is quite clear in several verses that so called natural laws can be suspended by "the ultimate reality" For example in Surah 2: 117: "when He (Allah) decreeth a matter, he saith to it: "Be," and it is."

The goal of the Sufi and all spiritual paths is nearness to God. In Sufism this is done via following the Sufi path and practices and struggling against the lower self or nafs. It is the lower self that keeps humans distant from God. Islam and Sufism is about surrendering to the will of God through following this path. Once near God, alterations of mechanistic laws may occur. This nearness to Allah is the explanation of so called miracles performed within all religious contexts of ancient times and today.

Rapid wound healing is a very impressive phenomenon to observe and experience, but Islam and Sufism teaches that one's heart is the center of one's being that becomes diseased (Surah: 5:52) and hardened (6:43) from wrong acts (sins). Sufism, however, offers healing for the heart as noted in the Quran: 10:57 "O mankind! There hath come to you a direction from your Lord and a healing for the (diseases) in your hearts – and for those who believe guidance and a Mercy." Thus, when the heart has been purified through spiritual struggle nearness and true healing will occur.

Conclusion

Sufism is a mystical tradition within Islam and is based on drawing nearer to Allah, through the spirituality of the Prophet Muhammad Masters of present Sufi schools trace their origins back to the Prophet through a chain of Masters. Sufism can be described as a path or way of attainment to Allah with its possible paranormal powers, knowledge, and healing. The psychology of Sufism is geared toward this attainment. The Sufi way involves following orthodox Islamic practices such as daily prayer, fasting, and some dietary prohibition, as well as frequent worshipful meditation. These practices may have not only spiritual purposes, but also many positive health implications. Although Sufism generally is focused on spiritual development, some Sufi schools have focused on healing. This healing is a blend of Sufi philosophy with other Islamic healing traditions. Paranormal Sufi healing abilities have been observed and explained on the basis of a spiritual link mediated through the Sufi Master back to the Prophet and Allah. Such phenomena from the Sufi way do not appear to be due to meditative or altered states of consciousness, but may perhaps be due to a higher consciousness.

The implications of Sufism for integrative health is that Western high tech medicine can be helpful for medical and surgical emergencies, but may not be as helpful for chronic non-life threatening conditions. What is needed today is a blending of high tech with "high touch". Sufism is one of the least studied approaches that offered an integration of Era I, II, and III. The Sufi way, universal path for spiritual traditions, includes: prayer, fasting, meditation, avoiding intoxicants, unhealthy lifestyles, and jihad, or battle against the lower self and ultimate attainment of nearness to "The God" (Allah). Finally, Roy challenges the belief that people reject spiritual types of healing because there is no "…generally accepted hypothesis in science that permits it." On the contrary he argues that: "…It is not the lack of an explanatory theory that retards this field so much as the innate resistance to changing one's world view."(Roy 2004). Perhaps further work will begin to change this state of affairs at our meetings.

References

Al-Jilani, A. A. Q., Al- Dargazelli, S. & Fatoohi, L. (2008). ***Purification of the mind***: Lunma Plena Publishing.

Angha, N. (1994). ***Principles of Sufism***. Fremont: Asian Humanities Press.

Armstrong, K. (1992). *Muhammad: A biography of the prophet*. San Francisco: Harper.

Benor, D. J. (2001). ***Spiritual healing: Scientific validation of a healing revolution***. Southfield: Vision Publications.

Chishti, H. (1991). ***The book of sufi healing***. Rochester: Inner Traditions International.

Coppens, P. (2010). *Mirin Dajo: Wonderman*. http://www.philipcoppens.com/mirindajo.html. Accessed August 5, 2010.

Cumming, H., & Leffler, K. (2007). *John of God: The Brazilian healer who's touched the lives of millions*. New York: Hillsboro.

Don, N. S., & Moura, G. (2000). Trance surgery in Brazil. *Alternative Therapies in Health and Medicine, 6*(4), 39–48.

Dossey, L. (1993). *Healing words: The power of prayer and the practice of medicine*. San Francisco: Harper.

Dossey, L. (1998). Deliberately caused bodily damage. *Alternative Therapies in Health and Medicine, 4*(5), 11–16. 103–111.

Dossey, L. (1999). Reponse to Peter Mulacz (Letter to the editor). *Joural of the Society for Psychical Research, 63*(856), 265–250.

Eisenberg, D. K. R., Foster, C., Norlock, F., Calkins, D., & Delbanco, T. (1993). Unconventional medicine in the United States. Prevalence, costs, and patterns of use. *The New England Journal of Medicine, 328*(4), 246–252.

Eisenberg, D. D. R., Ettner, S., Appel, S., Wilkey, S., Rompay, M., & Kessler, R. (1998). Trends in alternative medicine use in the United States, 1990–1997. *Journal of the American Medical Association, 280*, 1569–1575.

Fatoohi, L. (1999). Reponse to Peter Mulacz (Letter to the editor). *Joural of the Society for Psychical Research, 63*(855), 179–181.
Fatoohi, L. (2009). *Jihad in the qur'an: The truth from the source*. Birmingham: Luna Plena Publishing.
Fatoohi, L. (2010). *Jesus the muslim prophet*. Birmingham: Luna Plena Publishing.
Fuller, J. G. (1974). *Arigo: Surgeon of the rusty knife*. New York: Crowell.
Ginandes, C., et al. (2003). Can medical hypnosis accelerate post-surgical wound healing? Results of a clinical trial. *American Journal of Clinical Hypnosis, 45*(4), 333–351.
Green, E., & Green, A. (1977). *Beyond biofeedback*. New York: Delacorte Press.
Hall, H. (2000). Deliberately caused bodily damage: Metahypnotic phenomena? *Journal of the Society for Psychical Research, 64*(861), 211–223.
Hall, H., & Schwartz, G. (2004). Rapid wound healing: A sufi perspective. *Seminars in Integrative Medicine, 2*(3), 116–123.
Hall, H., et al. (2001). The scientific study of unusual rapid wound healing: A case report. *Advances in Mind-Body Medicine, 17*(3), 203–209.
Hussein, J., et al. (1994a). The deliberately caused bodily damage phenomena: Mind, body, energy or what? *International Journal of Alternative and Complementary Medicine, 12*(9), 9–11.
Hussein, J., et al. (1994b). The deliberately caused bodily damage phenomena: Mind, body, energy or what? *International Journal of Alternative and Complementary Medicine, 12*(10), 21–24.
Hussein, J., et al. (1994c). The deliberately caused bodily damage phenomena: Mind, body, energy or what? *International Journal of Alternative and Complementary Medicine, 12*(11), 25–28.
Hussein, J., et al. (1997). Deliberately caused bodily damage phenomena. *Journal of the Society for Psychical Research, 62*(849), 97–113.
Jahn, R., & Dunne, B. (1987). *Margins of reality: The role of consciousness in the physical world*. San Diego/New York/London: Harvest.
Khan, H. (1988). *The music of life*. New Lebanon: Omega Publications.
Kiecolt-Glaser, J. K., et al. (1995). Slowing of wound healing by psychological stress. *The Lancet, 346*(8984), 1194–1196.
Korotkov, K. (2002). *Human energy field: Study with GDV bioelectrography*. Fair Lawn: Backbone Publishing.
Korotkov, K. (2004). *Measuring energy fields: State-of-the-science*. Fair Lawn: Backbone Publishing.
Korotkov, K. G., Bundzen, P. V., Bronnikov, V. M., & Lognikova, L. U. (2005). Bioelectrographic correlates of the direct vision phenomenon. *Journal of Alternative and Complementary Medicine, 11*(5), 885–893.
Krucoff, M. W., et al. (2001). Integrative noetic therapies as adjuncts to percutaneous intervention during unstable coronary syndromes: Monitoring and Actualization of Noetic Training (MANTRA) feasibility pilot. *American Heart Journal, 142*(5), 760–769.
Krucoff, M. W., et al. (2005). Music, imagery, touch, and prayer as adjuncts to interventional cardiac care: The Monitoring and Actualisation of Noetic Trainings (MANTRA) II randomised study. *The Lancet, 366*(9481), 211–217.
Kuhn, T. S. (1970). *The structure of scientific revolutions*. Chicago: University of Chicago Press.
Micozzi, M. S. (2006). *Fundamentals of complementary and integrative medicine*. St. Louis: Saunders Elsevier.
Mulacz, W. (1998). Deliberately caused bodily damage (DCBD) phenomena: A different perspective. *Journal of the Society for Psychical Research, 62*(852), 434–444.
National Geographic (Producer). (2005). *Is it real? Superhuman powers* [Motion picture].
Nelson, R., et al. (1996). FieldREG anomilies in groups. *Journal of Scientific Exploration, 10*(1), 111–141.
Nelson, R., et al. (1998). FieldREG II: Consciousness field effects. Replication and explorations. *Journal of Scientific Exploration, 12*(3), 425–454.
Peper, E., et al. (2006). Tongue piercing by a Yogi: QEEG observations. *Applied Psychophysiology and Biofeedback, 31*(4), 331–338.

Posner, G. (1998). *Taking a stab at paranormal claim.* http://www.csicop.org/sb/show/taking_a_stab_at_a_paranormal_claim

Radin, D., & Yount, G. (2004). Effects of healing intention on cultured cells and truly random events. *International Journal of Alternative and Complementary Medicine, 10*(1), 103–112.

Roy, R. (2004). ***Science of whole person healing: Proceedings of the first interdisciplinary** international conference*. New York: iUniverse.

Schwartz, G. E., & Simon, W. L. (2007). *The energy healing experiments: Science reveals our natural power to heal.* New York: Atria Books.

Taft, R., et al. (2005). Time-lapse analysis of potential cellular responsiveness to Johrei, a Japanese healing technique. *BMC Complementary and Alternative Medicine, 5*, 2.

Yount, G., et **al. (2004). Biofield perception: A series of pilot studies with cultured human cells.** *International Journal of Alternative and Complementary Medicine, 10*(3), 463–467.

Zachariae, R., et al. (2005). The effect of spiritual healing on in vitro tumour cell proliferation and viability-an experimental study. *British Journal of Cancer, 93*(5), 538–543.

An Emerging New Model for Consciousness: The Consciousness Field Model

Robert K.C. Forman

Abstract A surprising level of agreement about the nature of consciousness emerged at a recent multi-disciplinary "meeting of experts" gathering on Neuroscience, Spirituality and Consciousness in Freiburg Germany, 2008. Contra the reigning hypothesis for consciousness, the epiphenomenal model, this hypothesis suggests: (i) Consciousness is a fundamental element of reality. (ii) Consciousness is mediated by the brain. That is, consciousness takes place with the help of the brain but is not within it. Brains are transducers of consciousness. (iii) Consciousness is independent of the brain. As the field of consciousness is experienced through consciousness-transducing brains, consciousness appears to exist independently, though it remains so far unobservable unless transduced by brains. (iv) The ability to be conscious of a connection with something larger may be the fundamental nature that distinguishes the human being. The ability to sense 'something larger' (which cannot itself be observed), a mystical ability, may be the skill distinguishing human beings from other hominids.

Asking the Question with *All* the Data

What is the nature of consciousness? This deceptively simple question is one of the most hotly debated and intractable questions of the modern sciences and humanities in the twenty-first century.

R.K.C. Forman (✉)
Jerusalem Institute of Advanced Studies, Jerusalem, Israel
e-mail: Forman@TheForge.org

H. Walach et al. (eds.), *Neuroscience, Consciousness and Spirituality*, Studies in Neuroscience, Consciousness and Spirituality 1,
DOI 10.1007/978-94-007-2079-4_17,

We know very little by way of an answer. Consciousness can be looked at from inside (through phenomenology, religion, spirituality, or generally introversion) or outside (through neuroscience, physics, biology, or generally "objective" studies).

We do not even know whether the answer will come from the sciences, the humanities or both. Nor whether it will look like a sentence or a formula.

The "prevailing view" of consciousness is the "epiphenomenal model", which grows out of the materialist worldview. It states:

- Consciousness is a product, some suggest an "excretion", of the brain. As Francis Crick put it:

> You,' your joys and your sorrows, your memories and your ambitions, your sense of personal identity and free will, are in fact no more than the behavior of a vast assembly of nerve cells and their associated molecules. Who you are is nothing but a pack of neurons. (Sharif, Schooler and Vohs 2008, p. 181, quoting Crick's The Astonishing Hypothesis, no reference)

- Consciousness can be, some would say has been, explained as having a materialistic or bodily cause, for *material* is the only and ultimate reality.
- Since consciousness has a bodily cause, the death of the body will mean the death of all subjective experience.

This is so, despite the fact that as of yet there is no scientific basis for such a causal claim. The thought that all mental processes are necessarily physical processes is at this point more a metaphysical assumption than proven scientific fact (Begley 2007, p. 132).

I will report on a new hypothesis of consciousness which emerged as a kind of majority report at the 2008 Freiburg, Germany "experts meeting" on "*Neuroscience, Consciousness and Spirituality*." It grew out of discussions of researchers in at least four disciplines: neuroscience and biology, psychology, religion and spirituality, physics and parapsychology.

Before beginning however, I would like to offer a thought about including spirituality, religion and parapsychology as well as neuroscience and biology in our article and our list of domains. While neurological studies of spirituality have become more accepted recently, it is still common for scientists to deny or debunk studies of parapsychology as well as reports of spiritual experiences. There are many well constructed and repeated studies with high degrees of deviation from chance which suggest a possible new model for both consciousness and more broadly the nature of reality. Nonetheless, studies of parapsychology, near death experiences (NDE), out of body experiences (OBE) and mystical claims are routinely dismissed, suppressed or ridiculed by 'hard headed' scientists and theoreticians. They do so primarily, it would appear, because their results *primae facie* challenge the reigning theories of a materialistic universe and materialistic causes of consciousness.

This is a dangerous maneuver, for two reasons. First to dismiss data coming from such fields without thorough investigation and exploration is to commit the logical fallacy of *petitio principii* (begging the question), i.e., to assume what is to be proven in order to prove it. The question before us is the following: Is the universe only material or might there be something more, something non material, in addition to the material? To claim that NDEs, parapsychology, or spiritual experiences of a non material field cannot be valid because one's physicalist assumptions suggest that it cannot be, is to assume an answer to the very question before us, and thus to commit a fallacy.

Second, to dismiss or suppress these studies because they challenge one's "common sense" or one's assumptions is bad science. It dismisses evidence as opposed to exploring it and examining its implications. Had the data of the double slit experiment been ignored because it did not fit with the reigning (Newtonian) model, we would not have quantum physics today. Had the theory of natural selection been denied because Darwin's evidence did not fit with the reigning models, we would not have the modern field of evolutionary biology. As Frederik van Eedlen put it,

> It is the greatest enemy of scientific progress if one rejects or refuses beforehand to study unknown and seemingly strange facts and findings. (quoted in the chapter by van Lommel, this book)

Parapsychology, NDE studies, spiritual phenomenology and other fields indeed do seem strange. But they may offer some of most revelatory and interesting data for investigating the nature of human consciousness. For this reason I have tapped some such studies.

The Consciousness Field Model

The following set of principles is not yet generally accepted. But the fact that such a hypothesis has emerged among thinkers in a wide array of disciplines makes it seem especially plausible. If any theory of consciousness is to be generally accepted it will have to be considered plausible in most or all of the fields touched by the phenomenon, and this one seems to be.

The hypothesis may be stated in the following four principles.

Consciousness is a Fundamental Element of Reality, like an Additional Dimension

- Jonathan Schooler (2008) suggests that consciousness may be "located in its own brane[1] further external to the dimensions of the physical world." He quotes Alan Wallace (2007) that

> Nothing in physics prevents us from adding a "space of elements of consciousness" to the natural world consisting of mass-energy, space time, and informational states… consciousness may exist by itself, even in the absence of matter, just like gravitational waves, excitation of space, may exist in the absence of protons and electrons (Wallace 2007)

[1] 'Brane' denotes in physics any dimensional or extended object in string theory.

- Mario Beauregard notes that some conclude from parallels between temporal lobe epilepsy and spiritual experiences that the latter are nothing but such epileptic episodes. Yet, he goes on to observe that most epileptics do not report spiritual experiences and most mystics are not epileptics. His conclusion is that they are not necessarily related and that consciousness may exist in itself as a "spiritual level of reality" and impact the brain. (see the chapter by Beauregard, this book)
- Of the 344 patients who underwent cardiac resuscitation that were interviewed by Dutch Cardiologist Pim van Lommel, some 62 (18%) reported memories of a near death experience (see the chapter by van Lommel, this book). His numbers of NDEs were similar to those of American Bruce Greyson (15.8% of 116 survivors had an NDE), British scientists Sam Parnia and Peter Fenwick (11% of 63 survivors had an NDE), and British Penny Sartori (23% of 39) (Greyson 2003; Parnia et al. 2001; Sartori 2006; see also Almeder 1992). Cardiologist Michael Sabom details a fascinating and veridical account of Pam Reynolds, who had a particularly vivid NDE. She underwent a surgical procedure for the removal of a basal artery aneurism and for that had to be put into a state close to physiological death with a cooling of brain temperature down to 15°C and a temporal shunting off of blood supply; putting her in a state near death. When resuscitated she reported vivid memories of conversations, unusual implements and the process used during the operation on her body, as well as memories of dead relatives interacting with her, all despite the fact that her brain activity had flat lined (Sabom 1998). These numbers, studies and reports suggest that human beings can be conscious and have memories, sometimes veridical, despite the absence of any significant brain activity to support it. This data implies that consciousness is independent of the brain, so, as van Lommel put it, consciousness "finds its origin in and is stored in a non-local space." (see the chapter by van Lommel, this book, see also Schwartz 2003)
- Mystics report experiencing a non-local level of reality, spread out and non-individuated, which stands independent of, and yet connected with the mystic him or herself and the rest of reality. As one mystic reports:

> Over the years, the interior silence that first became permanent many years ago has slowly changed. Gradually, imperceptibly, this sense of who I am, this silence inside, has grown as if quasi-physically larger. In the beginning it just seemed like I was silent inside. Then this sense of quietness has, as it were expanded to permeate my whole body. Some years later, it came to seem no longer even limited to my own body, but even wider, larger than my body. It's such a peculiar thing to describe! It's as if who I am, my very consciousness itself, has become bigger, wider, less localized. By now its as if I extend some distance beyond my body, as if I'm many feet wide. What is me is now this expanse, this silence, that spreads out (Forman 2011).

- The Upanishads and other Hindu Texts describe a hidden, expanded level of reality, *Brahman*, said to be the unformed principle within the formed world.
- Kabbalah authors suggest that the experience of the *Ein Sof* is that of "no end" or "unending." Kabbalah teachers also speak of an *Ein* (or Ayn), which is described as infinite no-thingness (see the chapter by Lancaster, this book).

- Meister Eckhart describes the *Gottheit*, the Godhead, which is an unmanifest infinite principle beneath all reality, of which God and Christ are merely representations, with which human beings may become co-terminus (Forman 1991). The evidence of such religious mystics point to an experience of a fundamental element of reality, consciousness, that is non local, or spread out.
- This hypothesis that consciousness is an independent field which transcends time may connect with quantum mechanical behavior in some way. It may help explain the human side of the quantum collapse in the Einstein-Podolsky-Rosen paradox or the double slit experiment collapse.
- Some individuals demonstrate the ability to perceive beyond the limits of their visual system. Pearce and Pratt ran experiments for remote viewing in which subjects were asked to visualize cards from either 100 or 250 yards away; of 1,850 calls, 370 were expected to be correct by chance, whereas 558 hits were scored ($p < 10^{-22}$) (Rhine and Pratt 1954). Archeological digs have taken place based on certain subjects' remote viewing abilities: The Egyptian city of Marea was accurately unearthed based on the unusual accuracy of two people with these abilities, a Mr. McMullen and Ms. Hammond (Schwartz 2000, see also McMoneagle 1997). These studies suggest that human consciousness has some non-local aspect, perhaps the kind of field described by some mystics, and that some human beings may have the ability to perceive by its means.

Consciousness is Mediated by the Brain, not Excreted by it

- Throughout the Freiburg conference we heard phrases suggesting that consciousness is a "transducer" or like a "radio receiver" or a "relay station."
- In his studies of NDEs, Pim van Lommel observed that consciousness can exist independently of brain activity. This implies that the mind is not excreted or produced by brain activities. This is why he stated that the brain is a conveyor of consciousness, not a producer.

> Could our brain be compared to the TV set, which receives electromagnetic waves and transforms them into image and sound? Could it as well be compared to the TV camera, which transforms image and sound into electromagnetic waves?... The function of the brain should be compared with a transceiver, a transmitter/receiver, or interface. Thus there are two complementary aspects of consciousness, which cannot be reduced one to the other, and the function of neuronal networks should be regarded as receivers and conveyors, not as retainers of consciousness and memories. (see the chapter by van Lommel, this book)

- Hartmann Römer suggested that there is a "non local correlation between consciousness and the brain." (see the chapter by Walach and Römer, this book)
- Arthur Deikman suggests that mystical and spiritual experiences of a non-local reality is fostered by training the mind to *cease* limiting its perception to what our cultures train us to recognize. Meditation may reduce the entrancement of automatized patterns and widen the range of our receptivity (Deikman 1972).

- Hinterberger et al (in this book) identify a high amplitude coherence in the beta range between frontal/prefrontal and parietal/occipital areas during meditation that positively correlates with years of meditative experience ($r=0.61$–0.68, $p<0.001$). Antoine Lutz suggested that in meditation we see greater coherence across large brain areas, and high "amplitude gamma synchronicity" (Lutz 2008; Lutz et al. 2004, see also Austin 1999, 2006, p. 53; Aftanas and Golocheikine 2001; Hankey 2006; Orme-Johnson and Haynes 1981; Travis and Arenander 2006; Hebert et al. 2005). These changes seen by Lutz, Hinterberger and others may represent the result of *de*automatizing our mental processes to greater and greater degrees of stimulus.
- In Hinduism, Brahman, the non dual, panentheistic principle, is a core notion. It is said to exsit independently of the cosomos and of each person, and is at the ontological basis of all beings. Like the consciousness of our hypothesis, Brahman exists as its own domain of reality, or in a dimension all on its own. But, again like the field of consciousness in our hypothesis, it somehow comes into or manifests as form, thus coming to be "formed Brahman," much as consciousness comes to be embodied as an individual. Thus formless Brahman, like the consciousness of our hypothesis, exists simultaneously with, yet is hidden within its formed aspect: The "formless within form." The human being is enlivened through contact with this other sort of reality. We receive it, as a TV might receive radio waves. Yet according to Hindu doctrine, our Atman, or consciousness, can come to experience its true nature as Brahman, much as one might experience the independent domain of consciousness in certain anomalous experiences.

This hypothesis is in alignment with that of William James, Huxley and Bergson. After perusing the range of religious and spiritual experiences, James concluded that the brain may serve a transmissive function as opposed to a productive one. He drew an analogy to a prism, which does not produce light but refracts it to produce a spectrum (James 1902). Following James, Aldous Huxley and Henri Bergson suggested that the brain does not produce consciousness but rather acts as a kind of filter, blocking out much input and allowing through only a narrow band of perceivable reality. (see the chapter by Beauregard, this book).

Consciousness is Independent of Brain Processes. (This Stems from Principle 1 & 2)

- If the field of consciousness is indeed experienced through consciousness-transducing brains, then it must exist independently of the brain, though it will remain unobservable unless transduced by brains.
- Near Death Experiences in general, which sometimes involve veridical knowledge, clearly suggests that people can have experiences even when there is effectively no brain activity. If we hypothesize that there must be *some* minimal brain activity during these events, it is certainly not the kind of higher level functioning

that should, under the reigning theory, give rise to such complex experiences and memories.

- Telepathy and remote viewing point to the possibility that consciousness is connected to a region or field that is independent of the brain. Distant viewing and predictive knowledge, as in the case of the discovery of Marea, Egypt especially suggest such a connection (Schwartz 2000).
- Mystical experiences sometimes carry the sense of being non local or spread out beyond the body. This suggests that human beings can have experiences of a field beyond, and thus independent of, the body and the brain. This suggests (though does not prove) that such people are experiencing the field of consciousness directly. My own research suggests this (Forman 1998). Barendregt argues that in Buddhist meditation one arrives at a state which is "a kind of 'emptiness…also called selflessness." (see the chapter by Barendregt, this book)
- Beauregard argues that though there is brain activity during mystical events, "it is conceivable that the neural correlates of [such events] reflect the actual connection of the brain with a spiritual level of reality." (see the chapter by Beauregard, this book)

Our Ability to Connect with Something Larger may be the Basic Nature of the Human Being

- Matt Rossano argues that about two hundred thousand years ago, Shamans brought people to health by having them focus their attention. This may have led to selection for that ability. And, he continued, our ability to connect with that which is larger may have been one of the distinguishing capacities of human beings, as one of the original capacities that gave early human beings an advantage over other species (see the chapter by Rossano, this book).
- In order to become enculturated, human beings must be taught what there is in the world (a *Weltanschauung*) and, along with it, what *not* to notice. English speakers do not tend to notice different snow types that, because of their linguistic training, Eskimos are trained to perceive. Enculturation is a limiting as well as a training process. To open oneself up to a wider range of impulses, perceptions and ways of thinking into which our cultures entrain us may be the natural state of human beings. The enculturation process may help form the specific neural pathway we see in brain patterns, even without causing consciousness per se.
- Meditation may serve to de-automatize the enculturation processes, making it a way to recover that which is more fundamentally ours (Deikman 1972). The first step of moving beyond automatized processes towards spirituality may be "to let go of all worldviews," helping us recover our lost sense of a connection to something larger. According to Brian Lancaster, the Kabbalah teaches that we should "untie the knots from the self, step out of the structure we've created" (see the chapter by Lancaster, this book).

Conclusions and Further Questions

These four principles provide a framework for a new hypothesis of consciousness which I call the Consciousness Field Model (CFM). If it becomes generally accepted, like any good theory it opens up many unanswered questions. Some are:

- How might the body or brain receive or transduce consciousness?
- Can we verify the existence of consciousness outside of the brain/person? If so in what sense? What would it take to affirm or deny such an independent field of consciousness?
- Might there be any way to measure consciousness outside the brain, i.e. a "consciousness-o-meter?"
- From where might the domain of consciousness come?

This CFM model could potentially tie together discoveries from East and West, spirituality and science, brain research and introspective experience, the humanities and the sciences, and may account for anomalous experiences. As Jeffrey Schwartz remarked at the conclusion of the gathering, the implications here, political as well as scientific, are enormous. And they are important.

Acknowledgments I am grateful to the Jerusalem Institute of Advanced Studies and to Philip Wexler and Yoni Garb for support in writing this review article.

References

Aftanas, L. I., & Golocheikine, S. A. (2001). Human anterior and frontal midline theta and lower alpha reflect emotionally positive state and internalized attention: High-resolution EEG investigation of meditation. *Neuroscience Letters, 310*(1), 57–60.

Almeder, R. (1992). *Death and personal survival: The evidence for life after death.* Lanham: Littlefield.

Austin, J. (1999). *Zen and the brain: Toward an understanding of meditation and consciousness.* Cambridge: The MIT Press.

Austin, J. (2006). *Zen brain reflections.* Cambridge: The MIT Press.

Barendregt, H. (2011). Mindfulness meditation: Deconditioning and changing view. In H. Walach, S. Schmidt, & W.B. Jonas (Eds.), *Neuroscience, consciousness and spirituality.* New York: Springer.

Beauregard, M. (2011). Neuroscience and spirituality: Findings and consequences. In H. Walach, S. Schmidt, & W.B. Jonas (Eds.), *Neuroscience, consciousness and spirituality.* New York: Springer.

Begley, S. (2007). *Train your mind, change your brain: How a new science reveals our extraordinary potential to transform ourselves.* New York: Ballentine Books.

Deikman, A. (1972). Deautomatization and the mystical experience. In C. T. Tart (Ed.), *Altered states of consciousness.* New York: Anchor/Doubleday. Available at http://www.deikman.com/deautomat.html

Forman, R. (1991). *Meister Eckhart: Mystic as theologian.* Rockport: Element Books.

Forman, R. (1998). What does mysticism have to tell us about consciousness? *Journal of Consciousness Studies, 5*(2), 185–201.

Forman, R. (2011). *Enlightenment ain't what it's cracked up to be: A journey of discovery, snow and jazz in the soul*, Washington, DC: O-Books.

Greyson, B. (2003). Incidence and correlates of near death experiences in a cardiac care unit. *General Hospital Psychiatry, 24*, 269–276.

Hankey, A. (2006). Studies of advanced stages of meditation in the tibetan buddhist and vedic traditions. I: A comparison of general changes. *Evidence Based Complimentary and Alternative Medicine, 3*(4), 513–521.

Hebert, R., Lehman, D., Tan, G., Travis, F., & Arenandar, A. (2005). Enhanced EEG alpha time-domain phase synchrony during Transcendental Meditation: Implications for cortical implication theory. *Signal Processing, 85*(11), 2213–2232.

Hinterberger, T., Kohls, N., Kamei, T., & Walach, H. (2011). Neurophysiological correlates to psychological trait variables in experienced meditative practitioners. In H. Walach, S. Schmidt, & W.B. Jonas (Eds.), *Neuroscience, consciousness and spirituality*. New York: Springer.

James, W. (1902). *Varieties of religious experience*. New York: Longmans, Green, 1916.

Lancaster, B. L. (2011). The hard problem revisited: From cognitive neuroscience to Kabbalah and back again. In H. Walach, S. Schmidt, & W.B. Jonas (Eds.), *Neuroscience, consciousness and spirituality*. New York: Springer.

Lutz, A. (2008, July). *The relation between mental training/inner transformation in the contemplative traditions and neuroplasticity*. Paper presented at the expert meeting Neuroscience, Concisousness and Spirituality, Freiburg, Germany.

Lutz, A., Greischar, L., Rawlings, N., Ricard, M., & Davidson, R. (2004). Long-term meditators self-induce high-amplitude gamma synchrony during mental practice. *Proceedings of the National Academy of Sciences of the United States of America, 101*(46), 16369–16373.

McMoneagle, J. (1997). *Mind Trek: Exploring consciousness, time and space through remote viewing*. Charlottesville: Hampton Roads Publishing Co.

Metzinger, T. (2008, July). *Science and spirituality: No self, no Soul, no Faith*. Paper presented at the expert meeting Neuroscience, Concisousness and Spirituality, Freiburg, Germany.

Orme-Johnson, D. W., & Haynes, C. (1981). EEG phase coherence, pure consciousness, creativity, and Tm—Sidhi experiences. *International Journal of Neuroscience, 13*(4), 211–217.

Parnia, S., Waller, D. G., Fenwick, P., et al. (2001). A qualitative and quantitative study of the incidence, features and aetiology of near death experience in cardiac arrest survivors. *Resuscitation, 48*, 149–156.

Radin, D. (1997). *The conscious universe: The scientific truth of psychic phenomena*. San Francisco: HarperOne.

Rhine, J. B., & Pratt, G. A. (1954). Review of the Pearce-Pratt distance series of Esp tests. *Journal of Parapsychology, 18*, 165–177.

Römer, H., & Walach, H. (2011). Complementarity of phenomenal and physiological observables: A primer on generalised quantum theory and its scope for neuroscience and consciousness studies. In H. Walach, S. Schmidt, & W.B. Jonas (Eds.), *Neuroscience, consciousness and spirituality*. New York: Springer.

Rossano, M. (2011). Setting our own terms: How we used ritual to become human. In H. Walach, S. Schmidt, & W.B. Jonas (Eds.), *Neuroscience, consciousness and spirituality*. New York: Springer.

Russell, R. (2007). *The journey of Robert Monroe: From out-of-body explorer to consciousness Pioneer*. Charlottesville: Hampton Roads Publishing Co.

Sabom, M. (1998). *Light and death: One doctor's fascinating account of near death experiences*. Grand Rapids: Zondervan Publishing.

Sartori, P. (2006). The incidence and phenomenology of near death experiences. *Network Review, 90*, 23–25.

Schooler, J. A. (2008, July). *Cognitive neuroscience view of consciousness and spirituality*. Paper presented at the expert meeting Neuroscience, Concisousness and Spirituality, Freiburg, Germany.

Schwartz, S. (2000). *The location and reconstruction of a Byzantine structure in Marea, Egypt, including a comparison of electronic remote sensing and remote viewing*. http://www.stephanaschwartz.com/PDF/Marea.pdf. Last accessed August 12, 2009

Schwartz, G.E. (2003). *The afterlife experiments: Breakthrough scientific evidence of life after life*. New York: Atria Books.
Sharif, A., Schooler, J., & Vohs, K. (2008). The hazards of claiming to have solved the problem of free will. In J. Baer, J. Kaufman, & R. Baumeister (Eds.), *Are we free? Psychology and free will* (pp. 181–204). Oxford: Oxford University Press.
Smithies, O. (2005). Many little things: One geneticist's view of complex diseases. *Nature Reviews Genetics, 6*, 419–425.
Stace, W. (1987). *Mysticism and philosophy*. New York: Tarcher.
Tart, C. (2009). *The end of materialism: How evidence of the paranormal is bringing science and spirit together*. Oakland: Harbinger Books.
Travis, F. T., & Arenander, A. (2006). Cross-sectional and longitudinal study of effects of transcendental meditation practice on frontal power asymmetry and frontal coherence. *International Journal of Neuroscience, 116*(11), 1519–1538.
van Lommel, P. (2011). Endless consciousness: A concept based on scientific studies on near death experiences. In H. Walach, S. Schmidt, & W.B. Jonas (Eds.), *Neuroscience, consciousness and spirituality*. New York: Springer.
Wallace, B. A. (2007). *Hidden dimensions: The unification of physics and consciousness* (Columbia series in science and religion). New York: Columbia University Press.
William James, *The Varieties of Religious Experience*, New York: Longmans, Green, 1916. Originally published in 1902.

Index

H. Walach et al. (eds.), *Neuroscience, Consciousness and Spirituality*, Studies in Neuroscience, Consciousness and Spirituality 1,
DOI 10.1007/978-94-007-2079-4, © Springer Science+Business Media B.V. 2011